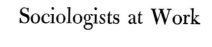

Sociologists at Work

SOCIOLOGISTS

ESSAYS ON THE CRAFT OF SOCIAL RESEARCH

Basic Books, Inc.

Publishers

NEW YORK • LONDON

AT WORK

EDITED BY

Phillip E. Hammond

Robert N. Bellah

Peter M. Blau

James S. Coleman

Melville Dalton

James A. Davis

Reneé C. Fox

Blanche Geer

Herbert H. Hyman

Seymour Martin Lipset

David Riesman

Stanley H. Udy, Jr.

Jeanne Watson

Charles R. Wright

To S. N. H. *and* J. M. H.

The Authors

ROBERT N. BELLAH is Associate Professor of Sociology and Regional Studies at Harvard University. He is currently engaged in research on modern Japan which should eventually lead to a sequel to *Tokugawa Religion*.

PETER M. BLAU is Professor of Sociology at The University of Chicago. Trained at Columbia, his latest book is *Exchange and Power in Social Life*.

JAMES S. COLEMAN is Professor of Social Relations at The Johns Hopkins University.

MELVILLE DALTON is Professor of Sociology and Research Sociologist in the Institute of Industrial Relations, University of California, Los Angeles. His graduate training was at The University of Chicago.

JAMES A. DAVIS is Associate Professor of Sociology and Senior Study Director at National Opinion Research Center, The University of Chicago.

RENÉE C. FOX is Associate Professor of Sociology at Barnard College, Columbia University. Her interest and research in the sociology of medicine and science have not only carried her to Belgium, the land described in her contribution to this volume, but have opened new vistas of investigation in the Congo, where she is currently spending a sabbatical year.

BLANCHE GEER is Associate Professor of Sociology and Education and Research Associate at the Youth Development Center, Syracuse University. She is a co-author of *Boys in White*.

PHILLIP E. HAMMOND is Assistant Professor of Sociology, Yale University. He is currently engaged in studies in the sociology of religion.

HERBERT H. HYMAN is Professor of Sociology and Associate Director of the Bureau of Applied Social Research, Columbia University.

SEYMOUR MARTIN LIPSET is Professor of Sociology and Director of the Institute of International Studies, University of California, Berkeley. He is the author of *The First New Nation, Political Man,* (with James S. Coleman and Martin Trow) *Union Democracy,* (with Reinhard Bendix) *Social Mobility in Industrial Society, Agrarian Socialism. Political Man* won the MacIver Award for 1962. He is a Fellow of the American Academy of Arts and Sciences and served as Henry Ford II Research Professor, Yale University, 1960-1961.

DAVID RIESMAN is Henry Ford II Professor of the Social Sciences at Harvard University. He attended Harvard College and Harvard Law School, taught and practiced law, and from 1946 to 1958 was at the University of Chicago as a member of the Sociology Department, of the Committee on Human Development, and the Social Sciences Staff of the College. His present research is in the field of higher education.

STANLEY H. UDY, JR., is Associate Professor of Sociology, Yale University. Trained at Princeton, he is presently working on comparative analyses of selected American industrial firms.

JEANNE WATSON (Ph.D., 1952, Social Psychology, University of Michigan) worked at the Survey Research Center and the Research Center for Group Dynamics at the University of Michigan prior to her position at the University of Chicago.

CHARLES R. WRIGHT is Associate Professor of Sociology, UCLA. He was formerly on the faculty of Columbia University and on the staff of its Bureau of Applied Social Research. In 1963 he served as OAS Professor to Latin America (Chile). His publications include *Mass Communications: A Sociological Perspective, Public Leadership, Applications of Methods of Evaluation.*

Contents

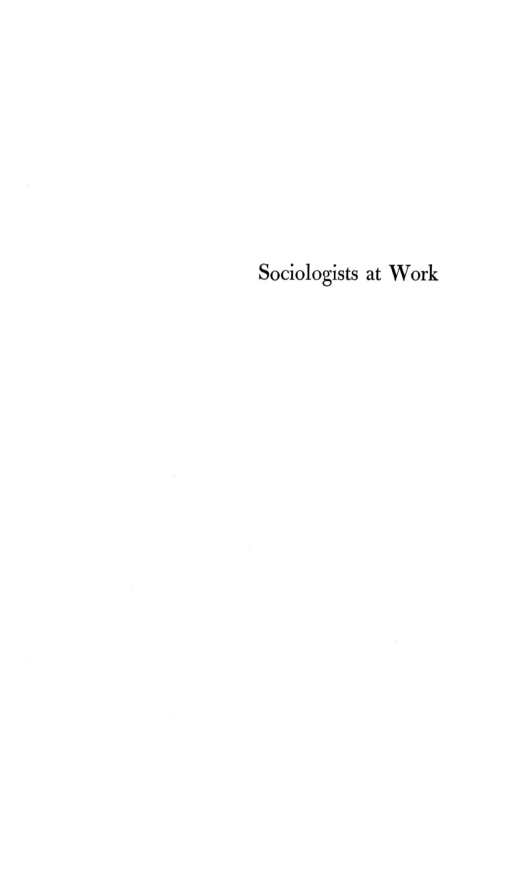

Sociologists at Work

1 Introduction

PHILLIP E. HAMMOND

In 1831, shortly after de Tocqueville and his companion, Beaumont, arrived in America, they were guests in the home of "a great landowner" who lived up the Hudson River from New York City. Interested in the consequences as well as the antecedents of democracy in America, de Tocqueville inquired of his host what chief drawbacks were contained in democratic society. The answer he received—that inheritance laws equalized the portions of the heirs—was to plague the observer from France for the remainder of his American tour. For apparently it was from this remark that de Tocqueville developed a view which attributed to inheritance laws the effects of leveling the population, restricting the development of class boundaries, small houses, instability of the social fabric, and so forth.

Not until near the end of his journey did he have conversations correcting his point of view. Despite the months he spent visiting stately homes, touring prisons, riding horseback through Michigan with Indians, being a guest of mayors, governors, and congressmen, of laborers, tenant farmers, and rivermen, too late did he come to realize that the inheritance laws did not *require* equal portions, that subdivision was likely to be of money and not property, and that, in any event, American *commercial* success could not be passed on as readily as French *territorial* success to the next generation.

As George Wilson Pierson, historian of de Tocqueville and Beaumont's American odyssey, states:

> [The pattern of inheritance was] a very inadequate explanation of the origins of social equality and political democracy in the United States. Unfortunately they entertained no doubts and made no investigation. As a result, before the two friends were to discover their error, the thought about the influence of the inheritance laws had obtained

1

such a grip on their minds that they could not shake it off. The conviction . . . was therefore to warp their judgment, again and again to lead their interpretations astray, and finally one day to reappear in the pages of Tocqueville's book.[1]

To the social observer and researcher, Pierson's "biography" of de Tocqueville's *Democracy in America* makes as rewarding reading as its subject. There exists no clearer example, from no more classic a work, of the remark by the philosopher of science, Hanson: "People, not their eyes, see. Cameras, and eye-balls, are blind." [2] The chronicling, month by month, of the investigations by these two "sociologists before their day" conveys as do few other documents the role of the circumstantial, the irrational, and nonrational, as well as the logical and systematic, nature of social research. Not only are the "research methods" made obvious (e.g., de Tocqueville typically asked many different persons the same questions, he sought out documents from government bodies and officials, he regularly corresponded with his father in France to determine the comparability of American and French institutions),[3] but also the role of "controlling ideas," of evanescent theory, of the fortuitous comment, emerges in a way that is seldom indicated in traditional research-reporting.

For very few social scientific reports, it appears, with all their discussion of methods, contain accounts of the "method" by which they came about. There are almost no chronicles of social research. And yet this missing component is an important one, as seen in the frequency with which existing chronicles are cited as well as by the number of times eminent and experienced researchers have called for more such accounts.[4]

Edward Shils, for example, in a quasi-autobiographical essay several years ago, described how he entertained the notion of "primary group" from student days on. He concluded the piece by saying that his purpose was to show "a more just conception of the collaboration of research and theory" which is "a disorderly movement . . . full of instances of things known and overlooked, unexpected emergencies, and rediscoveries of long known facts and hypotheses which in the time of their original discovery had no fitting articulation and which found such articulation only after a considerable time." [5]

And Robert K. Merton, noting the huge literature on scientific methods, remarks that "this literature is concerned with . . . ways in which scientists ought to think, feel, and act. It does not necessarily describe, in needed detail, the ways in which scientists actually do think, feel, and act." [6]

This need for a fuller accounting of investigative activity is felt not alone in the *social* sciences. Conant, speaking both as chemist and educator in his essays, *On Understanding Science,* declares, "To my mind there has been too little said in the popular accounts of science about both the dynamic quality of the enterprise and the fact that it is concerned with evolving conceptual schemes rather than the classification of facts." [7]

Beveridge concurs, somewhat elliptically, in his comment that "no one believes an hypothesis except its originator but everyone believes an experiment except the experimenter." [8]

This component of research—Conant calls it the " 'feel' for the Tactics and Strategy of Science"—is admittedly elusive and hard to transmit. Some scientists would insist that it is not possible to give formal instruction in how to do research, and the history of science is full of the physicist's or mathematician's reference to his discoveries as "aesthetic" experiences, presumably because he indeed finds his creative researches difficult to describe. But though at some reaches the origins of ideas, the theoretical breakthroughs during the course of research, are to be described only as poetic, as mysterious or awesome, at some other level they can be understood, or at least described, as very human activities. Taken together, the essays of this volume illustrate the point.

It would be an error, however, to expect of these essays a logical program of "how to do social research." They deal more with the "context of discovery" than with the "context of justification," and the latter only is amenable to logical analysis.[9] But because most research-methods texts, like research monographs, deal exclusively with the context of justification, matters related to the context of discovery are seldom found in print. As a partial correction, then, the essays are offered.

The authors were asked to portray their own research activity as it was experienced during some *specific* investigation. Their instructions were to particularize, not generalize. The suggestion was also made that, in format, the essays be chronologic or ideologic, that is, organized around the sequence of events in time or the sequence of ideas in the mind of the researcher. Happily both principles are observable, frequently within a single chronicle. Finally—and in defense against any who may hold that even a variant of scientific reporting must necessarily be impersonal and dispassionate—the authors were encouraged to write in the first person if they chose, to indicate freely where changes would be made if the research were to be repeated, and generally to let the reader in on the *sub rosa* phases of contemporary social research.

Beyond the intrinsic entertainment that comes with making public any erstwhile private activity, the resulting set of chronicles exhibits several messages or themes seldom made explicit in discussions on research method. These themes might appropriately be flagged in this Introduction. First is the view of "theory versus research" contained in the essays. And second is the identification of various ways in which social research is a social activity, thereby involving institutionalized rules and behavior.

THEORY VERSUS RESEARCH

In what sense are theory and research dissimilar? Of course, the interviewer ringing doorbells or the observer behind the one-way mirror is behaving in a way different from the thoughtful reader sitting behind a desk. But is it correct to call the one a researcher and the other a theorist? Or, perhaps more accurately, is not a good deal of the meaning of "research" lost by such a characterization? For how can a question be asked or an observation be made without first having been thought about, conceptualized, or structured in terms of meaningful categories? And does not *that* structuring or conceptualization stem from reading books or some equivalent? Conversely, how can the thoughtful reader behind a desk be other than a researcher—that is, a systematic handler of observations—if his theory is to be good?

In the commoner view of research and theory, where researchers collect data and theorists do something with data, the former is too narrowly depicted and the latter too often portrayed as naïve. This view suggests that research is *only* the classification of facts when, in reality, as science it is concerned with "evolving conceptual schemes." Indeed, research by induction is patently *not* what scientific discovery typically involves but rather what has been called abduction, or "leading away," that is, theorizing.[10] In this sense, then, dichotomizing into research and theory fails to do justice to either activity. The old worn dichotomies, e.g., speculator versus systematizer, guesser versus accumulator, romantic versus classicist, armchair theorist versus head-counter, Aristotelian versus Baconian, and more recently, ruminant versus factitioner or cloud-hopper versus clodhopper—all these and more, though perhaps descriptive of something, are not accurate or fair portrayals of contemporary social inquiry. Good specimens involve theory and research bound up as one.

These essays illustrate the point very well. Consider the common experience of the following researchers as they have chronicled it:

> The usual textbook notion of social research is that one forms a hypothesis and then proceeds to gather data to confirm or negate it. . . . But . . . in the field of comparative and historical sociology, the researcher often finds himself with an abundance of data, and the problem is how to make sense of it. — B E L L A H

> [The] intellectual challenge in survey analysis is in ordering and synthesizing the diverse information. . . . There are so many questions which might be asked . . . that a thousand different studies could come out of the same data. . . . The real job of the study director is to select and integrate. — D A V I S

> Some of the hypotheses advanced . . . early . . . were later abandoned; others were supported by empirical observations, but even these were often modified and refined in the course of research. . . . The double aim is always to develop and refine theoretical insights which explain reality . . . and to discriminate between the correct and the false explanatory principles. — B L A U

Dalton is referring to this theoretical sensitivity demanded during, and not simply before, the research process when he warns against being a slave to one's hypotheses. And Coleman remarks that too frequently neglected in the examination of research activity is the "synthesis," as well as the analysis, necessary to make research coherent. Udy, faced with the decision of how many societies from around the world he would investigate, decided "to use a large number of cases . . . and . . . try to discover 'what is going on here.'"

Riesman and Watson, in their candid chronicle, make it quite clear that underlying part of the Sociability Project's pangs was the lack of success in finding appropriate synthesizing principles for ordering their observations. And Geer, in the kind of report all too infrequent in social-science literature, describes the birth of an "integrating principle" as it occurred in her exploratory field work.

A common experience, then, of these social researchers is the sense of struggling with data so that conceptual schemes can be imposed. It is this imposition of conceptual order that distinguishes research from cataloguing. And imposing conceptual order is what the thoughtful reader sitting behind a desk is also trying to do.

SOCIAL RESEARCH AS A SOCIAL ACTIVITY

The imposition of conceptual order onto data is a social activity.[11] As such, it exhibits characteristics common to other social behavior, and the essays of this volume supply several especially good illustrations. First is the point that even the lone investigator is involved in a group activity. Second, the ethics of social inquiry—ethics not only of relationships between researchers and subjects but also among researchers —is of central concern to contemporary social research. And third, the fact that social scientists investigate the behavior of their fellows has not only ethical but also intellectual consequences for their research activity.

Every scientist is aware that his research is in some way the product of others' prior work and the precursor of subsequent work. Indeed, as has been said, of all parts of culture perhaps only science is truly cumulative. And to ensure its cumulative character, its continuity, science has instituted norms by which scientists are to behave: proper acknowledgment of one's intellectual benefactors, full citation of similar work conducted elsewhere, obligation to publicize one's results, and so forth.[12]

Such norms, however, expose primarily the more formal links among segments of a scientific community. There are also informal links which, though less visible, may be equally important in understanding the cumulative nature of science, the teacher–student and colleague–colleague links being obvious examples. Each of the chronicles illustrates this cumulative phenomenon in social science by revealing both source and content of the influences. Bellah, for example, indicates clearly not only how Weber ignited his interest in the religious forces in Japan's industrialization but also how Parsons, via course assignment and personal communication, helped shape his search for variables and a mode of analysis. Likewise, Udy mentions the help, both formal and informal, both substantive and technical, he received from Murdock. In Udy's case also there is the example, in his reading of nineteenth-century European economic historians, of the rediscovery of conceptual distinctions made and then forgotten. Lipset grew up in the home of a typographer; Dalton was trained as a chemist. Blau shared a Marxian concern for the impact of industrial organization on the worker; Fox could share the apprehensions of her friends, Belgian medical researchers studying in this country, who were faced with prospects of returning to the research setting in their homeland. Some *sources* of influence then, be they men, situations, or intellectual traditions, are set forth in these chronicles.

The *content* of influences, however, is also documented. For some researchers, the focus of the study is given, and variables must be sought. Wright and Hyman were supplied with a situation where the client defined the "problem"; they describe the task of locating proper variables. For others, the situation is reversed. Davis' chronicle, also involving a client, describes how, after confronting the client's problem, he was left with many variables but no ready-made focus. His search was for a global idea. Similar is the experience, so ingenuously described by Coleman, of a sophisticated researcher who can come to a point in his analysis where, despite the presence of many variables, he simply does not know the next step. The overarching scheme is still missing.

Different yet is the research situation where both focus and variables may be present; what is sought is a mode of analysis. Blau's ingenuity at discovering indicators for conducting a functional analysis of bureaucracy is an example. He chronicles his search for these analytic procedures and reveals the social contexts in which the search took place.

Four of the chronicles, by coincidence, exemplify an extreme and clear case of informal linkage in the development of an analytic procedure. The case concerns the refinement of the "contextual-properties" or "structural-effects" notion as found in the essays of Lipset, Blau, Davis, and Coleman. The theoretical point that persons' behaviors are shaped by the groups of which they are members (and nonmembers) is, of course, a keystone in social science and has been known for centuries. The *idea,* that is, that group attributes emerge from the coalescence of individual attributes and in turn influence individual behavior is not new. But the development of a *mode of analysis* for researching the idea is a powerful innovation in social research. Especially in survey research, where mechanical sophistication had been great, discovery of a method for converting social psychological data into truly group data has permitted enormous increments in the growth of sociology. The four chronicles reveal a good deal of the development of this innovation and indicate jointly a way in which research is a social activity.

The applied uses of social research in consumer studies had, by the time of World War II, resulted in considerable knowledge about public-opinion polling or survey techniques. Sampling, constructing questionnaires, interviewing, and other aspects of survey research had become quite refined. The relevance of survey methods for traditional sociological theory remained limited, however, because, it was thought, in survey research the investigated unit is the *individual* whereas in theory sociology is interested in *groups*.

But with World War II came the Research Branch and its massive studies by Stouffer and others.[13] Having access to thousands of respondents who at the same time composed various "real" groups, the Research Branch found it possible to collect information from and about individuals, summarize that information for all persons forming various groups, and then characterize the same persons a second time with the summary information of their groups. In this way, statements could be made about groups and about the effect of group membership on individuals. No longer would survey research have only a respondent's *report* on the nature of his groups; now other members of those groups could also be questioned.

It is true that some such technique had been used before,[14] but the logic of its procedure had been little understood until the work of the Research Branch was published and commentary on the work followed.[15]

A few years later, survey researchers were incorporating the idea of "contextual" properties into a variety of investigations. One of the first of these, at Columbia University, was the Lipset, Trow, and Coleman study of the typographers union with its many small "chapels" or print shops. Chapels were sampled, their members interviewed, and the chapels then described by the aggregated answers of their members. In this way, it could be determined that, say, individuals in two different chapels, equally predisposed on the basis of their individual attributes to vote in union elections, might behave in different ways on voting day, depending on the composition of their respective chapels. And this could be true regardless of the person's perception of the make-up of his group.

Earlier, Blau had been investigating bureaucratic units and had discovered radical differences in the behavior of agents of different units, though outwardly their situations were quite similar. Blau's success then and later at The University of Chicago in stating his observation in methodical terms led to his formulation of "structural effects," which, as Davis relates below, got carried via students in Blau's courses to the National Opinion Research Center and Davis' project on Great Books and Small Groups. There the idea was further modified in both article and book.

Meanwhile Coleman, acquainted with the notion of contextual analysis at Columbia, had moved to Chicago and begun research in ten secondary schools. From its inception, this project was designed to capitalize on the experience of its predecessors. Again modifications were forthcoming.

By now, the survey researcher is heir to a number of monographs illustrating the notion of contextual properties or structural effects, and he also has available a series of methodological papers outlining the logic of the idea.[16] It would be difficult to assign priority of discovery in this instance; many contributions occurred almost simultaneously in a period of fifteen years. Probably to Lazarsfeld, with his penchant for formalizing the method of social inquiry and with his influence on a generation of students, should go much of the credit for the dissemination of the idea. But it is the fact of the dissemination, not the assignment of priority, that is the focus here. The four chronicles together make it eminently apparent that each researcher was involved in a cumulative venture and had contact, direct or indirect, with the others. However typical the instance, it illustrates one way in which social research is a social activity.

A second way in which social research can be seen to be a social activity is found in the several chronicles that discuss the ethics of social inquiry. Dalton, especially, devotes considerable space to ethical considerations in participant observation, while Riesman and Watson raise the correlative question of the impact these considerations had on their research team. Geer discusses the proper behavior of the field researcher cast into a situation with few guidelines.

Hired in industrial organization for one purpose but taking the post in order to accomplish another purpose, Dalton is in a position to discuss the rationale behind the "deception" sometimes found in social research. Opinion within professional sociology is by no means uniform on this ethical question,[17] and the public can be assumed to hold an even greater variety of opinion. The essential point here is that, participant observation being a social activity, a series of role relationships is assumed during and after the investigation by the researcher, and the ethics of the situation is discussed in terms of how well these obligations are met. Dalton's comments on the secretary in love, on "obstructive allegiances," and so on make his discussion more than an exhortation in absolutes.

Riesman and Watson confront a variant of this ethical question: what is the effect on researchers when they engage in an activity heretofore defined by them as deceitful? Part of a broader conflict within the total staff over differing modes of research, the ethical issue became an added encumbrance for the team's operation. In their chronicle, which surely must be one of the least deceitful documents ever written, Riesman and Watson agonizingly appraise the impact on the social relations of the staff of being "dishonest."

Blanche Geer, in still another variation on the ethical theme, also sees the problem as one of the quality of social relationships, not as a question answered in advance as "good" or "bad." Indeed, her discussion of this point is not even cast in the language of ethics. Instead, the researcher, having elicited the cooperation of subjects, is then confronted with the obligation to enact his role faithfully. Failure to do so, she implies, is not only less efficacious but deceptive in another way and thus unethical. How one elicits cooperation or establishes rapport is not so crucial a test as continuing a relationship entered into in good faith by both sides. The example of exploratory field worker and hundreds of college "previewers" is a poignant one.

Still a third way in which social research is a social activity is seen in every chronicle: the researcher may stand in a special relationship with his subject. That social science combines the methods of science with the subject matter of the humanities is only grossly the context for this point, however, since natural scientists may also exhibit this attachment to subject, just as social scientists, of course, may not. The impact is not only ethical, as when a social scientist must decide whether deceiving his subject is permitted or a physicist imagines the outcome made possible by his weapons discovery. The impact is also intellectual; it calls for modifications and adjustments in his method, his strategy, his theory. Not only an ethical stance, in other words, but a cognitive stance as well must be taken by the researcher in his relationship with subject.

And this aspect is no doubt intensified in social research. The opportunity to empathize with the investigated objects, if not unique to social science, is certainly greater there.[18] And though empathic identification does not, by itself, constitute empirical knowledge, such a social phenomenon built into social research does have certain consequences for its research strategy, some of which are revealed in these chronicles. The most obvious of these consequences is the heuristic value of empathy; feelings may be identified, and hypotheses about them may then be investigated. Though all the chronicles implicitly contain this element, Dalton's and Geer's are especially sensitive at this point, describing particular conversations, expressions, and contexts that led to eventual insights and hypotheses.

A second intellectual consequence for research strategy of the social relationship between investigator and subject is found in the logic of measurement. Again, this can be true in physical science, but social-science measurement probably involves a larger share of "classificatory" concepts where "states of being" are imputed to persons or collectivities by inference from objective manifestations.[19] Few, indeed, are the standardized, agreed-on measures of such concepts appropriate for multiple situations. Many measures by which subjects are to be characterized have only approximately the same meaning to those subjects. Democracy must be assessed differently when investigated in the context of a labor union, a young people's camp, Belgian politics, or the parlors of private homes where sociable interaction occurs. Social rank means quite different things in industrial bureaucracies, midwestern communities, and adolescent social structures; social rank, therefore, must be measured each time in a way appropriate to the research site. Blau, for example, indicates how bureaucratic record-keeping had multiple meanings, some clearly understood by supervisors and workers but others only vaguely intuited by them. His use of records as measures, therefore, differed from what might have been, had number of client interviews and number of placements been taken to mean only the quanta of activity. Similarly, Davis mentions the task of settling differences among the persons who were party to the planning of that research—differences arising in part from various possible meanings almost any measure of "effectiveness of a Great Books Program" might have.

As these and several other chronicles illustrate especially well, the fact that researcher and subject are socially related contributes both to the creation of the measurement problem and to its solution: the researcher will understand and want to characterize correctly the units he investigates; he will know and understand the limitations intrinsic to many of his measures; and he will, in reporting his findings, supply full enough exposition to allow readers to know and understand the meaning of his measures.[20]

A third intellectual consequence for social research that stems from the researcher's relationship with his subject matter is a variant of the second: the particularities of the subject studied may be as important to the researcher as its generalities. Reminiscent of the older distinction between "ideographic" and "nomothetic"—narrative or theoretic—the third point is simply that social research is frequently both. In this sense, social research proceeds much as research in geography or natural history, where prior theory is used to understand the present instance and the present instance is used to extend theory. One may, of course, argue logically that any general law can be observed only in its instances, but that is to miss the very relevant point of the researcher's *orientation* toward his given instance.

Thus, in her investigation of the institutional context of medical research in Belgium, Fox was both testing theory with a set of observations and applying theory by making sense of the Belgian case. Lipset not only set out to discover structural sources of political democracy in general but also sought to identify the reason for democracy in the International Typographical Union. Bellah was building on the Weberian thesis linking religion and a rational economy; he was also seeking to understand a particular period in the history of Japan. Coleman, launching an investigation of alternative mobility structures, also studied the institutionalized variety and similarity in American secondary schools. Conversely, Wright and Hyman, asked to evaluate the effectiveness of a particular teaching program, studied character education generally and how to evaluate it. One reason for this dual orientation, these chronicles imply, is the investigator's social relationship with his subject, prompting his interest in *it* as well as the abstract qualities it exemplifies.[21]

READING CHRONICLES OF SOCIAL RESEARCH

Several themes to appear in the following pages have been identified in advance. If the set of chronicles achieves its purpose, still more will be suggested to the reader. With many different methods represented, conducted by social scientists from many different "schools," these essays are meant to give body to the research skeleton conveyed in research-reporting and methods texts.

If it may be an error to follow dogmatically the didactics of methods texts with their codification of the logic involved in the "context of justification," it is certainly an error to expect of these essays on the "context of discovery" a set of rules to follow in social research. Everyone, if he thinks about it, knows that in scientific inquiry (1) chance plays a part, (2) hypotheses can be too restrictive, (3) imagination is of great importance, and (4) uncritical reading of previous work can have adverse effects. Everyone also knows, however, that (1) one must not take chances willy-nilly, (2) the hypothesis remains the most important mental technique of the investigator, (3) the thinker must be very critical of his own images and visions as they arise, and (4) even reading what is true so far as it goes can have the same adverse effect.[22]

The response to such aphorisms by persons trying to convey how research really is conducted is to relate anecdotes. Chronicles are series of anecdotes. We would hope that these chronicles can therefore be read for pleasure, telling as they do something of the "personal" lives of social researchers. We would also hope, however, that these essays, read in conjunction with the monographs they chronicle, will, by example if not by precept, serve to enlarge and enrich the reader's understanding of and appreciation for contemporary social research. An exciting endeavor, social research in its chronicling, too, has instructive as well as enjoyable parts.

• NOTES

1. George W. Pierson, *Tocqueville in America*, abridged by D. C. Lunt (Garden City, L.I.: Anchor Books, 1959), p. 85. See also pp. 76, 84, 243-244, 290.

2. Norwood R. Hanson, *Patterns of Discovery* (New York: Cambridge University Press, 1958), p .6.

3. There is even a full discussion of the two travelers' growing tired of observing penitentiaries, the manifest object of their tour and the basis on which the French Government supported them, as they became increasingly intrigued with "democratic institutions." The more cynical will see the modern analogue here, too: social researchers requesting agency funds for an "applied" investigation when their real interests are "pure" and theoretical. Davis, below, includes candid comments on this phenomenon in research.

4. Perhaps the best-known chronicle in modern sociology is William F. Whyte's Appendix in the second edition of *Street Corner Society* (Chicago, Ill.: The University of Chicago Press, 1955). Others include: C. Wright Mills, "On Intellectual Craftsmanship," in *The Sociological Imagination* (New York: Oxford University Press, 1959); Alvin Gouldner, Appendix, in *Patterns of Industrial Bureaucracy* (Glencoe, Ill.: The Free Press, 1954); Edward Shils, "Primordial, Personal, Sacred and Civil Ties," *British Journal of Sociology*, VIII (1957), 130-145. Among others, Paul F. Lazarsfeld has long called for more such commentaries or biographies of research. This writer happily acknowledges his debt to Prof. Lazarsfeld for sensitizing him to this feature of social research.

5. Shils, *op. cit.*, p. 144.

6. Robert K. Merton, Foreword, in Bernard Barber, *Science and the Social Order* (rev. ed.; New York: Collier Books, 1962), p. 19.

7. James B. Conant, *On Understanding Science* (New Haven, Conn.: Yale University Press, 1947), p. 116.

8. W. I. B. Beveridge, *The Art of Scientific Investigation* (3rd ed.; New York: W. W. Norton and Company, 1957), p. 47.

9. Hans Reichenbach, *The Rise of Scientific Philosophy* (Berkeley and Los Angeles: University of California Press, 1951), p. 231.

10. Hanson, *op. cit.*, p. 71.

11. Conant, *op. cit.*, and Beveridge, *op. cit.*, make this a central point.

12. A growing literature is available on this matter. See, for example, Barber, *op. cit.*; Bernard Barber and Walter Hirsch, eds., *The Sociology of Science* (Glencoe, Ill.: The Free Press, 1962), especially parts 4 and 5. Several papers by Robert K. Merton are precisely to this point. See "Priorities in Scientific Discovery," *American Sociological Review*, XXII (1957), 635-659; "Singletons and Multiples in Scientific Discovery," publication of the Bureau of Applied Social Research, Columbia University, No. A-336; "The Ambivalence of Scientists," *Bulletin of The Johns Hopkins Hospital*, CXII (1963), 77-97.

13. S. A. Stouffer *et al.*, eds., *The American Soldier* (2 vols.; Princeton, N.J.: Princeton University Press, 1949).

14. H. C. Selvin, "Durkheim's *Suicide* and Problems of Empirical Research," *American Journal of Sociology*, LXIII (1958), 607-619.

15. P. L. Kendall and P. F. Lazarsfeld, "Problems of Survey Analysis," in R. K. Merton and P. F. Lazarsfeld, eds., *Continuities in Social Research: Studies in the Scope and Method of "The American Soldier"* (Glencoe, Ill.: The Free Press, 1950).

16. For example, Peter Blau, "Structural Effects," *American Sociological Review*, XXV (1960), 178-193; James A. Davis, J. L. Spaeth, and C. Huson, "Analyzing Effects of Group Composition," *American Sociological Review*, XXVI (1961), 215-225; P. F. Lazarsfeld and Herbert Menzel, "On the Relation between Individual and Collective Properties," in A. Etzioni, ed.,

Complex Organizations (New York: Holt, Rinehart and Winston, 1961); H. C. Selvin and W. O. Hagstrom, "The Empirical Classification of Groups," *American Sociological Review*, XXVIII (1963), 399-411.

17. For a position different from one contained in these chronicles, see Edward Shils, "Social Inquiry and Autonomy of the Individual," in Daniel Lerner, ed., *The Human Meaning of the Social Sciences* (New York: Meridian Books, 1959).

18. Ernest Nagel, *The Structure of Science* (New York: Harcourt, 1961), pp. 473-485, is a very clear discussion of the logical status of this "subjective nature of social subject matter."

19. P. F. Lazarsfeld, "Problems in Methodology," in R. K. Merton, L. Broom, and L. S. Cottrell, Jr., eds., *Sociology Today* (New York: Basic Books, 1959), p. 46.

20. *Ibid.*, pp. 47-67.

21. A similar point, with its implications for applied social research, is made in a paper too little attended by sociologists: Max F. Millikan, "Inquiry and Policy: The Relation of Knowledge to Action," in Daniel Lerner, *op. cit.* Note that the point is *not* that human behavior, unlike behavior of things, is too variegated for systematic observation. It is *not* a matter of social researchers' attending to "all the rich kaleidoscopic detail" instead of "counting," as the issue is sometimes, and disdainfully, put. It *is* the point that social research frequently is interested in the particular qualities of the subject matter as much as the general qualities.

22. Beveridge, *op. cit*; R. K. Merton, "The Ambivalence of Scientists," *op. cit.*, pp. 78-79.

2 The Research Process

in the Study of

*The Dynamics of Bureaucracy**

PETER M. BLAU

When I entered graduate school at Columbia University shortly after having been discharged from the United States Army at the end of 1945, I clipped two pictures from magazines and put them on the wall of my room: one of Karl Marx and one of Sigmund Freud. This was not a bad choice of intellectual heroes, though hardly an original one. Many social scientists in the thirties, on this side of the Atlantic as well as on the other where I had grown up, were attracted to Marx's theory or to Freud's, and quite a few were influenced by both and attempted to reconcile and integrate them. My interest in these ideas had been kindled by Frederick W. Henssler, my sociology instructor in a small midwestern college, whose stimulating discussions also led to my decision to become a sociologist. But while I considered myself a sociologist when I entered Columbia, I had not yet formed a clear conception of the distinctive nature of sociological inquiry, and the pictures on my wall show that I had no sociological hero.

My orientation changed considerably while I was in graduate school. One important source of influence was the classical tradition in sociology: the great theories of social structure of Max Weber, Émile Durkheim, and their contemporaries. The most direct and greatest influence on my sociological thinking, however, was exerted by Robert K. Merton.

* Peter M. Blau, *The Dynamics of Bureaucracy* (Chicago, Ill.: The University of Chicago Press, 1955). A revised edition was published in 1963 which contains as its last two chapters a somewhat altered form of this essay.

16

An interest in theories of social structure may take many forms, and mine was largely shaped by Merton's approach. Compare, for example, the way in which Weber's and Durkheim's theories are utilized and extended in Talcott Parsons' *The Structure of Social Action,* on the one hand, and in Merton's essays "Bureaucratic Structure and Personality" and "Social Structure and Anomie," [1] on the other. (I deliberately select monographs that were already published when I was in graduate school.) Parsons is concerned with the interrelations between broad institutional complexes and, specifically, between abstract analytical aspects of institutional systems. Merton, in contrast, focuses attention on the influences exerted by social structures on patterns of conduct; the core concept is that of structural constraints. To be sure, Merton also analyzes how social regularities in behavior become institutionalized and modify the social structure. But in either case, with Merton the emphasis is on the relations between some elements of social structure and an observable pattern of conduct rather than directly on the relations between various abstract elements of social structure, as it is in Parsons. Later writings reveal parallel differences. Thus, Parsons' abstract conception of functionalism contrasts sharply with Merton's functional paradigm with its down-to-earth concepts and emphasis on observed consequences. I was Merton's research assistant while he was framing his essay on functionalism, and the opportunity I had of working with him while he developed this new theoretical framework undoubtedly contributed to the great impression it made on me.

My interest in the study of bureaucracy emerged naturally as the result of these influences on my orientation. If one's concern is with the compelling force the social structure exerts on human behavior, a logical place to study it is in a bureaucracy, where this force is so apparent and pronounced and yet often produces quite unexpected results, as Merton's analysis of the subject suggests. Besides, the implicit functional analysis in Weber's theory of bureaucracy raised the intriguing possibility that new insights could be gained from explicitly applying Merton's revised functional framework to the subject. Finally, to a person who continued to have an interest in social reform, the threat posed by bureaucracy to democratic institutions supplied an additional incentive for studying it.

The atmosphere at Columbia in those days, when Merton was becoming increasingly involved in empirical research and had already started his fruitful collaboration with Paul F. Lazarsfeld, tended to destroy the preconception, which most of us students initially shared, that a social theorist is not concerned with systematic empirical investigations. My new interest in the integration of theory and research motivated me to take all the courses and seminars on research method Lazarsfeld offered and to decide to do an empirical investigation. Since the emphasis at Columbia was on sampling surveys, I even considered briefly an interviewing survey of bureaucratic officials, but I decided against it because this procedure would not enable me to deal with the problems of bureaucratic structure that interested me most.

An alternative research procedure began to crystallize in my mind as I reflected on what I had learned in Conrad M. Arensberg's course on the sociology of industry. There we had analyzed in detail studies of industrial work groups. Why not do a similar study of groups of officials in a bureaucracy? Arensberg explained to us the quantitative method of recording social interaction that he and E. D. Chapple had worked out,[2] and, while I argued with him against the social behaviorism implicit in the approach, I became convinced in the course of these discussions of the utility of the method, quite independent of the behavioristic approach from which it derived. I had found a quantitative method for studying at least one aspect of social structure—the patterns of social association in a group. One fact that helped to silence my initial doubts about the positivistic implications of this method was that it had been used in one of the most imaginative studies of group structure I had read: William F. Whyte's *Street Corner Society*.[3]

My decision was now apparent: I planned a quantitative study of work groups in bureaucracy. This was my way of integrating social theory and social research. It would be an empirical study of work groups in the manner of the research on industrial work groups, but in contrast to these investigations it would be guided by Weber's theory of bureaucracy. I would attempt to collect quantitative data on bureaucratic work groups and to apply Merton's functional framework in the analysis of these data.

PROBLEMS OF FIELD WORK IN A BUREAUCRACY

How does one conduct a theoretically oriented case study that uses quantitative as well as qualitative information? Not enough systematic knowledge on the distinctive problems of the case study has been accumulated to answer this question in general terms. I want to describe here how I carried out a case study of officials in two bureaucracies: the decisions I made in designing the study, in collecting the data in the field, and in analyzing the findings. Irrational elements often entered into these decisions, and some were obvious blunders, but it is precisely from these that future researchers may learn most.

Of course, I shall not deal with all decisions I made or even all I recorded. Although it might be of interest, I have not endeavored to make a quantitative analysis of my field notes and memos, which occupy nearly three file cabinets. This is merely a case study of a case study; a few cases are selected to illustrate various types of decisions. In this first part, we shall examine the design of the study and various problems encountered during field observation in a bureaucracy. In the second part, we shall trace the history of some of the ideas and problems that are analyzed in the book, noting how these concepts originated, what clues furnished insights that led to their modification or refinement, and what the limitations of the analysis are.

Initial Design

During my last semester in graduate school in the spring of 1948, I wrote many lengthy memos outlining a multitude of problems about bureaucratic work groups that could be studied. As one reads these scores of pages, two basic themes stand out: first, how the formal organization influences the informal organization of work groups and how the latter, in turn, influences the performance of duties; and second, a functional analysis of the informal organization, exploring the distinctive functions of the informal group structure in a bureaucracy. My ideas ranged from broad problems to be explored to quite specific hypotheses. For example, I stated that the various forms taken by the displacement of goals should be ascertained, as should the specific processes through which displacement comes about; I suggested that a basic function of the informal organization in a bureaucracy is to mitigate bureaucratic impersonality; more specifically, I hypothesized that informal relations among colleagues serve the function of relieving the tensions that arise in the contacts of officials with the public.

Some of the hypotheses advanced at this early stage were later abandoned; others were supported by empirical observations, but even these were often modified and refined in the course of research. This process of selection and modification indicates that no claim can be made that hypotheses have been subjected to a rigorous test in such a case study. But the idea that research methods can be classified neatly into hypothesis-testing and insight-supplying ones is grossly misleading, since these are polar types that appear in actual investigations in various admixtures. The double aim is always to develop and refine theoretical insights which explain reality—for instance, group structures in bureaucracy—and to discriminate between the correct and the false explanatory principles.

It is all too easy to obtain some impressionistic evidence for our broad theoretical speculations. Such evidence, therefore, helps us little in discriminating between diverse or even conflicting theoretical principles. My endeavor to stipulate hypotheses, some in advance of the empirical research and some in the course of it, and to collect at least some quantitative data served the purpose of furnishing a screening device for insights. Those ideas that survived this screening test, while still only hypotheses (since a case study cannot validate general principles of bureaucratic structure), were more apt to be correct than the original speculations. Moreover, the quantitative analysis of specific relations between variables often produced unexpected findings that challenged the imagination and led to refinements of theoretical conceptions.

The significance of a quantitative case study, then, is (1) that it stimulates the kind of theoretical insights that can be derived only from quantitative analysis as well as the kind that results from close observation of an empirical situation and (2) that it provides more severe checks on these insights than an impressionistic study and thus somewhat increases the probable validity of the conclusions. It is only in retrospect, however, that I have arrived at this general formulation. While I was preparing my study, I was concerned simply in exploiting pertinent theories and translating their insights into specific research problems. This helped me in deciding what kind of bureaucratic group would be most suitable for the study. It was also necessary to prepare a proposal for the Social Science Research Council, which later awarded me a predoctoral fellowship. Last, though not least, the extensive outline of problems and hypotheses became the basis for the development of the research procedures for my case study.

On the suggestion of Merton, from whose criticism and advice I benefited greatly at this stage, I modified my plan and decided to compare groups in two bureaucracies—a public and a private one. I re-formulated my hypotheses in terms of this comparison between government agency and large private firm and spent much of the summer trying to obtain permission for study from one organization of each type. The immediate cause for abandoning this plan was that I was unable to obtain permission for doing this research from any suitable private firm, but by the time I had to make this decision there were other reasons for it, too. My experience with the first agency made me realize that, since I had so much more to learn about social patterns in government agencies, a comparison of groups in two government agencies which differ in some ways but are not too dissimilar might be most fruitful.

In the weeks before I started observation in the first government agency, I designed a detailed schedule of research procedures. This was completed after I had entered the organization and become acquainted with the actual setup. I had earlier decided to use three basic methods—direct observation, interviewing, and analysis of official records—and to employ various quantitative as well as qualitative research techniques under each. Thus, I planned not only to observe whatever I could notice about the relations among colleagues in the office but also to obtain a systematic record of all their social interaction for a specified period; not only to interview selected members of the organization on specific issues but also to administer a semistructured interview to all members of the work groups intensively studied; not only to read the procedure manual but also to abstract some quantitative information from it.

I put each research problem I had outlined on 5 × 8 slips, making several carbons, which were cross-classified by the various research procedures to be used to investigate this problem. For example, the slip that dealt with contacts between different sections in the organization was classified under written regulations, interview with management representative, active observation (by which I meant that I would ask questions to clarify what I observed), quantitative observation, self-recording, and interview with officials. The resulting file indicated the different substantive problems that might be studied with any given research procedure.

This schedule served as the basic guide for my study, but it was not intended to be rigid or fixed. As new ideas or new investigation techniques occurred to me in the course of observation, I added them to the file. And when it became apparent that a certain problem could not be studied with a given procedure, or not at all in this situation, I modified the schedule accordingly. I repeatedly went over the file in the early phase of observation, revising it and abstracting lists of problems under the various research procedures, which I carried with me into the field. There I explored how well the various procedures would actually lend themselves to studying the different problems. The revised lists resulting from this exploration were regularly consulted, and they directed my research activities.

I found such a schedule of procedures, somewhat flexible yet exerting some control over the research process, very useful. To be sure, it is possible that I missed some exciting leads because I was too concerned with following a research schedule. It may also be that I failed to collect systematic data I could have obtained because I was too easily intrigued by new possibilities and diverted from following the predesigned procedure. But, whatever the wisdom of my specific decisions, I think that the general principle of using such a research schedule in a case study is sound.

In a bureaucratic field situation, the needs of the organization and not those of the observer determine what occurs and thus what can be studied at any given time. While some events occur regularly, enabling the researcher to determine when to study them, others do not. A change in regulations is introduced, a conference is called, two officials have an argument, and the observer must be ready to turn his attention to these incidents, which may bear on some of his central research problems. The researcher who compulsively insists on following his predesigned plan will miss these rare opportunities. Conversely, the one who is seduced by every new lead will find that he has failed to collect information on the theoretical problems that prompted his research. A research schedule that is recurrently revised but quite closely followed guards against these dangers. During the last weeks of observation, I went over my lists carefully, and, while my design was so ambitious that I could not possibly carry out all my plans, I was able to select those items for completion that I considered most important. At the very least, my research schedule prevented me from inadvertently forgetting to obtain some information that was essential in terms of my theoretical framework.

Chronology

While there is a rough chronological order in this paper, it is often violated to more fully present the development of an idea. In the second part, moreover, each section presents the development of some conceptual problem from beginning to end. Since the sequence of our discussion does not coincide with the chronology of events in the actual research, it might be helpful to outline this chronology briefly.

The federal agency of law enforcement, although it is considered in the second part of the book, was studied first. The observation period started on August 31, 1948, and lasted for about three months. After an initial period of observation and an examination of official records, a section of eighteen officials (which I later called Department Y) was selected for intensive study, but the investigation also encompassed several special groups, such as reviewers and stenographers. I observed officials in the office, accompanied them on their field visits, had lunch with them, even went to an American Legion party, held scores of informal interviews and hundreds of briefer conversations, and abstracted various data from official records. I took notes of my observations and interviews either on the spot or as soon afterward as possible. In the evening, I typed all my field notes in duplicate. Since these were written hastily and I relied much on memory when I typed them up, I tried hard to finish typing each day's notes on the same day, but I was not always successful, even though I often worked a sixteen-hour day. Week ends were devoted to catching up on those field notes, doing some cursory analysis of the data to get some feel of what I had found, and going over the procedure schedule to plan the work for the coming week. It was a full three months.

At the beginning of December, I took time off to conduct a preliminary analysis of my data, after which I returned to the agency briefly to complete my observation, particularly to seek to answer some of the questions the analysis had raised. Even though I had stopped going regularly to the office for observation, by Christmas I took advantage of the opportunity of attending the office Christmas party.

I worked further on the preliminary analysis, which served as the basis for preparing an interview schedule. In January and February, I conducted interviews with the members of Department Y, and students from a methodology course at Columbia University conducted additional interviews with clerical personnel for me.

I started the study of the state employment agency on March 2, 1949, and followed the same general schedule. The period of observation was about three months. In May, I interrupted the process of data collection for a week to begin the preliminary analysis and then returned to the agency to complete my observations. In June, I worked further on the preliminary analysis and prepared interview guides not only for the members of what I referred to as Department X, which had been intensively studied, but also for the members of other departments which I planned to study for comparative purposes. These interviews were conducted during July and August.

In the fall of 1949, I started to teach at Wayne State University. My twelve-hour teaching load occupied me fully during the first semester, and while I was able to devote some time to the study subsequently, most of the work on it had to be done during vacations. Hence, it took me an entire year, from the beginning of 1950 to the beginning of 1951, to complete the basic analysis, and it was not until the fall of 1951 that the first draft of the manuscript was completed. On the basis of Merton's detailed and helpful criticisms of this manuscript, I made extensive revisions before I submitted the final draft of my dissertation to Columbia University in March 1952. Afterward, I continued to make further revisions, and another three years elapsed before *The Dynamics of Bureaucracy* was finally published in the spring of 1955—seven years after I had started work on it.

Entry and Orientation

An interviewing survey of a sample of New Yorkers can be conducted without official permission from anybody, but a field study of a bureaucracy cannot be executed without the explicit permission of management. This poses special problems. While not all respondents selected for a sampling survey agree to be interviewed and those who do are typically not representative of those who do not, there are known procedures for correcting this self-selection bias. But the problem of self-selection is far more extreme in research on organizations, and it is not easily possible to correct for the bias it introduces. Suppose someone wants to study bureaucratic rigidities and fear of innovation. The very fact that the management gives him permission to conduct his investigation in its organization indicates that it is not resistant to trying something new. It may well be no accident that all old-established bureaucracies I approached refused permission for the study and that both organizations that opened the way were relatively young ones, founded during the New Deal. Perhaps self-selection makes it inevitable that the organizations we study are those in which bureaucratic rigidities are least pronounced.

The members of the organization know, of course, that management has given permission to conduct the study, and this creates another problem. The observer cannot escape an initial identification with management, since the assumption is that management must have a direct interest in the study. In the federal agency, I was suspected of being a representative of the Hoover Commission, which at that time carried out investigations in various branches of the federal government. These suspicions compound the problem posed by the sheer presence of the observer, since people become self-conscious if somebody sits in a corner and watches them, even if he tries to do so unobtrusively. Ultimately, I overcame these difficulties, but not until blunders I committed owing to my inexperience had first intensified them.

On my first day, the district commissioner introduced me to his top assistants, told me that one of them would serve as my guide during my period of orientation, and assigned an office to me where I could read in private the extensive rules and regulations that governed the agency's operations. During the first week, the official who acted as my guide explained the operating procedure to me in lengthy interviews and answered the questions I had after reading the books of rules and regulations. He also introduced me to the supervisors of the various departments, and I had occasion to interview them. During the second week, I further explored the operations of various departments, discovered the quantitative operational records that were kept, abstracted information from them, and revised my research design by adapting it to the concrete circumstances in this agency. I still spent much time in the private office, and only supervisors and some selected officials had as yet met me; most agents had not, although they had undoubtedly noticed my presence.

This thorough initial orientation was a good preparation for the observation period. The basic knowledge I acquired about the organization and its operations enabled me to translate my general research procedures into specific operational terms appropriate for this field situation, and it helped me to select a suitable department for intensive study. But spending two weeks on becoming oriented to the agency's operating procedures also had serious disadvantages.

At the very start of the intensive observation of Department Y, I was introduced to all its members by a senior official at a meeting and given an opportunity to explain the study briefly. I realized that an observer must explain at the outset who he is and what he wants to do. But I failed to realize that what I defined as the beginning of the actual observation was not the beginning for these agents. I had been seen around for two weeks, and my failure to explicitly clarify my identity earlier gave rumors about me that much time to circulate. The private office and my preoccupation with becoming familiar with a complex bureaucratic structure had blinded me to the fact that I was already being observed by these agents, even though I had not yet started observing them.

I learned from this experience. Six months later, in the study of the state employment agency, I explained the study in a brief talk to the members of Department X the day I set foot in their local office. (It was possible to arrange the preceding one-week orientation in a different location.) It seemed to be easier to establish rapport here than in the federal agency, but whether this was due to the lecture or the experience I had acquired as an observer is a moot point. In any case, the lecture was far less effective than it could have been, as I found out when I gave another lecture at the end of the observation period in the employment agency. The object of this farewell lecture was to illustrate what such a study of social relations seeks to discover; I presented Whyte's analysis of the relation between informal status in a gang and bowling score, plus some of my own preliminary findings on consultations among officials and their discussions about clients. This lecture was much more successful than the first one. Officials were interested, asked numerous questions, and made some revealing comments afterward. One agency interviewer said that there should have been a meeting in which I explained my study at the beginning, not just at the end; apparently she had entirely forgotten that there had been such a meeting. Another interviewer, who had not forgotten the earlier talk, mentioned how much more interesting the second one was and how much more relaxed I appeared when giving it.

In the first lecture, I described the objective of the study in general terms, covering such points as the importance of human relations and the need for firsthand knowledge of government agencies. In the second lecture, I illustrated the study's objective with concrete findings. It is evident from the reactions that the second topic would have been a better way to introduce the study than the first. There are a number of reasons: a discussion of actual findings is much more interesting than a mere description of formal objectives, creating some interest at the outset in what the observer has to say and wants to do. Besides, the concrete examples of sociological analysis, judiciously chosen, demonstrate more effectively than hollow-sounding explicit disavowals that the observer is really not an efficiency expert who wants to check up on the work of officials. The evidence the observer furnishes of what he and other social scientists can do, finally, not only affirms his professional identity but also helps to command respect for his research and to motivate respondents to cooperate with it.

THE ROLE OF THE OBSERVER

After I had been introduced to the members of Department Y in the federal agency, I spent most of my time in the room where they and the members of another department were located. I took every opportunity to become acquainted with these agents, asking them about their work, going to lunch with little groups, slowly beginning to establish some rapport. But I soon became impatient with the slow progress I made and decided that, since I was sitting in this room and watching what was going on anyway, I could use my time more economically by making the quantitative record I had planned of the social interaction among agents. I started this record of all social contacts in the department within a week from the day I had been introduced to its members. This was a serious mistake.

I had just begun to overcome the suspicion and resistance aroused by my entry, and now I employed a technique of observation that the agents found very objectionable and that increased their resistance to the study again. Of course, I explained that I simply wanted to get a systematic record of the social contacts among officials, but this did not overcome the objections of the agents. Even those who apparently believed my explanations considered such a record ridiculous and emphasized that I could not gain an understanding of the agent's job by sitting in the office but must go into the field where the most important work was being done. Others suspected that I was really trying to check on how much time they wasted, as exemplified by the mocking comment one whispered to me when he left the room: "I'm going to the washroom; will be back in two minutes." The continual observation to which keeping such a record subjects respondents makes them self-conscious and is irritating. Evidently I should have waited until my rapport was much better before using this technique (as I did, of course, in the second study). Why did I make the blunder of using it prematurely?

I think the answer is not simply that lack of experience prevented me from knowing how much resistance this method of observing inter-action would create in a group not yet fully reconciled to my presence. Common sense should have told me so, had not irrational factors pre-vented me from realizing it. I was a lone observer in the midst of an integrated group of officials who were initially suspicious of and even somewhat hostile to me and my research. While they were part of the bureaucratic structure, my position was not anchored in it. My anxiety engendered by this insecure position was undoubtedly intensified by the pressure I felt to progress with my observations, since I was not sure whether I could achieve my research aims in the limited time available. It seems (and I use the tentative wording advisedly because I am now reconstructing mental processes of which I was then not fully aware) that I tried to cope with this anxiety by imposing a rigid struc-ture on my research activities. This emotional reaction may have prompted my decision to turn so early from more exploratory observa-tions to the precisely circumscribed and fairly routine task of recording interaction frequencies.

The feelings of insecurity that the bureaucratic field situation tends to evoke in the observer, particularly the inexperienced one, are gener-ally a major source of blunders. This is the fundamental pragmatic rea-son, quite aside from considerations of professional ethics, why the ob-server, in my opinion, should not resort to concealment and deception. It is difficult to simulate a role successfully over long periods of time, and if concern over detection adds to the observer's other worries he is not likely to be effective in discharging his research responsibilities. I explained quite openly what I was doing and the aims of my study to any respondent who was interested, the major exception being that I never called it a study of "bureaucracy." (Refraining from deception does not imply, of course, revealing one's hypotheses to respondents, since proper research procedure may require these to remain con-cealed.)

A few times I did try to conceal something, against my better judgment, and this typically turned out to be a mistake. For example, the negative reaction of agents to the recording of social interaction made me worry about how senior officials would react to my use of this technique. Once, when a managerial official passed my desk, I inadvertently placed my hand over the recording sheet; and another time, when the district commissioner stopped by my desk and asked me about the research, my concern over his reaction, which proved to be quite unjustified, led me to give a vague and confused answer. Here were two silly blunders that resulted from unnecessary attempts to conceal my research activities.

I was at first suspected of masking my true identity and pretending to be an outside observer while really being a representative of some government commission. My trump card in establishing rapport, therefore, was actually being what I pretended to be, and chances were good that this would become apparent in continuing social interaction. I could not have played this trump if I had in fact practiced deception, for doing so entailed permitting my natural behavior in social intercourse to reveal the kind of person I am. There is little doubt that my success in overcoming the strong initial resistance to the study by some union members stemmed from the fact that they perceived my genuine sympathy with their viewpoint, even though I never explicitly expressed my political opinions. As our informal discussions revealed my familiarity with university life and the sociological literature, my claim of being a social researcher was validated.

But the most convincing evidence that I really was not part of the government service was supplied by me inadvertently, often by mistakes I made. When agents, apparently not believing that I was an outsider, came to me for advice on a problem in their cases, my genuine ignorance of the complex official regulations became quite apparent. When I talked too freely to interviewers in the state agency during working hours, having become used to this practice in the federal agency, a supervisor several times told me, politely but firmly, not to interfere with their work. I had made a mistake in not being more careful, but these incidents were to my advantage, since they demonstrated to the interviewers that I was only a tolerated outsider and really not part of the management hierarchy.

As one would expect, I found some officials more easily approachable than others, more interested in the research, and more willing to furnish information. A few seemed quite eager to talk to me and volunteered sensitive information on topics that others were most reluctant to discuss at all; for example, they freely criticized agency procedure and even their colleagues. It is my impression that the best informants in the early weeks tended to be officials who occupied marginal positions in the work group or the organization. Being not fully integrated among colleagues or somewhat alienated from the bureaucratic system may have made these officials more critical of their social environment, less restrained by feelings of loyalty from sharing their criticism with an outsider, and more interested in the approval of the observer than were those who received much social support and approval within the organization.

The marginal position of the observer in the bureaucratic field situation complements the marginal position of these informants, and this entails a danger. The observer may be tempted to rely too much on those officials who make themselves easily accessible to him. If he yields to this temptation, he will obtain a distorted picture of the organization and the group structure. Moreover, if he becomes identified with deviant individuals or cliques, his ability to establish rapport with the majority and his effectiveness as an impartial observer will be impeded.

It is not possible, however, to avoid entirely the self-selection of informants. The observer is no neutral machine that selects respondents purely at random and devotes exactly the same amount of time to each. He can hardly be expected to reject the overtures some officials make to him, not only because his insecure position creates a need for social acceptance but also because his research responsibilities demand that he take advantage of these opportunities for obtaining information. Somewhat marginal officials proved to be an invaluable source of new insights, particularly about dysfunctions of various institutions and practices. Their incomplete integration in the existing social structure made them perceptive observers of it and its shortcomings, and their concern with commanding the observer's respect gave them incentives to share their most interesting observations. The solution to the problem of self-selection of informants is not to ignore this important source of information but to supplement it with quantitative data based on responses from the entire group or a representative sample.

Interest in earning the observer's respect may well be an informant's major motivating force for supplying information and explanations. Respondents make a contribution to the research in exchange for the respect they win by doing so, provided they care about being respected by the observer. The most competent officials can command respect by demonstrating their superior knowledge and skills, but the less competent ones cannot, and this leads some of them to seek to earn the observer's respect by acting as secondhand participant-observers and sharing their inside knowledge and insights with him.

In addition to this difference between respondents variously located in the social structure, there is a difference in the time when they tend to be good informants. During the later phases of the field work, my best informants were no longer largely officials who occupied marginal positions but included some of the most competent and highly respected ones. I think what happened was that once I became accepted as a social scientist my prestige rose, increasing the value of my approval and respect for my respondents. As long as my respect was not worth much, only the marginal officials who commanded little respect in the organization were interested in earning it; but later, when my respect came to be worth more, other officials, too, became interested in earning it.

Finally, a few remarks are in order about the interviews administered subsequent to the observation period. On my last day in the office, many officials whom I asked for interview appointments assumed that the interview would take place during office hours. When informed that the interview would be during off-hours, some were reluctant to devote their own time to it, but group pressures soon came to operate in my favor. (I had this experience in both agencies.) As these potential refusals, who had said they wanted to think it over and that I should call them back, noticed that most of their colleagues made interview appointments with me, a few approached me and said they had decided they could make an appointment now, and most of the others readily made one later. Only two of the thirty-eight respondents in the two departments intensively studied were not interviewed. (Even these two never overtly refused but postponed appointments so often that I finally gave up.)

The question that evoked most resistance was the one on party preference in the last election, notwithstanding my repeated assurance that answers were anonymous and confidential. That this was not due simply to distrust of me is indicated by the fact that one agent said he would be glad to tell me if I would only promise not to use his answer in the research. (I refused to make such a promise.) Since politics is a sensitive area in civil service, agents did not want my report to describe the political preferences of civil servants (particularly in 1948, when Wallace was a presidential candidate), and this, rather than fear that I would identify individual responses, probably accounted for their reluctance to express their voting preferences. The anonymity of our research reports protects only individuals from exposure and not the groups they are identified with, for our monographs do reveal the characteristics of various groups—civil servants, Negroes, Jews, and others. In a situation where the group with which responses will be associated in the final report is apparent, persons identified with this group will not be moved by our assurance of anonymity to reveal characteristics they do not want attributed to their group.

THE PROCESS OF CONCEPTUAL REFINEMENT

In this part, I want to trace the development of some of the ideas about bureaucratic work groups that are analyzed in *The Dynamics of Bureaucracy*. One possible procedure for doing so would be to start with all ideas and hypotheses I derived in advance from the theoretical literature and examine which ones were discarded and which ones found their way into the book. Then, one could investigate which other ideas originated at later stages in the research, during the field observation or interviewing, or still later in the course of the analysis. But this procedure is actually not feasible. Before I went into the field, I wrote page after page of research problems and hypotheses, many of which were necessarily discarded because I did not have an opportunity to investigate them, as others appeared more promising or because no meaningful pattern could be discerned. For example, I analyzed at length the extensive statistical records in the federal agency to derive a measure of performance quality, but I could not construct a valid one and therefore did not even present this analysis. It would be tedious and hardly worthwhile to describe all the ideas I later discarded.

An alternative procedure, which I shall adopt, is to select a few of the main conceptual problems analyzed in the book and follow their history from the first spark of the idea through the modifications and refinements that occurred in the course of the research to the final analysis and interpretation of the data. For most of these problems, as we shall see, we can find an early trace, be it ever so faint, but the prefield-work conception typically underwent fundamental changes as the result of the research experience. The field situation is ripe with serendipity: new insights are gained, and they can often be corroborated with empirical evidence; yet these insights can typically be traced back to earlier theoretical conceptions and thus serve to refine them.

Consultation

One of the fundamental general principles that governed my study from its inception was that informal relations influence the performance of duties. As I explored the implications of this proposition in the summer of 1948, I specified the hypothesis that officials who have frequent informal contacts (for example, who regularly lunch together) tend to ask one another for information they need rather than some other official whom they might be officially expected to ask. Here, I seem to have predicted the existence of the consultation pattern I later observed in the federal agency, but what I predicted was actually only a small element of what I was to find. I had in mind simple requests for information and did not anticipate at all the extensive practice that prevailed among agents of giving advice on complex problems. Nor did I realize the important implications of the consultation pattern for the group structure.

During my period of orientation in the federal agency, I asked a supervisor whether agents sometimes cooperated with each other. I did not yet know the operating procedures well and wanted to find out whether their duties sometimes required agents to work together on a case. The supervisor's answer, however, was not in terms of what I had had in mind. He said: "They are not permitted to consult other agents. If they have a problem, they must take it up with me." Unexpectedly, I had obtained my first clue to the practice of consulting colleagues.

At the very beginning of the intensive observation in Department Y, I could not fail to notice that agents frequently discussed problems of their cases with other agents and asked their advice. I was impressed right away by the significance of this pattern. Here was an officially prohibited practice that clearly had important implications for official operations, providing the kind of link between informal work group and bureaucratic operations in which I was primarily interested. Besides, this practice offered an opportunity for analyzing an aspect of the content of the informal relations among officials which would supplement the analysis I had already planned of the quantitative form of their social interaction. Exactly a week after I had entered Department Y, I started a tally of all the consultations I could observe, recording on a matrix which agent asked which colleague for advice. Two weeks later, when I went over my procedure schedule, I noted that I should ask in the interview sociometric and other questions about consultation as well as continue the record of consultations in the office.

A section on consultation was included in the preliminary analysis I did at the beginning of December near the end of the field observation. In brief outline, I pointed out that agents who needed advice on difficult problems were put under cross pressure by the requirement to consult only the supervisor and noted their reluctance to reveal their difficulties to him lest it affect his evaluation of them. The unofficial practice of consulting colleagues that resulted from this cross pressure served to cement the informal relations in the work group. I suggested also that the social esteem for the consultant strengthened his position among colleagues and that one dysfunction of the pattern might be that it weakens the esteem for and position of the supervisor. This discussion of consultation was the central topic of my first report to the Social Science Research Council, which I prepared later that month.

The skeleton of the final analysis of the consultation pattern can already be discerned in this early exploration of it. Numerous specific elements were still to be added, of course. For example, there yet were no data on the actual network of consultation nor any discussion of the tendency to consult not only experts but often colleagues whose competence is not superior to one's own. Most important, however, was the fact that the two crucial insights for an understanding of this pattern were still missing: the principle that the informal institution of unofficial consultation improves performance even when no advice is obtained and the conception of consultation as an exchange process.

The interviews furnished clues for these insights. Several agents explained that they often discussed their cases with colleagues not in order to ask advice but because they wanted to share an interesting experience or have an opportunity to think out loud. At first, I interpreted these statements as defensive reactions through which agents wanted to convince me that they really did not need advice as often as I had seen them discuss problems with colleagues. The fact that the most expert agent in the department made such a statement, however, was not compatible with this interpretation. As I tried to think of a better one, the expression "thinking out loud" gave me the idea of what I called consultation in disguise. Analyzing a problem in the presence of a fellow expert can be considered consultation in disguise, since the nonverbal communications from listener to speaker indicate to the latter whether he is on the right track and thus serve the same function as explicit requests for confirmation of one's judgment. Talking out loud reduces the anxiety engendered by difficult problems and thereby improves the ability to solve them.

This interpretation of a special case of consultation suggested a principle for explaining consultations in general. The experience of being able to obtain advice from colleagues tends to relieve anxiety over decision-making, and the experience of being consulted by colleagues tends to increase self-confidence. Through these two processes, the pattern of consultation improves the ability of agents to make correct decisions on their own without consulting anyone. (I would have liked to obtain empirical evidence to confirm this conclusion, but only an experiment or a panel study could furnish the data necessary for this purpose. Although the data do show, for instance, that the agents most often consulted were the most expert decision-makers, this was, of course, largely due to the fact that experts were particularly attractive consultants. Whether it was also true that being consulted often improved decision-making ability, as hypothesized, could be ascertained only by observing changes in competence through time.)

The interview responses indicated that agents enjoyed being consulted, and the explanation one agent volunteered, "It's flattering, I suppose," gave me the idea of conceptualizing consultations as an exchange of value. By asking a colleague for help with solving problems, an agent implicitly paid his respect to the other's superior competence in exchange for the advice he received. When I first had this crucial insight in the winter of 1949, it was crudely formulated and embedded in a discussion that required much subsequent refinement; it later became the cornerstone of the analysis of consultation.

Two important implications of this principle were perceived only later in the course of the intensive analysis of my data. Taking a hint from Whyte's discussion of the relationship between incurring obligations and the status hierarchy in *Street Corner Society*, I came to realize that the process of consultation may be the basic mechanism through which informal status became differentiated among agents. Some agents acquired superior status in the group in exchange for helping others with their work, just as the gang leader's status entailed doing more favors for the other members than they did for him. Expert knowledge was necessary, but not sufficient for achieving high status among colleagues. Those agents who possessed expert knowledge but did not freely give advice to others did not achieve high standing in the group.

Another implication of the concept of exchange was that the "marginal principle" might be applied to the analysis of social processes. As experts were increasingly consulted, the value (utility) of the respect implicit in being asked once more for advice declined, and the value of the cost in time of giving another piece of advice increased. This could explain why popular consultants, although they enjoyed giving advice, did not like it when others consulted them very frequently. It also explained why agents often consulted partners on their own level of competence rather than superior experts; by doing so with their less serious problems, they did not exhaust the supply of advice that superior experts were willing to give them.

This is as far as I went in the book. In closing, let me point out some limitations of this analysis—and how I could have gone further. My application of the marginal principle was very primitive. As a matter of fact, I simply used the common-sense notion that the more you have of something desirable, the less eager you will be to get still more of it. It was George C. Homans who, first in an article and later in his book *Social Behavior*,[4] explicitly called attention to the significance of principles of economic theory for the analysis of the group I studied and social behavior in general. The idea of using suitable aspects of economic theory, properly adapted, in the analysis of the exchange processes that occur in social life opens up a new and promising approach. The explicit recognition of this principle would have helped to advance the analysis of consultation further. The tendency to consult in mutual partnerships instead of always asking experts for advice, for example, could have been explained more adequately by applying the marginal principle not only to the consultant but also to the agent who consults. As an agent often pays respect for advice, the cost in utility of once more subordinating himself by asking for the expert's help increases, making him willing to settle for inferior advice if he can get it at a lesser cost. By consulting a partner who recurrently consults him, such an agent pays for advice not with respect, which he can ill afford, but with consultation time devoted to the partner's problems, which he enjoys since he is rarely consulted.

I might also have better conceptualized the analysis of status differentiation. Seduced by the phrase "paying respect," I did not fully exploit Whyte's insight that obligations are at the root of informally generated status differences. If agents regularly consult an expert colleague, they become dependent on and obligated to him, and their obligations and dependence constrain them to defer to his wishes. It is this deference, not merely respect, that the others pay the consultant for his expert advice, and that is the basis of his superior status in the group. I do not really know why I did not carry the analysis of consultation that far, since these are the very concepts I used in the analysis of supervisory authority.

Supervisory Authority

The initial design of my study did not particularly focus on supervision. My interest in the subject was stimulated by a change in supervisor in Department Y that occurred about a month after the beginning of the field work in the federal agency. The former supervisor was a blunt first sergeant; the new one was a smooth operator.

In his first departmental meeting, the new supervisor explained his operating procedures to agents. His discussion was replete with references to his identification with the agents and the benefits they would derive from the efforts he would make in their behalf. He also expressed repeatedly his disidentification with management, to which he typically referred as "the brass hats in the front office." When he announced a minor change in procedure, for example, he added: "I don't want you to think that these are my bright ideas." These deliberate attempts to curry favor aroused my curiosity.

At the beginning of the next departmental meeting two weeks later, the supervisor, apparently trying to prove himself a regular fellow, explicitly told agents that he would overlook violations of a minor office rule. He explained that the assistant district commissioner had complained to him that his supervisory report indicated no tardiness in Department Y, whereas some of its members had been seen coming late to the office. The supervisor ended with the admonition: "We have to watch this from now on. Don't come in late more than once a month. If you do, sneak up the back stairs or something." Later during the meeting, he explained how he had persuaded the district commissioner to fly to Washington to try to have put into effect a prospective promotion for most agents earlier than had been planned.

The tactics of this new supervisor suggested to me an explanation of the processes through which supervisory authority becomes established. The essential principle is that the supervisor yields to group pressure, permits subordinates to violate some rules, and exerts efforts in their behalf because these practices obligate them to him. Their obligations to him and continuing dependence on him for benefits constrain subordinates to defer to his wishes and comply with his requests, thus extending his influence beyond the limits of his formal authority. The social agreement that tends to develop among the group of subordinates concerning their common obligations to the supervisor specifies and legitimates the range of his influence and thereby transforms it from power into socially validated authority over them. The new supervisor, who must initially establish his effective authority, is under special pressure to find ways to obligate subordinates to himself.

Most new supervisors do not operate like the one I observed. Yet these ideas, which I assume to be applicable to supervisors in general, would undoubtedly not have come to me had I not had an opportunity to watch this man or someone like him. The self-conscious and manipulative attempts of this supervisor to create feelings of obligation among his subordinates merely placed in high relief processes that occur usually in more subdued and subtle form. That new supervisors whose approach is matter-of-fact also make special efforts to win the good will of their subordinates is suggested by the only bit of quantitative evidence I was able to collect on this topic. All three supervisors who had been transferred during my period of observation were more favorable in their evaluation of the work of subordinates in their new departments than they had been in their old departments. Indeed, the one supervisor's exaggerated claims of how much he did for his subordinates were designed merely to further increase their obligations to him, but these obligations fundamentally rested on the things he actually did to benefit them. This important insight was supplied by an agent shortly after the new supervisor had been assigned to Department Y. In a discussion with me, this highly competent and conscientious official criticized the new supervisor not for making false claims of helping agents, as I would have suspected, but for being actually too ready to help them get by in order to assure that they would help him out in turn. "It's the proposition: you scratch my back, I scratch your back."

This agent suggested the principle that underlies my conception of the establishment of supervisory authority. It is the same principle that underlies my analysis of consultation as an exchange process, but I never made the explicit connection between the two. I did not use the concept of exchange in the analysis of supervision, although it is implicit in the discussion, and I used only in passing the concept of obligation in the analysis of consultation. Nevertheless, it may well have been that the train of thought stimulated by such comments about the new supervisor and watching him operate led to my conception of social exchange, even though I explicitly used the conception not in this analysis but in that of consultation.

Statistical Records

I did not anticipate the investigation of statistical records of performance. This was not merely an oversight. I wrote in a memo in the spring of 1948 that production records, which were a basic source of data in the famous study of work groups reported in *Management and the Worker* by F. J. Roethlisberger and W. J. Dickson,[5] would not be available for white-collar workers in public or private bureaucracies. My ignorance on this point reflected the state of the literature. There was no published analysis using quantitative performance records in white-collar offices, and the mere mention of their existence was so rare that I had not yet come across it.

In my first week of orientation, I learned about the detailed and varied quantitative records kept in the federal agency on operations and on the performance of every agent. My interest was immediately aroused, and I started scrutinizing these records and abstracting information from them. But at that time I had no idea of analyzing the significance of this quantitative evaluation procedure for operations in the bureaucracy. My interest in them was confined to employing them as a source of information about the performance of officials.

My observation in Department Y made it evident that supervisors used these production measures extensively as standards in terms of which to criticize the performance of agents and as incentives to spur them to exert more effort. The supervisor would tell agents that they must turn in more cases; he would explain to them that they found legal violations in so small a proportion of their cases that chances were they overlooked some of them; or he might discuss numerous measures of very specific aspects of operations to pinpoint shortcomings in performance. When the supervisor gave agents the periodic civil service efficiency rating, he would often justify it by referring to their performance record. Despite the complexity and variety of the measures, primarily three factors were emphasized. This suggested, as I commented in my field notes in November, that production records constitute a control mechanism that is quite different from the prototype of bureaucratic control and more akin to the profit principle. The agent is judged by a few objective quantitative criteria of effective law enforcement, just as the salesman is judged by his sales.

I did not follow this line of thought further at this time, however, because my attention was drawn to the dysfunctions of this evaluation system. Agents complained about the use of statistical criteria to put pressure on them, and several explained to me that such pressure is most likely to lead to silly practices. For example, it would motivate agents to consider as legal violations bookkeeping mistakes they found in the firms they investigated, just to improve their statistical record of successful cases. This was a beautiful illustration of the displacement of goals in a form quite different from the one Merton had examined. Bureaucratic emphasis on statistical records of operations, designed as a means to improve performance, induced officials to view making a good showing on the record as an end in itself.

My field work in the federal agency, then, had alerted me to the significance of statistical records in a bureaucracy. I had already thought of them as a control device, although it had not as yet occurred to me that they serve as a tool in the hands of the supervisor and also as a direct mechanism of control that partly substitutes for him. My attention had been particularly drawn to some dysfunctional consequences of such a quantitative system of evaluating performance. This was the background with which I began the observation in the employment agency, where I found simpler statistical records in which specific influences could be more readily ascertained.

At the beginning of the second week of observation in Department X of the state employment agency, an interviewer spontaneously asked me: "Were you told that we are working on production like in a factory? That's what we don't like. The main emphasis is on the number of placements." I had already noticed some strange practices. The desk assigned to me, because it happened to be unoccupied, directly faced that of one of the most competitive interviewers, and I had seen her keep slips of job openings on her desk and even push them under a pad to conceal them instead of putting them into the file box where they belonged. I attributed these practices at first to a desire to save the best jobs for one's own clients, but it soon became apparent that such hoarding of job slips was motivated primarily by a concern with making many placements.

In the following weeks, I had opportunities to observe much competitive hoarding of and vying for job slips. Interviewers told me that the statistical performance records, which had been introduced in this department a little more than a year earlier, were responsible for these competitive tendencies, and some supplied many illustrations of illicit practices encouraged by competition. The monthly statistical records, which were passed around to all members of the department, indicated how many interviews each official had held and how many referrals and placements he had made, both in raw numbers and in percentages.

This comparatively simple performance record invited quantitative analysis, particularly since the emphasis in the department on placing clients in jobs suggested that placements would constitute a fairly valid measure of performance, that is, production. At the end of the second month, I took time out to analyze these statistical records and the accumulated job slips from which they were derived. I saw that initials on the slip showed which interviewer had taken the job order over the phone from an employer and which one or ones had referred clients to this job. Since I had learned that the best opportunity for competitive hoarding of job orders occurred when they were first taken, before other interviewers even knew about them, I realized that the proportion of referrals made by an interviewer to jobs he himself had received over the phone would furnish a fairly reliable index of competitiveness. This was a fortunate discovery, for I could not possibly have obtained a reliable measure of concealed competitive practices, either through observation or through interviewing. I analyzed the data and obtained the basic findings for two sections in the department: placement productivity and competitiveness were directly related in the generally more competitive section but not in the less competitive section, and the total productivity of the more competitive section was lower than that of the less competitive one.

In this case, I had derived quantitative findings while I was still in the field and before I had come to understand fully the pattern to which the data pertained. I had explored various dysfunctions of performance records but had as yet no clear ideas about their functions.

My talks with officials during the last month of observation furnished two important clues that directed my attention to their functions. When I asked a supervisor whether he discussed the performance records with interviewers he first answered "Yes" and then corrected himself, saying that he had formerly discussed performance, but now that the records were available, he just sent them around and let them speak for themselves. Another time, the interviewer with the best performance record explained to me how each month she compared her performance on every single index with that of her colleagues, and if she was behind on any one of them, even though ahead on others, she modified her practices in an attempt to catch up. If I remember correctly, these illustrations of the direct influence exerted by performance records were what helped me to place in proper perspective various other pertinent data I had obtained earlier.

In the preliminary analysis I carried out the following summer, I conceived of the use of statistical records for evaluating performance as a bureaucratic mechanism of control that enables managerial officials far removed from operating employees to exercise a direct influence over operations. For, by furnishing each official as well as his superiors with precise knowledge of how his performance compares with that of others, the statistical records designed by management motivate officials to improve their performance in order to receive a good rating. Their direct influence on performance greatly facilitates the supervisor's job. I went on to note that statistical records were introduced by management in Department X to correct some deficiencies in placement operations, that they influenced operations, and that they promoted competition. In conclusion, I presented the correlations between productivity and competition previously mentioned, but I still was not able to offer any interpretation for these findings.

At this point, I had developed a functional analysis of statistical performance records as a bureaucratic control mechanism, but I had no systematic evidence to support my interpretative scheme. I had found some quantitative relationships between placement productivity and competitiveness, but I had no systematic interpretation for them. Evidently there was a need for two kinds of complementation, and I endeavored to meet this need in the subsequent full-scale analysis.

The hypothesis that the performance records in Department X served the function of furthering the objectives specified by management implies that the sheer existence of these records should improve placement productivity as well as the other phases of operations included as objectives. Of course, I had no way of telling what, say, the proportion of each interviewer's referrals resulting in placements had been before any performance records were kept. As I pored over the various statistical data on operations in the employment agency, however, I discovered that quantitative records on the placement activities of various *departments* had been kept before the records on the performance of *individual interviewers* had been introduced. This made it possible to check whether the productivity in Department X had, indeed, increased when the records on individual performance were introduced. It had! The findings supported the hypothesis.

I started my interpretation of the data on competitiveness and productivity by suggesting that a positive correlation between the two in one group but not in another may account for the greater competitiveness of the former group. Competitive hoarding was apparently an effective way of improving an interviewer's placement productivity only in the first group. This required an explanation of the difference in competitiveness between the two sections. I searched my field notes carefully and found three differences in the experience of these two groups that might be responsible: the members of the less competitive group had had more employment security when the records had been introduced, they were more professionally oriented toward their work, and their supervisor placed less emphasis on the statistical records than the supervisor of the more competitive section. How about the intriguing paradox that the competitive *group* was less productive than the other group, but the competitive *individual* in it was more productive than other individuals? A possible explanation finally occurred to me. Group cohesion makes it possible to discourage competitive tendencies, and it reduces status anxieties which impede effective performance. Relieving status anxieties by competitive striving is an alternative which is open to the individual in the noncohesive group. Since both group cohesion, which involves little competition, and individual competitiveness serve to lessen status anxieties, these apparently opposite factors have the same influence on improving placement operations. I even found some quantitative data that indirectly supported this argument.

In closing, let me again indicate how the analysis could be carried further. A minor point first: I wish now I had made more explicit that the direct influence on operations exerted by statistical records is due to their making *visible* the differences, and particularly the deficiencies, in performance. They make precisely apparent to every official and his colleagues, as well as to his superiors, what he does poorly. It is this visible evidence of his shortcomings that sets in motion attempts to overcome them. A more important point is that the paradoxical relationships between competitiveness and productivity are pregnant with implications that can be more fully exploited.

The finding that two factors are related directly for individuals (at least, in one group) but inversely for groups parallels a finding Stouffer and his colleagues reported in *The American Soldier:* [6] soldiers who had once been promoted tended to have more favorable attitudes to the army promotion system than privates, but these attitudes were generally less favorable in an outfit containing many promoted soldiers than in another containing few. I said so in a footnote, but it was only years later when I developed the concept of structural effects that I fully realized the significance of such findings. Group attributes, as Patricia L. Kendall and Paul Lazarsfeld have pointed out in their discussion[7] of *The American Soldier,* typically have corresponding individual attributes because we tend to characterize groups by measures that summarize characteristics of their members, such as their average competitiveness or the opinions that prevail among them.

The introduction of both a group measure and the corresponding individual measure into a multivariate analysis has some interesting implications for the study of structural constraints. It can answer the question, to use a concrete example, of whether the prevalence of a service orientation in a group of officials exerts structural constraints on them to treat clients more considerately or whether any correlation between service orientation and considerate treatment is merely due to the fact that those individuals who have a more favorable orientation to clients also treat them more considerately than do others, which is hardly surprising. The structural effect would be demonstrated by showing that the members of groups, whatever their individual orientation to clients, where a service orientation prevails, tend to treat clients more considerately than the members of other groups. There are more complex instances of structural effects which can combine with the effects of the corresponding individual attributes in various ways, and these more complex combinations of structural and individual influences are illustrated by Stouffer's findings and by mine.

Organization and Functional Framework

The last case of conceptual clarification I want to report does not pertain to a specific problem but to the over-all organization of the analysis in terms of a functional framework.

After completing the collection of data in the summer of 1949, I outlined the various data to be analyzed quantitatively and the different areas for qualitative analysis. This was helpful when I started the analysis half a year later. As I completed each part of the quantitative analysis—for instance, that of the interview responses I had put on McBee cards—I wrote one or a few memos about it. My field notes had been filed in folders roughly arranged by topic. I had typed two copies for cross-filing since I had found that most observations were pertinent to more than one topic. I read the notes in each folder and put my analysis of them into one or several memos. I moved back and forth between quantitative and qualitative analyses, trying to cover related problems consecutively and sometimes writing memos integrating several previous ones. I analyzed all data on the federal agency before I started on those on the state agency. I ended up with over three hundred memos, organized in folders by topics. After several reorganizations, these folders became the chapters of the monograph. I had too much material, however, and a good part was later discarded as less interesting or less relevant to the main theme than the rest (for example, two chapters on careers).

The functional framework was a general guide throughout, although it was by no means prominent in every part of the analysis. The question arises of how I decided to trace only certain unanticipated consequences of a pattern, but this question is as difficult to answer as the question of how one decided to investigate some patterns and not others. The functional paradigm and its specific applications to bureaucratic work groups in terms of Weber's theory made me sensitive to particular kinds of matters in the agencies observed. Generally, I did not try to remember to look for functions and dysfunctions while in the field, although once my attention had become centered on a given pattern, I did deliberately search for all its consequences that I could discover. Some connections, however, occurred to me only in the course of the analysis. I realized, for example, only after I had completed the quantitative analysis which showed that there was some discrimination at the reception desk but that interviewers did not give preference to whites in their referrals, that the pressure created by statistical records might have the latent function of discouraging discrimination against Negroes. (A recent replication by Harry Cohen of the employment-agency study suggests a refinement of this conclusion;[8] specifically, the effect of statistical records on discrimination among employment interviewers may be contingent on whether the employers in the industry served discriminate against Negroes.)

It is not easy to say what nonrational or irrational factors influenced uawares my decisions to investigate the problems I did, but one might be mentioned. From the topics in the book, it would seem that illicit practices aroused my interest, perhaps not only because I was explicitly concerned with informal patterns but also because the illicit stimulates curiosity. I discuss illegitimate competitive practices, illegal discrimination, prohibited consultations, and offers of bribes. The strange thing is that I am quite wary of—biased against, if you will—the study of intrinsically interesting subject matters where a journalistic fascination with the engrossing substantive issues easily diverts the analyst from focusing on problems of theoretical significance. I chose an unglamorous bureaucracy for study, not the United Nations or the Pentagon, and my analysis ignores the sensational facets of problems in favor of those aspects I consider of theoretical significance; for example, my discussion of bribe offers is concerned simply with the enforcement of group norms, not with the sensational issue of bribery. Notwithstanding this orientation and my focus in the subsequent analysis, the enticing allurement of the illicit may have drawn my attention to certain problems.

As I approached the end of my analysis, I began to wonder how to integrate the separate elements into a whole. To be sure, the functional framework gave a common theme to my discussion of different topics, but there were as yet no apparent direct connections between them. In the fall of 1950, before I had completed the analysis, I had an idea that helped me to make these connections.

It occurred to me that the concept of emergent need supplies a crucial link between the dysfunctions of some social patterns and the functions of others, at least in a bureaucracy. For if dysfunctions lessen adjustment and functions further it, as Merton has specified, one would expect that the emergent need for adjustment created by dysfunctions of one pattern often gives rise to another which operates to meet this need. This is particularly apt to happen in a bureaucracy, where officials on various levels are held responsible for maintaining the adjustment necessary to achieve precisely formulated objectives. Since social practices have multiple consequences, one instituted to meet a given organizational need may have unanticipated dysfunctions producing new needs, in response to which still other readjustments may occur in the social structure, and so forth.

This reconceptualization of Merton's paradigm gave me a framework for organizing the analysis and interrelating its various parts. The scheme was especially well suited for the investigation of bureaucratic organizations, which are, of course, much more integrated social structures than, say, societies and where formally stipulated objectives provide more precise criteria for defining functions and dysfunctions than the general concept of adjustment. Moreover, this scheme centered attention on the dynamic processes of change in organizations, which are always in danger of being ignored by the investigator concerned, as I was, with ascertaining the interdependence between elements in a social structure.

The explicit application of the scheme enabled me to see connections between various segments of the analysis that I had not previously noticed. It became apparent that the statistical records in the employment agency furnished a focus for interrelating a large part of my material on this agency. Performance records had been introduced to meet existing problems of operations. They served this purpose; but they also engendered competition, which had the dysfunctions of interfering with the work of a group of specialists and of impeding placement service in general. The two emergent needs for adjustment were met, respectively, by practices of specialists that elicited the cooperation of regular interviewers despite their competitive tendencies and by cooperative norms that developed in one group of interviewers to stem competitive tendencies. Another dysfunction of production records was to motivate interviewers, in the interest of making many placements, to engage in practices that led to conflicts with clients. The need to relieve the consequent tensions was met through informal discussions among colleagues, which served to restore equanimity after conflict.

The good fit of the data on the employment agency to the scheme guided my decisions on how to present my material in the monograph. I had not been sure whether to discuss simultaneously the data from both agencies that pertained to a particular problem or to present first one entire case and then the other. The latter course was now clearly indicated, since otherwise I could not trace the dynamic interrelations between elements in a given bureaucratic structure. While I applied the scheme also to the data from the federal agency, not all the problems I analyzed there could be connected in its terms. The fact that my general conceptual framework could be best illustrated with the material from the employment agency prompted me to present it first, although it had been collected and analyzed second.

CONCLUSIONS

As I come to the end of this chronicle and pause for a moment to hold a mirror up to it just as it is a mirror to *The Dynamics of Bureaucracy*, I realize that I cannot tell what stimulated my decisions to organize the discussion in certain ways. Why did I, for example, present my observations of the supervisor in departmental meetings first and only then start to indicate what ideas they gave me, while I explained from the very beginning what various observations meant to me in the other sections? I do not know. Was it because the section on supervisory authority is shorter? The detailed outline I made before I started writing does not indicate this difference. The decision to organize differently the section on the supervisor occurred to me while I was writing, and I cannot tell what suggested it. If I am unable to say what stimulated such a decision a few days ago, I could not possibly recollect the specific occasion that gave rise to an insight more than a decade ago. Fortunately, my field notes contained many comments which have enabled me sometimes to trace how certain observations suggested some new ideas, but there are also many gaps between an idea and its later refinement. By and large, such a chronicle can describe only the process of the development of research ideas in broad outline, and this is what I have tried to do with as much precision as I could muster.

Guided by Weber's theory and Merton's functional conception, I conceived of my investigation from its inception as a case study of bureaucratic work groups. It was to be an analysis of quantitative as well as qualitative empirical data, and this is what it turned out to be. As I have indicated, however, my specific initial ideas underwent fundamental modifications, sometimes beyond recognition, in the course of the research, and even the functional framework was revised shortly before completing the analysis.

• NOTES

1. Talcott Parsons, *The Structure of Social Action* (New York: McGraw-Hill, 1937); R. K. Merton, *Social Theory and Social Structure* (Glencoe, Ill.: The Free Press, 1949).

2. Eliot D. Chapple, "Measuring Human Relations: An Introduction to the Study of the Interaction of Individuals," *Genetic Psychology Monographs,* XXII (1940), 3-147.

3. William F. Whyte, *Street Corner Society* (Chicago, Ill.: The University of Chicago Press, 1943).

4. George C. Homans, *Social Behavior: Its Elementary Forms* (New York: Harcourt, Brace and World, 1961).

5. F. J. Roethlisberger and W. J. Dickson, *Management and the Worker* (Cambridge, Mass.: Harvard University Press, 1939).

6. S. A. Stouffer *et al.*, eds., *The American Soldier* (2 vols.; Princeton: Princeton University Press, 1949).

7. In R. K. Merton and P. F. Lazarsfeld, eds., *Continuities in Social Research: Studies in the Scope and Method of "The American Soldier"* (Glencoe, Ill.: The Free Press, 1950).

8. Harry Cohen, "The Demonics of Bureaucracy: A Study of a Government Employment Agency," Ph.D. dissertation, University of Illinois, 1962. A summary of portions of Cohen's work is contained in Peter M. Blau, *The Dynamics of Bureaucracy* (Chicago, Ill.: The University of Chicago Press, 1963).

3 Preconceptions

and Methods in

Men Who Manage*

MELVILLE DALTON

I

Preconceptions

Coming to the introspective task of describing the play of one's preferences on one's alleged detachment in making a study brings to mind as models not the great researchers but the confessors. Plagued with the usual mix of conventional and unorthodox sentiments about method in the social sciences, I suffer additionally from experience as a chemist. Hence, I have doubts about the appropriate method for discussing method. It seems logical that the more one professes to know about a subject, the more one is committed to profess that one knows how to know. But in the clash between what ought to be and the professional concern with what is, should one adopt the contrite mood of an Augustine or John Woolman on the one hand (and tremble about sins of technique) or follow the cynical tell-all manner of a Casanova on the other? Probably both biases are useful if seasoned with the self-reporting temper of another confessor—John Stuart Mill.

Whatever recital I make, I am aware of my bedevilment among the "truth" theories of coherence, correspondence, and pragmatism. I find relief in thinking that the only truth is "socially consistent sensation" among competent judges of "sound mind" and that insistence on a social fact may subtly alter the fact. It seems that I should recognize

* Melville Dalton, *Men Who Manage* (New York: John Wiley, and Sons, 1959).

50

William James's "knowledge about" but lean toward his "knowledge of" as I clearly did in *Men Who Manage* (from hereon referred to as MWM). Whatever dogmas I oppose or truths I support, I fear I may inadvertently dogmatize somewhat. To be supremely objective would require a distractingly subjective preoccupation—one that could interfere with the research. In discussing my intellectual homages, I shall work to limit the polemic vein inherent in preferences.

Inventory of Biases

Before describing methods used in MWM, I shall itemize conscious biases that colored the research by creating problems and forcing decisions for me that might not have troubled students with different devotions. I do not claim a monopoly on the biases, for they are all held in some degree by different students and are old. The fact that they are old does not mean that they have been answered or are unworthy of discussion. To shrug off an issue cavalierly as "an old problem" may imply a hunger for novelties or a pragmatic flight from intellectual trouble, but it solves nothing.

By attempting a statement of personal warps, with an effort to explain or even to "justify" them, I may be able to expose unwitting slants. This is not to exhort others to share my biases, nor to boast of them, but to present them in enough detail to reveal the nuances between them and the similar predilections of others who may be sustained by a different cluster of crotchets in making the same points. The different sources of similar biases are also worthy of study.

My major investigative colorations can be discussed under the topics of (1) conceptions of scientific method, (2) hypothesis as a symbol, (3) quantification as an end, (4) explicitness, and (5) the ethics of research.

Conceptions of Scientific Method

The dominant attitude in the social sciences today is that *science* implies a method, a procedure characterized by the invariant steps of hypothesis, observation, testing, and confirmation. It is presumed that (1) following these steps is a simple matter confirmed, and agreed to, by all scientists; (2) personal feelings must not and do not enter into the pure inquiry—except to check such feelings; (3) this is the only route to valid knowledge; that, in short, there is just one scientific method. If doubters request that the method be distinguished from the methods used in other branches of knowledge, they are told that the methods of science are polar to those of art, literature, music, and related activities.

Many of us support this scheme and accept it as something the natural scientists themselves have developed. My own bias is that (1) the scheme has uses, but (2) there is more to discovery and confirmation than these steps indicate, and (3) the natural scientists themselves do not defer to the model as we do. Yet from the volume of talking, reading, and writing that we do about science and the methods we attribute to the natural scientists, I infer that they are implicit gods for many of us. If we feel we must follow fixed methods, which I think can be defeating, we may have to choose other preceptors than the natural scientists. If a choice were possible, I would naturally prefer simple, rapid, and infallible methods. If I could find such methods, I would avoid the time-consuming, difficult, and suspect variants of "participant observation" with which I have become associated.

But many eminent physicists, chemists, and mathematicians question whether there is a reproducible method that all investigators could or should follow, and they have shown in their research that they take diverse, and often unascertainable, steps in discovery and in solving problems. To avoid errors of paraphrasing, I shall quote briefly from some of these specialists. (In quoting them, I am not, of course, suggesting that they should teach us the substance of sociology.) Their remarks are extracted with minimum cruelty to context.

P. W. Bridgman, Nobel Prize–winning physicist, says, "There is no scientific method as such. . . . The most vital feature of the scientist's procedure has been merely to do his utmost with his mind, *no holds barred.* . . . The so-called scientific method is merely a special case of the method of intelligence, and any apparently unique characteristics are to be explained by the nature of the subject matter rather than ascribed to the nature of the method itself." [1]

Of similar occupation and honors, Polycarp Kusch similarly denies the existence of a scientific method and holds that what we call by that name can be outlined only for simple problems.

Michael Polanyi, physical chemist recently turned social scientist, says, "The scientist's procedure is of course methodical. But his methods are but the maxims of an art which he applies in his own original way to the problem of his own choice." [2] Polanyi plays on this theme of *personal* influence on objectivity. One is objective only "by striving passionately to fulfil his personal obligations to universal standards." [3] He denies that science can be a set of statements determined only by observation and contends that conventional reports omit many of the steps in discovery and verification because of adhering to a false ideal of method and objectivity "as an automatic process depending on the speed of piling up evidence for hypotheses chosen at random." [4] He cites the reaction of adherents to that ideal after Einstein had inched his way to the theory of relativity. "When Einstein discovered rationality in nature, unaided by any observation that had not been available for at least fifty years before, our positivistic textbooks covered up the scandal by an appropriately embellished account of his discovery." [5]

Einstein himself and other physicists, chemists, physiologists, and biologists speak in a similar vein and cite many cases of undesigned discovery, of the role of "hunches" and "serendipity" in research, and of scientific advance with little benefit from formal conceptual tools. [6]

The vagaries of discovery have led some professional methodologists and philosophers of science to consign the thought processes of discovery to psychology and to settle on verification and proof as the hub of scientific method. [7] However, this may not be the way out, at least not just yet. For psychologists themselves currently say that "some very important parts of the scientific process do not lend themselves to mathematical, logical, or any formal treatment. . . . No one knows better than we how little can at the moment be said about . . . how the mind works, how problems arise, how hypotheses are formed, deductions made, and crucial experiments designed." [8] Another psychologist holds that discoveries result from "puzzlement—a state of mind which would not lend itself to any accurate verbal description; . . . there are researchers who do not work on a verbal plane, who cannot put into words what they are doing—whose thinking functions in terms of experiences, subconscious observations—who don't know what they have been after until they actually arrive at their discoveries." [9] The extremes of cultural (Kuhn and Merton) and psychological factors in discovery should be reconciled.

Hypothesis

No explicit hypotheses were formulated in MWM, chiefly for three reasons: (1) I never feel sure what is relevant for hypothesizing until

I have some intimacy with the situation—I think of a hypothesis as a well-founded conjecture; (2) once uttered, a hypothesis becomes obligatory to a degree; (3) there is danger that the hypothesis will become esteemed for itself and work as an abused symbol of science.

Throughout the research, I had, of course, hunches which served me as less exalted guides. As stated in the Introduction, the questions raised about ongoing and interconnecting events led the research. These questions might have been formalized as hypotheses. But concomitantly with my growing knowledge of events, I dropped many hunches and followed others. Some of these are discussed in the second part of this essay. To include all such changes of outlook and to report them as explicit hypotheses formulated, tested, and dropped when the process never attained such clarity seemed to me false and pedantic and, even if true, as entries that might be thought to encumber more than they would enlighten the report. Caution in hypothesizing is a bias dating from at least the time of Bacon's *Novum Organum*. His point remains sound: hypotheses should be made discreetly because, influenced by the "idols," they may be more selective and obstructive then helpful. More of a practicing scientist than Bacon, the physicist Johnson similarly fears "the tendency for hypothesis to degenerate into frozen prejudice." [10]

On the second point, a prematurely publicized hypothesis may bind both one's conscience and vanity. We are professionally committed to understatement rather than overstatement. In either case, my guiding crotchet is that a hypothetical statement is also one's attempt to be original. Having once made such a statement, one more easily overlooks negative findings, or, on having them pointed out, one's emotional freight often limits creativity to ingenious counterarguments, as in Goethe's long research and brilliant but empty theorizing in opposition to Newton's superior hypothesis on the nature of color. Darwin's magnanimous surrender to Agassiz' abler hypothesis about the structure of Glen Roy is not typical.

In the subject matter of MWM, the area of action lacks the relative uniformity of physics. To me, setting up a condition enabling a "crucial experiment," so that a series of hypotheses might be eliminated, leaving one in charge of the field, seemed out of the question. The number of possible "causes" seemed so large—and shifting—that eliminating one or a dozen hypotheses still left no sense of nailing down a specific "cause." To me, *hypothesis* implies a universal condition, or one that is true most of the time. I could, of course, have *posited* a series of constants, but this seemed beside the point. Considering the modern world's continuing refinements of every "truth" or the dropping of one truth in favor of another, I think that probably all is hypothesis save the one absolute that we must continue to hypothesize ourselves to new positions from which to make more helpful hypotheses. This is to say that I see little, if any, final certainty and that much of our knowledge is fallible. But I think that successive hypotheses, seriously developed and significantly applied, *can* move us from larger to smaller failures. Hence, rather than hypothesizing, I resorted to a kind of type analysis to uncover recurring processes and events. Almost inevitably this meant some implicitness of terms and assumptions and, naturally, reduced precision. The blind alleys I encountered on this route were less obstructive to me than the potential pitfalls in other avenues.

On the third point, throughout the research I was so concerned not to hypothesize trivially that even my hypothesis that it is often unwise to hypothesize remained implicit. At the other pole, one may give a sham dignity to one's efforts at research by "scientificizing"—glibly hypothesizing about matters easily verifiable or known by everyone. I may, for example, hypothesize that I have a letter in my box, test the hypothesis by looking, and then profess that I have been scientific. Possibly I err, but to me this is a pompous exercise of professional vocabulary that degrades what we all honor as science.

So, before framing hypotheses, I first sought intimacy with the area of study to raise questions worthy of hypothetical phrasing. Failure to phrase will not disturb those who get the questions, but embellished hypothesizing about some matters is wasted time and potentially embarrassing. "The essence of scientific method [is] to select for verification hypotheses having a *high* chance of being true. . . . It is travesty of the scientific method to conceive of it as a process which depends on the speed of accumulating evidence presenting itself automatically in respect to hypotheses selected at random." [11]

Quantification

In doing the research, I was interested in quantifying data that readily lent themselves to the technique but not concerned to quantify for its own sake. Many data that are not quantified could have been set up in tables by determined ingenuity. I have shared this attitude, but in the present research I found too many instances where perforce

I would have ignored widespread problems if I had quantified parts of the data in a way convincing to me. Where the problem was inextricably intertwined with others, I felt that too much injustice would be done to the whole to wrench it out for the sake of sampling and scaling theory. In such cases, one might objectively relate the mutilated part to a subjectively established criterion and in doing so inflate the part out of all proportion to the interlinked parts discarded because of quantitative inadequacy. For the present research, the price in some cases would have been blocked action and invalid results stemming from (1) evasive responses to direct questions in the touchy areas covered in MWM, chapters 3 through 7; (2) unavoidable disclosure of an inquiry unacceptable to some people, which would have ended parts of the study or jeopardized the whole; (3) offense to those who were able and willing to help me as long as I did not involve them in trouble.

My preference for idea over number has probably been buttressed by my unsystematic observation, in and out of the academic world, that idea is usually supreme over number as an influence in thought and behavior. Idea also dominates those with the idea that number is more important than idea. This appears most obvious where interest groups have compiled charts and columns of statistics to support a position; where, for example, scientists use statistics to explain the presence or absence of a tie between lung cancer and smoking. Repeatedly the idea, which the numbers are supposed to support, determines whether or not readers will study the report or reject it unread.

Supplementing my personal bias against quantifying at all costs is the disenchanting attitude of the mathematicians and physicists themselves. As with the concept of scientific method, these creators and refiners of the quantitative technique are far from endowing it with the infallibility and splendor that we often do. Their ideas and remarks throw doubt on mathematics as *the* matchless and impersonal tool, even when it is relevant to use. Let me crib from them again. K. Gödel, the mathematician, in 1931 made a reverberating case for the mathematician's inability to demonstrate whether the axioms of any deductive system involving arithmetic (such as that of the Whitehead-Russell *Principia Mathematica*) are consistent or mutually contradictory.[12] Like Heisenberg's principle of indeterminacy or Skolem's theorem, this will have little or no practical effect on our work, but it is theoretically disillusioning. And Gödel's remark that the development of mathematics along quantitative lines is a historical accident [13] is also disturbing. Russell's comment adds nothing to my certainty: "Physics is mathematical not because we know so much about the physical world, but because we know so little: it is only its mathematical properties that we can discover." [14] Possibly with tongue in cheek, Russell has also stated: "Mathematics may be defined as the subject in which we never know what we are talking about, nor whether what we are saying is true." [15]

In the spirit of Russell's first remark, Einstein adds: "As far as the laws of mathematics refer to reality, they are not certain; and as far as they are certain, they do not refer to reality." [16]

Qualification of Lord Kelvin's nineteenth-century dictum that the only knowledge we have is knowledge that is measurable enabled me to consider all problems I encountered without the distracting prior preoccupation of rejecting this or that item of behavior because it did not promise quantifiability. As I saw it, I was free to consider more qualities than quantity.

Explicitness in Research

Possibly from our wish to say all and the inability to do so with a definition of science that excludes such stopgap terms as "insight" or "intuition," we are led to elaborate on the need of being explicit. MWM lacks the ideal explicitness, especially on concepts or principles of any discipline. This results in part from the wish to address a general audience, but also from a bias against "explicity."

In "unfolding the meaning" of a point, one *exploits*[17] one's statements for more than they are worth, since there are no accepted absolutes. Feeling sure, but lacking proof that can be verbalized, we are moved to deck out our reports. Carried too far, our disinterested role shifts to the interested one of coercing agreement. But in seeking to win acceptance of an idea, we may unnecessarily handicap ourselves. Our pluperfect definition of science prevents us, unlike the natural scientists and mathematicians, from falling back on resource words suggestive of the method of art to fill in our demonstrations.[18]

Explication also fails when utterance is so drawn out that the reader's imagination has no play and the writer's fear of incompleteness leads to an overdrawn picture. One thinks of Santayana's definition of the fanatic as a person who, losing sight of his goals, redoubles his effort, or possibly of W. I. Thomas' "four wishes," which were not overdrawn so much as they were, paradoxically, explicit and incomplete. Their explicitness led others to what Thomas later regarded as an overemphasis on the importance of the concept.

In MWM, where I was convinced that a condition was true but was unable to prove it, I was all too brief and often guilty of inexplicitly fleeing or taking refuge in an et cetera in lieu of admitting ignorance or professing to have said all. In wearing this implicit garb, I have been somewhat comforted by the natural scientists and the sorcerers —not the apprentices—of quantification. Some of them are quite frank about the implicit bridge between knowing and proving. Their remarks are as important for method in general as for our present aspect of the topic.

For example, H. A. Kramers, the Dutch theoretical physicist, in analyzing Bridgman's view of the ambiguities of method, confesses: "My own pet notion is that in the world of human thought, and in physical science particularly, the most important and most fruitful concepts are those to which it is impossible to attach a well-defined meaning." [19] He cites the concept *energy* as an example. Paradoxically (at least for us), if energy *is* hard to define, it is still easy to measure; the distinction is worthy of thought, for the empirical support of some clearly defined concepts is hard to measure. Among the mathematicians, G. Polya first implicitly establishes his truth. His explicit advice is: "When you have satisfied yourself that the theorem is true, you start proving it." [20] K. F. Gauss, one parent of the normal curve, similarly confesses to the force of implicit method in his work but acknowledges the need to dress it formally. He has been quoted in many places for a century as saying: "I have had my solutions for a long time, but I do not yet know how I am to arrive at them."

The clearest possible statement of problems, concepts, and findings is essential for communication and learning. But explicitness may reach a point of diminishing returns. One man's explicitness may to another be "failure to communicate." When a student has given all explanations that his peers can seize and evaluate and he is still short, let him ape the humility of a Kramers or a Polya. He may aspire to be more explicit than they, but he need not be ashamed if he fails. He will be a disgrace if he goes so far as to try—with nothing but words and wishes—to drug the reader's imagination and beat him into intellectual submission.

Ethics in Research

Since much of the research in MWM was done "from the inside," ethical questions are implicit about relations with informants and collection of data. Some social scientists fear that such research is scientifically suspect; that, if concealed, it hurts the dignity and rights of research subjects; that it gives science a bad name; and that it is likely to be unethical.[21] What is unethical varies considerably with students. Commonly their codes are a mixture of eternal truisms with a thin knowledge of how the many codes of work groups are fashioned and function.

In an industrial firm, for example, each department, each division, as an interested in-group has an unprinted code about fealty due it. Codes stress loyalty to the unit in maintaining appearances, preserving secrets, and dealing with problems peculiar to its place in the service. Complexity of organization, general instability, diverse interests and backgrounds of personnel, and the gap between work life and private life make impossible one simple static code. This is not confined to business and industry but functions in any formal organization in which division of labor is linked to responsibility. The researcher himself, immodestly cutting across all parts of the organization, can easily be seen as immorally undercutting every department's implicit ethics as well as flouting the ethics of his professional group.

My interpretive bias is that the ethical precepts of a given department grow out of its relative position in the organization, its function and importance in relation to other departments, the character of its personnel, technology, and so on. Hence, inside the organization there are recurring practices—essential as well as not—that some members support and others censure as unethical. It seems clear that all persons recognize some norm, however amorphous or splintered, for all involved take some pains to conceal what would be denounced. In some cases, part of the official code is a dead letter except as it is used politically to control others. However this may be, the researcher who is obliged to get at all relevant behavior may obviously offend some persons in the organization. If there is reason to think that visible behavior is minor and misleading, he must get at the unknowns essential to complete the picture without, of course, damaging the research, the persons studied, and the profession. But in doing this, the researcher, and not a remote part-time ethicist who cannot say where his personal

code comes from, must size up the moral issues peculiar to his problem and bear responsibility for reconciling the diverse moral commitments he assumes in and out of his office. Naturally, he may involve
himself and others in trouble, but the promise of his work may be
worth the risk. Where would anatomy and surgery and dependent
specialties be if Mondino, Leonardo, Vesalius, and others had entirely
honored the absolutes of their day instead of haunting cemeteries and
gibbets in their search for cadavers?

Given the stated questions, my research called for attention to the
obscure as well as open aspects of organization. The predilections I
carried into the research were built on the following assumptions.

(1) Science and ethics differ in that (*a*) science is an *is* world, a set
of *facts* growing out of consensus among small groups of competent
people, while (*b*) ethics is a set of *wishes*, an *ought* world, built up by
an uncertain majority over an indefinite period. It is not, as with science, a subject for international agreement.

(2) Though society may reject—or destroy—a researcher and his
methods and findings, it may also change as a result of his findings[22]
and tolerate, even honor, him.

(3) Morality "is not a mere expression of the so-called 'mores' of
any particular race, nation, or state, or of any historical era. It is not
a code of repression or a mere dream of perfection. It arises from the
universal human situation, in which man finds himself confronted by
the necessity of reconciling conflicting interests." [23] Behavior may be
moral though it falls short of the perfection aspired to by a Rousseau
or Kant.[24] For morality is both a means and an end. It starts as a means
but may conclude as a static end, out of touch with many direct supporters as well as the larger world. As an axis between the concrete
means and mutable ends of groups, it is workable. But when fixed
ends are allowed to order all the means in a changing society, morality
functions either as an intellectual toy or as the exclusive possession
of some group. Morality as a means reconciles new ends with old, but
as an *end*, morality justifies any means to perpetuate itself, until recurring reform re-establishes the tie between abstract and concrete
good.

(4) Deception and distortion are restricted among members of the
in-group but are tacitly allowable with strangers. In a well-integrated
department, the department is an in-group of the organization. There
may, of course, be conflicts in the department and more than one in-

group. Except where there are strong crossties, *other* departments in an organization are first-circle strangers; those outside the organization, or beyond the neighborhood, are second-circle strangers, and so forth.

In terms of the usual categories (dogmatic idealism, ethical nihilism, ethical pluralism) into which ethicists classify men according to their *ought* worlds, the methods used in MWM would label me as an "ethical pluralist."[25] In being an "impartial observer," I do not therefore pretend that I entirely avoided opposing one good to another. Only a godlike student may align himself with the *one* good in research conduct. I can only reject such authorities to call on others, particularly those who hold the view that wisdom consists in following precedent, unless other equally worthy rules suffer too much by it, and in remembering related codes and using them where relevant.[26] This is not to condone expedient search for a moral principle in which one does not believe, to support commission of acts contrary to one's dominant code.

In the conflicts and compromises among the personnel, as well as in my failures and reorientations in studying them, I was struck by their relatively minor *adjustments* to codes as compared with their *inventiveness* in interpreting the code or in settling issues that could not be agreed on as coming under the code. The loose-leaf and ever-changing manuals of union, employers, and experts for guidance and interpretation were, in this respect, reminiscent of the different versions of extant world scriptures and the range of commentaries about their meanings.

To summarize, the social investigator must sort his values and obligations and weigh them repeatedly throughout the research process. In a democratic society, he cannot impose one fixed code on multiple conflicting codes. But he is committed to give as clear a picture of what exists as his limitations allow. Some of his research subjects will oppose him. In research, as in his daily nonprofessional activities, he may have to feel his way, protect his values while reconciling them with those of others. In his compromises he may lose ground occasionally, but in his research life he should counter with sustained drives, as in his community life. Ethics is a guideline which many are struggling to control. Whatever the storm, the researcher must move toward his ends and dispute the right of way with those unwilling to compromise to preserve the uncompromisable. Living at all—in or out of research situations—is dangerous. He damages *absolute* truth when he accepts conduct he dislikes in exchange for action he approves; and all of us do this consistently in order to live more significantly.

The essential thing is that one be neither an absolutist nor a nihilist in one's research ethics. The only absolute tolerated in a democracy is that one must weigh and consider in holding differences to a workable level. In the process, as in the legislative arena, there will be specific questions about what is ethical and specific name-calling, but there will be general agreements that constrain the ethical turmoil and move it in ethical directions. The lone "inside" researcher, cleaving to his value of extracting the essentials of the process from both those who favor and those who oppose his purposes, must suffer the universal defect of having his absolute ethic reduced to a relatively high purity—one might wryly equate it with that of *The United States Pharmacopoeia's* standard for drugs and chemicals—say, 98 per cent purity.

II

Specific Methods

From what was said in Part I, it might be thought that I used no formal methods in the research. I did use such means, but I often modified or supplemented them with devices adapted to situations and persons involved.

To get at the conditioning situations, let me first answer a question that several students have asked me, "Did you first plan the research and then contact the firms you studied, or were you a full-time employee doing research on company time when you were being paid to do your official job?" I was employed at Milo when I conceived the idea of studying that firm. I had earlier worked at Fruhling and returned there to renew old acquaintances, as reported below. I collected data from Milo both while employed there and later. I did not compromise my official duties, and in doing the research I received encouragement and aid from officers in responsible posts whom I did not coerce and who revealed no qualms about aiding me, though they did not wish to be identified in the report. Though I fitted some research into my formal duties, most of my interviewing was done after working hours both in the plant and in homes; when in the plant, I did not disrupt the formal duties of informants. For many months, all my days off were spent in some of the firms. The section in Part I on "ethics in research" should clarify my views on this question. Points of procedure and priority and the significance of connections among the companies for concepts and methods will come out in the discussion.

The research was not preceded by consciously elaborated hypotheses but grew vaguely out of my confusion and irritations. As a member of a staff group, I was given specific instructions about the nature of my work, how I was to conduct myself with operation groups, and to whom I was responsible. I was lectured on dos and don'ts for collaboration with line departments and other staffs. The order and neatness of instruction sharpened my sensitivity to unexpected events that occurred. I was repeatedly puzzled by the gaps between official and unofficial ways of doing things and by the open and covert name-calling.[27] Some associates were literalists in their interpretation of procedure, while some were paraphrasers.[28] Their disputes involved not only work methods but the meaning of our function. And this last was often belittled by the operating departments. At the same time, some members of my staff group thought there was a measure of logic in the criticisms of us raised among those subject to the consequences of our work.

My reaction to the confusion was not to distract myself with developing simple explicit hypotheses as guides but more expediently to frame questions about specific events in the disorder visible behind the official calm and to fill out partial answers I thought I had. This was a kind of implicit hypothesizing that gave me more freedom of thought and rapid movement from hunch to hunch than initial preoccupation with formal hypothesizing about limited facts would have allowed.

The first guiding question I raised was, why should there be conflict between experts (staff personnel) and administrators (Chapter 4)? I recalled that there were clashes at Fruhling, where I had worked on a different kind of staff. I visited old associates there and found similar processes of discord, though precipitating events might differ. Other questions developed from necessary contacts with a range of production departments. For example, as I saw—and was caught in— the contentions between union and management, I wondered at the way in which grievers and managers formed cross cliques (Chapter 5). To my logic, they were natural enemies who, when disturbances broke out, should turn to the contract for guidance rather than use it only as the last resort.

Piqued by what I regarded as improprieties, I raised silent questions

about other areas. Why was there disruptive conflict between Maintenance and Operation (Chapter 3)? If people were awarded posts because of specific fitness, why the disparity between their given and
exercised influence (chapters 3 and 6)? What was the meaning of the
double talk about success as dependent on knowing people rather
than on possessing administrative skills (Chapter 6)? What was behind the contradictory policy and practices associated with the use of
company materials and services (Chapter 7)? Why and how were
"control" staffs and official guardians variously compromised (chapters
4 and 7)? Why, among executives on the same formal level, were some
distressed and some not (chapters 8 and 9)? And why were there such
sharp differences in viewpoint and moral concern about given events
(Chapter 9)?

In time, a general question embraced all others: what orders the
schisms and ties between official and unofficial action? I was concurrently groping for a frame of meaning close to the disorder I was experiencing and grouping my impressions into tentative centers. In
other words, (1) I struggled among the mixtures of helpful and inadequate concepts I brought in from my academic training, as (2)
I also searched for more flexible and—as it seemed to me—more relevant concepts. From this mulling, I settled on such unconventional
specimens as "axis of reward-right," "out-of-role," "role-front," "theft
intelligencer," and so forth.

Utilizing excellent contacts (below), I extended my inquiries to
Attica and Rambeau with the hope of establishing similarities—or
uniquenesses—to reinforce or reshape my impressions. In some cases,
I was able to see allied problems more closely, or I found conditions
in one plant that stimulated further inquiry in the others. The concept of informal reward was sharpened by findings from Rambeau
and the drug chain. Data from Attica reinforced and broadened my
thinking about data from Milo and Fruhling on problems in staff–line
and union–management relations and less so on internal struggles for
control. The distorted personnel files of Attica, lack of access to the
files of Fruhling, and unrepresentative—though similar—informal data
from that firm forced me to limit to Milo the study of careers. However, I felt that the career consciousness and shared understandings
at work in altering Attica's employee records were cloth from a bolt
of the same weave, though I did not know the specific pattern. In all
cases and for all the problems I was thinking of at the time, I felt that

my hunch of ongoing interplay between official and unauthorized ways of dealing with problems was emerging from a postulated to a real "constant." And though my concern with numbers embraced only certain things, I felt a certainty akin to that in the remarks above of Gauss and Polya: the sureness of a firm hold on several qualities not caught in my trap for quantity.

Questions and hunches originating in the wider experiences at Milo and Fruhling were cross-fertilized by concurrent contacts at Attica and Rambeau. Since no simultaneous, systematic study could be made of all, and as Milo was the most accessible, that firm became the nucleus of inquiries and the continuing point of major effort.

From the concern[29] shown by some persons about revealing the unauthorized aspects of their behavior or that of others supplementing plant activities, I became convinced that formal approaches to these firms, if permitted, would reveal little more than official expectations as understood at various levels. And the chance of occasionally helpful slips of the tongue during formal interviews could not be followed fruitfully without departing from the announced mode of research. Hence, instead of this method I chose to accept and seek the aid of experienced, reliable, and, as nearly as possible, representative intimates[30] to supplement my inquiry into the more obscure conduct essential for the study. I felt that only in this way could I make the realistic first step of learning what the issues were—as defined by the conduct of personnel—and who was involved. Then I hoped more clearly to see the problems of recurring concern in terms of their extent and the ramifications of behavior that kept them alive.

We can more formally discuss the research under the topics of intimates, techniques and sources of data, and special problems.

Intimates

My duties at both Milo and Fruhling allowed much unquestioned movement about the firms and required considerable functional interaction with other departments. I did what I could to expand my existing circle of intimates and to develop confidential exchanges with acquaintances. To this end, I gave every legitimate service and possible courtesy and went beyond what was normal in giving personal aid.[31]

From this group of personnel I selected, over a period of about three years, as *intimates* those who (1) trusted me; (2) freely gave me information about their problems and fears and frankly tried to explain their own motivations; (3) had shown repeatedly that they could be counted on not to jeopardize the study; (4) accepted what I was able to tell them and refrained from prying into the information I was getting from others; and (5) gave me knowledge and aid (warnings, guidance, "tips") of a kind that, if known, would have endangered their careers.

These are criteria for the ideal informant. I made mistakes and was deceived in judging some persons accepted as intimates. This is part of the price the researcher must pay in studying those who have different interests from his and who may engage in their own unsystematic study of him and even systematically manipulate him to aid themselves and their department.

Though intimates knew of my general interest, I made no detailed statements of what I sought but indicated that I wanted broad information on "all kinds of personnel problems" from as many firms as possible. Carried too far, this course might insufficiently honor the plant under study. So I usually reinforced this method by asking informants questions about their earlier experiences there and elsewhere. The attempt to allay fear that their department or firm was of special interest sometimes stimulated informants to insist on giving local and current details and drawing parallels with their earlier experiences. I explored in detail any information or leads that came out.

The number and rank of intimates varied among the firms. Number seemed to me less important than the individual's position, reliability, and knowledge and the rapport achieved with him. More time was given to Milo and to cultivation of its personnel; hence the greater number of intimates there. Milo confidants totaled 81: 11 workmen, including 3 grievers; 24 first-line foremen; 14 general foremen; 6 line superintendents; 8 staff heads or assistants, and 18 staff supervisors or specialists; and 4 confidantes among secretaries should be included. By direct or nondirect methods, I interviewed 113 other persons in the Milo sample and 27 hourly paid foremen outside the sample.

One of the secretaries at Milo helped get the income data from that firm for me. The aid was given as part of a tacit exchange of favors. I approached her, as a person who had formerly been in my department and was now close to the payroll group, on the topic of managerial salaries. She indicated, as I knew, that these were confidential and that it would be dangerous to try to get them. Then, without saying whether she would try or not, she told me she had been wanting to talk with me about a specialist I knew and with whom she had had two dates. She "hoped he was interested" in her. But she felt that his superior education and more prominent family might be obstacles. Knowing of my training in sociology, she "had thought" I might counsel her. My training had not included courtship and marriage, but the possibility of getting data on salaries enabled me to adopt a counseling manner. I suggested some tactics and promised to learn what I could about the specialist, his family, and his plans for the future. After her next date, she called me for a conference and gave me the first group of salaries she had obtained. The conferences continued with the dating and the flow of data and after discontinuance of data brought on by a chance change of personnel. (Despite the counseling, the secretary married the specialist within a year.)

Reynolds of Attica was merely a speaking acquaintance. However, I had the confidence of one of his secretaries, two members of the grievance committee, and an engineer whom I had known earlier at Fruhling. I knew two of Reynolds' community intimates and five former Milo workmen in the plant who supplemented my information about Attica's racial problems. Reynolds' secretary was indispensable in aiding me to study personnel files. Since I could not examine the records in person in the plant without raising questions, she brought out six or eight folders each week end, over successive weeks, for me to study and return to her not later than Sunday evening so that they could be put in the files early Monday morning. She also indicated several alterations she was certain had been made in the files, but her knowledge was incomplete. Hence I have reported no details (MWM, p. 149).

The female secretaries mentioned above and several clerks were helpful in the research. A number of staff experts at Milo and Fruhling stimulated me to win the aid of such personnel. They observed that secretaries reflected the reactions of executives to staff reports. For example, some specialists, chiefly chemists and engineers, believed they were able to interpret how they "stood" with a given executive and how they were "doing" in comparison with rival experts who were submitting reports. They allegedly did this by observing the behavior of the secretaries toward them and noting how long their reports lay in the pile of papers awaiting executive perusal—and even to infer, not entirely a humorous matter, how many times a report had come to the top of the pile and been put back on the bottom, *by the secretary*, without having been read. The explanation was that the secretary hears talk—praise or blame—by the chief about a given expert and, wifelike, accepts this as correct. A specialist—for example, an engineer—coming up with original ideas or ideas welcome to the chief would be greeted warmly by the secretary. Those persons regarded as mediocre by the chief would be neglected by the secretary.

The potential contribution of female secretaries and clerks is usually unappreciated. Several things made them suitable to aid research of the kind I was doing. They are probably more status-conscious than males in the same roles, quicker to note symbols of rank and differences in influence and to spot certain of the factors involved. They are also more interested in events and social details and probably remember them better. Having this orientation and access to records and events, they are likely to possess considerable knowledge of unofficial activities and developing policy. Where his personal bent is a hindrance to clique participation or his clique ties are fouled and he feels isolated, a manager confides more to his secretaries than to others and even relies to a degree—sometimes great—on their judgment about some matters and persons. (Research on the marriages of secretaries to their bosses in all walks of life would probably show that much more than physical proximity and sex attraction is involved.) It is trite to note that women have long played secondary formal roles (with striking exceptions in France during the seventeenth and eighteenth centuries), have had to use indirect approaches to gain their ends, and have resented this condition. However, this may be relevant for the researcher, for where women secretaries are treated as intellectual menials they are disposed to be communicative with those who show awareness of their insights and knowledge of affairs.

At Rambeau, I had but three associates: Nevers, the husband of one of Reynolds' secretaries, and the secretary herself, who had formerly been employed there full time but then worked at Rambeau as an "extra" only on the evenings that the store was open. Her friendship with the bureaucratic female department head was most helpful (MWM, pp. 211-212). At Marathon Research Company (pp. 91-92), my only informant was the chemist whom I had known earlier at Milo. At Argo Transit (p. 203) the top manager and three persons in his office supplied me with more information than I needed. The X-ray technician (p. 203) was my sole contact at his hospital. Four people whom I had known earlier at Fruhling gave me details of their activities under piece-rate work at the unnamed cabinet factory (p. 197).

Initially, as a patron of the drugstore, I came to know the soda-fountain manager (pp. 212-213). She was troubled at times and went out of her way to serve me at the fountain and to tell me of her work experiences. Both the day and night pharmacists gave me a picture of her activities in the store, but both were biased against her and favored the preceding fountain manager who had allegedly been "much more reasonable" in the social use of the fountain and "working to make things easier for everybody." A nephew of the fountain manager also gave me supplementary details based on her reports to her family and on his own experience as an employee for the preceding fountain manager.

Some intimates were invaluable not only as sources of information but also for help in research situations. Especially at Milo and Fruhling, they occasionally served as what chemists would call "catalytic agents," because they accelerated reactions. In effect, they sometimes initiated and pushed uncontrolled experiments for me. In the staff groups, particularly, as well as in any situation where discussion was taking place and I knew in advance that I could be present and seemingly occupied with work, they introduced agreed-on topics and questions into the conversation. These stimuli were typically on issues in the problem areas. Usually "busy" over in a corner of the room, I was observing and taking notes on the remarks and other behavior of one or more people. Some intimates developed an interest in this kind of study and sketched events and conversations they thought relevant for me in my absence. Data collected by such persons obviously must be evaluated before use. Given the situations I have described, both the opportunity and inclination for personnel to distort are great. Here again, however, the possible returns outweigh the dangers and the labor of assessment.

Again I must "protest too much" by noting that some students may think the issue of relevant data here is minor compared with the *manner* in which it was collected. That is, such experimentation is regarded in some circles as "unethical." But, without the note-taking and careful assessment, this is a common and legitimate link in the fluid communication about events cutting across official and unofficial phases of organizational life. It seems no more unethical than the use of "visiting shoppers" (pp. 209-210) in business or "projective techniques" in psychological research.

Techniques and Sources of Data

As would be expected, my methods frequently overlapped, or they were irregular and incomplete as a blend of open and covert tactics—

open where I was sure of my ground, covert where I was uncertain of the obstacles and the time to act or where I feared defeat. Still, my major efforts can be discussed under formal interviewing, work diaries, and participant observation.

Formal Interviews

I did little formal interviewing because of the obvious problem of explaining what I was doing and the inadequacy of the approach for getting at unofficial activities, especially when other means were safer and more effective. However, in gathering career data on the managers (Chapter 6), I felt obliged to get formal statements as a key to official criteria both as a foil for, and as a possible lead to, unstated factors. My unspoken guiding questions were: Who was recruited and advanced? What were the bases on which people were chosen for preferment? What did "ability" mean, and how important was it in success as compared with seniority? How did people go about climbing in the ranks?

As a participant-observer—and thus possibly a half-specialist in self-deception—I framed my guiding questions, of course, on the basis of what I thought I already knew. But logical procedure demanded the corrective of checking my warped vision against other sources of information before returning for systematic soundings of several dozen intimates. The steps here then were, first, to interview formally several high officers with whom I was not intimate, with Milo presumably only one of several local firms in which all the best-informed managers would be asked the same question: "What are the things that enable men to rise here in the plant?" Next, I explored official statements in the supervisory manuals and handbooks. Only then did I turn to intimates for unofficial statements and seek to check accumulated data against those in the company files which a group of intimates were working to open for my study.

In formal interviewing, normally I did not take notes unless the informant wished me to (as did Cowper, p. 277). Instead, I listened with relaxed intentness and raised no questions as long as the interviewee was comfortable with his comments. If I had planned questions, I asked them as though they were spontaneous. But when unexpected leads appeared, I dropped agenda questions and pursued my object. Since I often feared that equal opportunities might not recur, my aim here was also to catch the background, seize side issues, and draw everything possible—then and in the future—from slips of the tongue. I did not, as I may imply, assume the manner of a cross-examiner. When note-taking was essential and I could not keep up, I used abbreviations, *ad hoc* symbols, trade terms, or jargon as quickly decipherable recall crutches. Once completed, whether notes were used or not, I reconstructed the interview as soon as possible. I first set up a skeleton of key items remembered—or noted down—and then expanded these, connected them by association with other parts, and reworked the whole until it seemed to be a facsimile.

I did this confidently, sustained by my suspicion that (1) recorded interviews lose much from the probably guarded comment of the subject (as a Rambeau employee said: "If it's important enough to be quizzed in this way, it's important enough to be careful about"); (2) the interviewee, in his talk as in his handwriting, would never twice say the same thing in precisely the same way; (3) without some knowledge of the informant's revealing visual behavior and unique intonations, the recorded interview is incomplete for a thorough analysis; and (4) though these subjective aspects can mislead the analyst, they can also reveal much to him, and they need to be assessed. Skilled judgment is indispensable and requires no apology. As part of this bias, I also sought to *recall* (even more dangerous than professing to *see* or *hear* all) and note down with the reconstructed interview the emphases made, facial expressions, marks of concern and relief, and other gestures as possible clues to more basic things. The restoration was relatively complete on topics covered and apparently accurate on grammar and speech mannerisms—judged by the occasional read-backs of some of our discussions I made to intimates and the checking of my notes with "catalysts" against what they believed had been said in group discussions involving people whose expression and idioms they knew better than I did.

Work Diaries

Through much of the period at Fruhling and Milo, I recorded events, biographical information, gossip, and initial hypotheses in loose-leaf notebooks. For example, I outlined and dated "unusual" incidents (which in a flow of deviations might include a return to "usual" practices); questions to ask certain people; excerpts from rule books and supervisory manuals; possible ties among events; signs of developing cliques; activities of cliques; out-of-role behavior; data on formal meetings—kind, occurrence, duration, and who was involved; additions to biographical data—and corrections turned up in unguarded talk; the search for and use of precedents by individuals and groups; contrary interpretations made of regulations, with notes on who made them and leads to follow; overheard remarks made by people about each other— complimentary and not; threats and accusations in arguments; joking remarks that seemed to bear on events; provocative statistics and ir- regularities in reports as leads to follow or as related to established data; and marginal notes by various departments and offices on work- analysis sheets and related reports.

This last item was more fruitful than it might seem. The notes were, in effect, questions about irrgularities, or interpretations, as defined by the note-maker, who had his own interest in knowing and usually was closer to many things than I. As such, the comments frequently gave me assurance that what I feared might be a fanciful hypothesis about what was occurring was worth further inquiry. The penciled explana- tions sometimes confirmed what I believed—and thus encouraged me— or suggested corrections or initiated a new line of inquiry for me.

Study of the notes also gave me insights into the ties between the official concerns of specific departments and the kind of comment they would make on the reports. Though vague, this was a step toward grasping the import of departmental consciousness and the necessary attendant secrecy on some matters. More explicitly analytical than line people and, of course, more responsible for analytical reports, the staff annotators were more pointed and voluble. Production Planning, for example, was concerned with irregularities in the "scheduling costs" of jobs, as well as with accelerated movement of materials when ware- house space was limited and, above all, with any irregularity that would lead to failures in meeting delivery deadlines. Industrial en- gineers were disturbed about the "too low" and "too high" perform- ances of certain individuals which naturally reflected on the staff as a group professing implicitly to motivate the diversely motivated and, like any group of experts, to justify their existence. Cost accountants and standard cost personnel were alarmed at any threat to their predic- tions. Routers from the Field Work Department (FWD) raised ques- tions about the irregularities of time charged to certain jobs and worried about the variations from predicted time the product would have to spend in given departments during the process of production. These comments and the necessity for analyzing them annoyed me but probably helped me by restraining quick certainties and glib explanations.

After coming to know some of the commentators better, I became aware that most of them were what I have called informalists. They were concerned to meet bureaucratic theory—passably—but they assessed situations without reverence for the theory. The formalists, on the other hand, were apparently either "above" believing that some officers could engage in unbureaucratic behavior or fearful of facing and interpreting the obscure aspects of organization. In any case, my experience was preparing me to see why departmental sustainers and face-savers were rewarded to a degree as persons not subject to standard controls.

As the diary grew, I studied it and raised questions. I incorporated tentative answers to the questions, rephrased or dropped some questions, and added new ones. As the problem areas became clearer, I detached some materials, labeled them, and classified them as logically as seemed possible. The silent but increasingly communicative document became both a confidant and a guide for the research. Accumulating evidence exposed errors of interpretation and emphasis, so that for a time I repeatedly sifted and reclassified the data. As I refined the categories which became chapters, the "miscellaneous" category waxed and waned. Concern to miss nothing significant and knowledge that I was missing some things obviously meant an accumulation of "irrelevant" data for the problems and categories I settled on as most meaningful. But under the stimulus of new questions from my daily prowlings, I frequently rechecked the "irrelevant" data. Sometimes I found positive items or a startling absence of something I was sure was or should have been there.

For example, among the extracts I had copied from manuals as general rules I thought "revealing" at the time, I expected to find entries on qualification for promotion. Finding nothing specific, I reread the manuals and found only vague references to the career merits of "application," "cooperation," having "a good record," and so forth. "Merit-rating" plans were mentioned and were in use by some of the staffs, but no sample plans or lists of essential traits were cited. The term "ability" was not defined in the manuals, though it often appeared in the comments of personnel. Their implicit use of the term set me to groping for their tacit facts. By making a new entry in the diary for noting down and comparing the contexts in which the term was used and pressing intimates for definitions, I concluded that the managers[32] could, if called on, agree generally on what they meant by the term. They used it to mean capacity to (1) maintain high production but low operating costs and a low rate of grievances (without illegal strikes) and of accidents; (2) make "good contributions" toward the solution of critical issues; (3) preserve "good relations" in the department and between departments; and (4) subordinate personal to organizational aims.

Participant Observation

Participant observation has long been used as a practical aid in legal prosecution, national defense, international relations, and among all contending interest groups. The practice, of course, moves from respectable to reprehensible shades. To some persons in the social sciences, it is both scientifically and ethically suspect. There are, indeed, several scientific shortcomings to the technique: (1) closeness to unique detail may limit attempts to classify data, to formulate problems, and to generalize; (2) the researcher's peculiar personality may attract him to unrepresentative informants or lead him to identify with some inconsequential subgroup; (3) his presence may disturb the very situation he is seeking to freeze for study, this hazard increasing when he must barter to enter the research arena—in effect, to promise help in solving problems; (4) where he works in a disguised role, he may give associates false clues, for their responses are directed to his simulated role and he may note them down as the *real* behavior without knowing that he was duped by unintended distortions; (5) when very friendly with his informants, the researcher may unwittingly communicate the answers he wishes. I sometimes considered the possibility that some of my intimates had me conditioned to the point where I was "acting like a trained seal"; certainly I had the experience of feeling that some informants had caught my unspoken interpretations and were speaking them back to me in my own words, so that at times I was reduced to thinking that "extrasensory perception" was at work; (6) if the researcher is not long in, and around, the area he is studying, he may mistake an unusual event for a typical one and overstress its importance.

Opponents of the technique charge that personal observation automatically has low validity. This is exaggerated. Admittedly some personal "observations" are more a projection than an observation. For evidence that they may also be valid we need but recall instances cited in Part I. Being what I am, I can say only that the merits of participant observation outweigh its defects, especially when the method is combined pragmatically with other methods as a supplement or as an equal or major research arm.

(1) The researcher is not bound by fixed, and sometimes crippling, research plans. He can adapt and reformulate the design as he sees (*a*) the insignificance of data he thought important or the need for those he had overlooked; (*b*) old entrees closed or new ones that must be dealt with differently; (*c*) that the problem is changing, is interlocked with others, needs redefining, and so forth. Regardless of how rigorously research is planned, these necessities usually arise where effort must be long and involved.

(2) The technique enables the inquirer to avoid pointless questions which often cause ridicule behind his back and injure the research in unconsidered ways.

(3) Greater intimacy allows the investigator more correctly to impute motives.

(4) He is also better able to get at the best-informed informants as he needs them later in the research. He has found the implicit analysts among his informants. On the basis of their tested leads he can return —as one who has himself been observed and tested—to assay any nuggets of fact they have panned from the stream of action and gossip around them. And, related to point (1), not only does he become more able to detect data irrelevant for his problem but he is better able to re-evaluate data he originally thought irrelevant. In short, he becomes increasingly able to make better judgments.

(5) The participant has a great advantage in getting at covert activity.

(6) He has time to build superior rapport before he asks disturbing questions.

(7) Also, since he is not committed to treating the always dissimilar informants uniformly, he can select uniquely equipped "specialists" in different areas of his problem.

(8) Finally, in many cases the established circulator is able to work his way to files and confidential data that the peripheral formalist usually never reaches.

Knowledge of weaknesses in the method obviously does not of itself remove the weaknesses. My consolatory bias was that knowledge leavened with judgment—born of mistakes caught early—allowed me to exploit the weaknesses and not to overplay the strengths of the technique.[33]

The freedom of movement I mentioned above allowed me to spot and, aping the line chiefs, draw on the challengeable offerings of malcontents. These included members among, first, my staff associates, then, Catholics, minority ethnic groups, first-line foremen in general, and various individuals wherever I met them. Such persons were disposed to trust me and to speak more freely than were other initially nonintimates. Their information on past and developing events was most helpful and constituted solid data, once it was corrected for such emotional distortions as the cry of Catholics at Milo that "95 per cent of management are Masons" and Sarto's charge that no Italian could enter the hierarchy. As fringe members of their groups, malcontents may distort from both ignorance and bias, but the concealed and subtle malcontent may be a full member and have access to many secrets— and he may also be a sporadic sufferer who tries the researcher's techniques and timing. Even the outspoken, rejected, and consistent grumbler can see *some* of the more objective events flowing from unknown decisions, and his resentment is likely to sharpen his perception to a point where the researcher's labor spent in skimming off the dross of distortions will be rewarded.

Participant observation allowed unquestioned cultivation of persons who sometimes gave me more help than either they or I recognized at the moment. Stein (pp. 156-157) was one of these. From his comments, I was better able to see the struggles among personnel to get "credit" for individual contributions. I then saw more clearly the importance of out-of-role activities, as well as the typical role inflation and exaggeration of credit due, that occurred. In the demand for *personal* recognition and reward, each job is inflated by its holder to impress others. Both the specific duties and the "importance" of functions are refined beyond measurement. Trifling services become vital operations as personnel ridicule first the triviality of one another's work, then that of other departments. The handbooks of pragmatic supervisors are often more important to them as symbols of their role than as working guides; when they lent me these books, they stressed the confidential content and set early deadlines for returning them.

Where the researcher also has a formal function, participation is, of course, indispensable for the research. In my own case, the carrying out of *line* assignments to find errors in inventories and in the labeling and classifying of raw materials; the attendance at safety, union–management (low-level), and other meetings; the study and compilation of reports—including those that involved "government jobs" and "graphite analyses," job descriptions, and personnel assessments; and the carrying out of unofficial assignments under an official title—all these were important in getting at the similar activities of others and in judging the interplay and interconnections of official and unofficial action, for gaining access to personnel files, and for learning what records were reliable as against those only "for the record." These functional intimacies further helped me to follow relations between Milo and the office after the demise of the FWD. Lacking personal communication with the office, I remained near the interplay by associating with the few visiting representatives that I could, some of the Milo personnel who had formerly been in the office and continued to communicate with friends there, a few Milo managers who made occasional trips to the office, and intimates who were critically involved in meeting expectations of the office.

I was led to study the grievance records in union–management relations by learning from the griever, Brady, that he and other members of the committee avoided formal settlements in favor of "gentlemen's agreements." I theorized that a comparison of the official reports might reveal a pattern, but I could not persuade other members of the Milo committee to aid me, and Beemer, Brady, and Spencer were unable to reinforce my efforts. The president of Fruhling's local compiled the data from his plant. Reynolds' secretary transcribed the Attica grievances and steps, let me study the minutes of meetings, and told me relevant decisions. The Attica committee confirmed the steps.

My access to Milo's personnel files grew out of disputes between two departments about the merits of certain tests given to incoming white-collar personnel. Near to one of the disputants, I argued, with imperfect detachment, that it might help to correlate statistically responses to the tests as a basis for a clearer appraisal. Since I was vague about the total information I wanted, it was decided that my academic training fitted me to extract "related information" from the files that a "clerk" might not do efficiently. Because some of the personnel who had taken the tests were managers and some were not, an influential intimate in one of the staffs supported my suggestion that we should also examine the management file for the records of those who had taken the tests. Knowing vaguely of my intent, he cut me short, when I started to explain what I personally wanted and why, with the statement: "I trust you to be discreet, so I don't care what you plan to do, and I don't want to know anything about it!"

This "method" of getting the occupational data for Chapter 6 was thus free, rather than designed. My experience with the Milo and Attica files led me to hypothesize that most personnel files are likely to be incomplete or distorted. Lower-ranking members give fuller information than higher-ups, who apparently stand on status rights and omit, or possibly withdraw, some information from the files as they gain status; or as criteria change, they make new entries. The lower-ranking personnel have less to give but naïvely give most of it; the higher-ups have more to give but give calculatingly and incompletely—or inflatedly.

In determining the unofficial influence of twenty-one Milo managers (MWM, p. 22) I was aided by fifteen intimates who served as judges. All were or had been close associates of the managers they were rating. Based on closeness to some subordinates of all grades from Taylor down, my participation was largely confined to challenging the ratings. The basis for ranking an individual was the relative deference of associates, superiors, and subordinates to his known attitudes, wishes, and informally expressed opinions and the concern to respect, prefer, or act on them. Scale analysis would have looked impressive here, but the judges were reluctant to submit to use of any formal device other than the chart.

There were minor disagreements about the placement of a few officers. For example, some of the judges who were line officers (typical of line attitudes toward staff people) objected to Rees's being regarded as more influential than Springer. But these same officers showed such fear of Rees that if their observable behavior alone was taken as a measure of Rees's influence, he should have been placed above Hardy. Two of the judges wished to put Peters below Taylor. These dissenters were general foremen who may have disliked Peters because he had been brought over from a staff organization by his sponsor, Blanke. The chart merely indicates that formal and actual authority differed. It is not presented as a measure of the difference.

The subject of Masonry was so touchy at Milo that even some intimates shrank from having a hand in establishing precise membership and the number of Catholics who had become Masons. What seemed like a simple thing to do aroused fears and alienated some of my fringe confidants whom I had mistakenly counted on for help and whom I now saw as themselves worthy of more study. These people now avoided me and aroused my fear that their concerns would lead to obstructive action against the research. I learned later that they "wanted" to aid me but feared that I would involve both myself and them in trouble, but they also feared not to aid me lest some of my intimates embarrass them. My own fears were idle, for two of these persons have since reported that their anxiety also checked discussion of it. Since the Masons were distributed among numerous lodges, to confirm membership I eventually had to submit lists of doubtful officers to seventeen intimates among the Masons. I felt obliged to give each a general explanation adapted to his personal tendencies as I saw them.

In the struggles at Milo between Catholics and Masons, the Catholics repeatedly charged that their group, allegedly a heavy majority in the surrounding community, was held to only 5 per cent of management. Since Magnesia's current and preceding mayors, as well as district congressmen, were all Catholic, I had no reason to doubt the rough accuracy of their statement about the city's population, but I wanted more detail to assess the ethnic make-up of the plant, and Magnesia lacked a directory of the city's religious groups. Catholic intimates cautioned me that I would do better to let them inquire of church officials and accept their communicated word. I agreed, but the figures, all showing a majority, were too different. Hence, I asked for introductions to senior priests. Here the wives of Catholic friends cleared me with the priests and arranged for interviews.

According to the husbands, their wives reported me as a "serious and obliging student with no axes to grind" and as one "who knows when to keep his mouth shut." One wife pictured me as a person who "wants to study problems that we're all interested in." I sought to fit this likeness in interviews with priests in different parishes. The first two priests estimated Magnesia's Catholic population to range from 59 to 85 per cent. The third priest, an austere rector converted from Protestantism, was as revered throughout the city as though he held the title of patriarch. His parish contained many of Milo's current and former managers. Therefore I felt obliged to visit him but, probably awed by his name, thought that I should develop some common ground with him before getting to the point. So I first spoke of his new church, then moved to church architecture in terms of styles, dimensions, and the background of European spirit and skills as against American. We talked of the rapid construction of St. Patrick's Cathedral and the National Shrine of the Immaculate Conception (then several years away from dedication) and compared them with the Washington Cathedral and that of St. John the Divine.

The rector was pleased at my interest and enthusiasm. He asked about my background and commented that I had an "Irish" name and that my ancestors were Catholics at one time. Possibly the prologue was helpful. He asked about the specific problems I was interested in. I indicated a wish to know more about the "local changing population," the "job patterns" in industry, and so forth. "Job patterns" was a red light. My explanation led to a discussion of social factors in success and whether "having an in" or "knowing the right people" could be a factor. I received some smiling but insufficiently verbalized admissions. Craving figures, I hinted at statistics I had collected about the community and quoted comments of unnamed Catholics in local industry. This brought the concession that the number of Catholic managers had been "declining for years." Like the first two priests, he offered no figures for Milo or Fruhling, but his estimate of the city's population, which his personality made more acceptable to me, was "about 70 per cent." [34]

Participant observation was vital for study of the FWD. I worked with this group for over a year. During that time, I heard gossip from members who had been in various levels and divisions of the hierarchy, received an orientation to events of the preceding five years, studied mechanics of the FWD operation, and heard members analyze the department's interplay with line executives as I concurrently followed developments in the shops. Later, as a staff representative in the shops, in a role apart from the FWD, I was able to study events more connectedly while maintaining unofficial contacts with the FWD to learn of defenses that line chiefs were presenting there—without, of course, reporting anything but trivia of either side to the other.[35] Here, however, I was made acutely aware of "departmental consciousness," something I knew obliquely before. Some of the personnel with whom I had been on close terms in the FWD now treated me as an outsider. I began to feel my way to a concept of "departmental identification" relevant for my experience.

Participant observation, in terms of what might be called in-plant and out-plant socializing, enabled me to study the ethnic make-up of personnel more effectively than I could with other methods. I was, of course, concerned to check the alleged exclusive selection by ethnic stock. I studied the national origins, surnames, and birthplaces of the managers, first by utilizing my personal knowledge, then by resorting to personnel records, next by checking with intimates to uncover name changes, and finally by free interviewing with doubtful persons. Socializing in the plant, especially on my days off,[36] allowed me to cultivate foreign-born and minority groups.

Since I had repeatedly heard their statistical lament that "you've got to be twice as smart to get half as far" and was aware that they saw me as one of the majority group, my initial approach was to talk with them about an unforgettable topic—their country of origin. Having a strong interest in European history and culture, I found this an easy way to establish rapport [37] with natives of Britain, Germany, Poland, Italy, Spain, Greece, and the Slovak and Baltic countries. Such ethnics were pleasantly surprised to find one of the usually condescending Americans interested in and having a small knowledge of the geography, show places, heroes, artists of various kinds, current political figures, interests, and accomplishments—including culinary skills—of their native countries and wanting to know more. These persons cleared me with suspicious first-generation Americans[38] who, as marginal men to a degree, were sometimes suspicious and resistive to my imperfectly concealed curiosity. It was but a step from such pleasant beginnings to knowledge of name-changing and associated attitudes, leads to internal plant events that I could not possibly cover in person, data not contained (or else distorted) in the personnel records, earlier events in the plant or the community or problems of their offspring employed in the plant that might bear on my present interests. Since many of these persons—workers and managers—at Milo and Fruhling were Catholics who had held a variety of jobs in the plant and elsewhere, they also supplied leads and valid information on the Catholic-Masonic struggle in Milo and its repercussions in the community. Some of the foreign-born introduced me to their local priests, who helped me with some details positively as well as by what they clearly avoided.

To get a rough view of Magnesia's total ethnic pattern, I went through the new city directory and randomly used surnames as the gauge of national origin. This device, of course, overlooked the extent to which names may have been changed [39] and the fact that migration weakens the criterion. Since nearly all of Magnesia's Negro population had Anglo-Saxon names, all such names from the area in which Negroes resided were counted as Negro, and all other names with their appropriate ethnic group. For example, names such as Purvinis or Tirilis would be regarded as Lithuanian or at least from the Baltic area originally. Census data supplemented those from the directory and showed that the foreign-born of Magnesia who were ethnically akin to the majority of Milo's leaders were less than 15 per cent of the city's population.

I learned the Milo membership in the yacht club by asking Geiger, its president, to let me study the list, which included both industrial and civic leaders of the community. Some socializing at the club enabled me to study their joint activities; and outside the companies, I developed closer relations with personnel during their periods of relaxation, attended the installation exercises for Masonic officers and personally verified some memberships, and continued contacts with people at Attica, Rambeau, and Fruhling.

Information on the political preferences of personnel seemed unworthy of the risk of a direct approach. Instead, I relied heavily on such indirect evidence of attitude as the facts that (1) having a choice of newspapers with different political shades, the managers chose to carry a famous Republican paper into the plant with them; (2) Democrats in management concealed their leaning; (3) all managers who served in public office did so as Republicans (shown on their occupational record). Naturally, I noted down the politics of anyone who gave me a clue. To this I added tallies of intimates and remembered the meaning of Haupt's reaction to the prank played on him by his subordinates to make him appear to be a Democrat (p. 91) and the demotion of anti-Republican Wilkins (p. 186).

As a participant, I found (what is probably trite to mention) that I could not interview in the conventional sense. At the beginning of the study, as a result of training that slighted my research situation, I was too steeped in the sacredness of formal approaches to function easily. Since my relations with intimates often were already structured, my straining toward a detached manner or persistently pursuing points defeated my purpose, until I learned better. My verbal interactions might simultaneously be both *ad hoc* and pointless for the research. As conversational rather than interview pieces, they moved rapidly about an intimate's view of things and so, in the contingencies of job demands, might be broken and incomplete. But if relevant for the research, I resumed exchanges at the same point, as far as the variables of mood, memory, and possible new concerns of the informant allowed. Hence, in some cases utterances over a series of exchanges are tied together as one statement. The characteristic thing about them is that they were precipitated by events involving the intimate; they were "situation-centered" and often unguarded. Though most reported utterances were prompted by a question from me or from an informal aide, some were parts of "noncatalytic" exchanges between others that I overheard in a shop or office. In other instances, comments were made as asides to me, in the shops, over an incident of the moment. Some statements, such as the long outpourings by Geiger (Chapter 4) and Evans (Chapter 6) were made at one time and without interruption by me. Geiger's remarks were in response to my question. "How did things go?" with reference to an exasperating cost meeting he had just left.

Usually expecting guarded talk, I sought when possible to catch men in or near critical situations and to learn in advance when important meetings were coming up and what bearing they would have on the unofficial aspects of various issues. Experiences with reneging informants prompted me to get comments or gestures of some kind from certain people before their feelings cooled so as to reduce the problem of reading between the *lyings*. In interviewing, I usually had in mind a schedule of points to follow. But when the respondent's talk uncovered events of seemingly greater importance, I omitted or adapted my prepared questions. Then, or at a later meeting, when I had exhausted the planned questions for that part of the research and was sure of the intimate, I asked loaded questions in various directions and followed promising responses.

Special Problems

Given the incorporated biases summarized in Part I, it was likely that I should involve myself in a series of theoretical and practical problems. I shall not try to establish the tie between my entering orientation and later difficulties, but the problems can be discussed under the topics of (1) polar actionists, (2) ethics of reward, (3) reneging informants, (4) objectivity in covert research, (5) knowing too much, and (6) obstructive allegiances.

Polar Actionists

In evaluating the occupational data for Chapter 6, I became aware that, in terms of their reactions to formalized procedures for advancement as well as for regulations in general, the personnel seemed to fall into two camps. These were the absolutists or formalists and the relativists or informalists. The first type accepted and even deferred to official modes of appraisal and were eager to work for bureaucratic refinements. The relativists used ladders that were helpful to them but were prepared to circumvent official ways and to devise expedient short cuts. Like any simple dichotomy, it is a sheep-or-goat concept of truth: it misses the differences within similarities and the similarities within differences. For my purpose, it had the obvious weaknesses that few personnel fit it well, that other undetected factors were probably at work, and that I had to rely on my judgment and that of imperfectly disinterested intimates to decide who did fit.

I used the theory, aware always that the occupational data (MWM, p. 159) could be interpreted to mean that there was no formal scheme for promotion at Milo, but that the informal data[40] (pp. 180-184) suggested that whatever plan did exist was qualified in practice by struggles between the relative relativists and absolutists. However, in finding peace through use of a loose concept, I became increasingly troubled about the problem of legal versus illegal actions and their ethical implications.

The Ethics of Reward

Administrative concern with what is *legal* may not overtly stress the body of rules governing interpersonal relations in the firm or the relations between a local plant and its central office, but the internal use of materials and services is usually a strongly legal matter. Heavily negative rules supplement the positive by specifying how materials and services are not to be used. Hence, when I saw them regularly employed in ways inconsistent with rules, I at first defined the practice as immoral, and I found personnel who agreed with me. But others disagreed; their arguments were not refined, although I became gradually aware that some functions not described or foreseen but that might *later* be recognized and formalized had to be done *now* to enable the department to survive.

In time, I saw that in terms of their thinking about misuse and proper use (theft and reward) of materials and services, personnel workably fitted into the rough typology of absolute–relative. I sought to avoid writing about the problem, for its larger meanings appeared endless. But I could not drop the question. Possibly my bias that historical data are a vital supplement to any study—that no period in man's history should be made an exclusive source for data on his behavior—was a factor in my obsession. In any case, I studied instances of bribery through history in relation to roles and events, and I reflected on the defenses and censures made in relation to persons and decisions at Milo, Rambeau, and the drug chain. The section on ethics in Part I is partly a result of my effort to deal with this problem.

But making a philosophical decision did not relieve methodological pain. I craved facts and constants, such as—to be pedantic—the relation between mercury and heat. Sulking for an absolute, I had to settle for half-operational definitions and examples of reward and theft and live with the dispute about their meaning in a flux of incompletely assessed situations. I might have gratified my metric impulse by counting unpleasant tasks done, or out-of-role contributions, or faces saved, or problems prevented by anticipation, or services as an informal liaison person, or personal sacrifices—but at the price of slighting attendant qualifying conditions. Possibly I should have developed and interfused both approaches. How to distinguish reward from theft in any "hard" way that leaves no borderline cases and that is not also arbitrary remains for me a problem.[41]

In some despair, at this point I departed from formalism and seized on the arguments of the physicists and mathematicians which justified frank recourse to intuition when other means fail. Having fled the problem in earlier writings, I felt that I must now face it, if only to talk of it cursorily, as in Chapter 7. I decided to ride formal technique as far as I could but no longer be obstructed by it.

The experience of a group of method-bound students at Rambeau confirmed me in the need to be wary of stretching hallowed techniques to fit unlike situations. Rambeau had repeatedly resisted retail unionism. A research organization volunteered to find out why, free of cost to Rambeau. Three-page questionnaires were distributed to all the highest- and lowest-performing salespeople. Many entries included salesmanship ("selling") or related occupations. Respondents were asked to rank occupations in terms of interest to them in different circumstances. Over 80 per cent chose selling (or a cognate). Yet informants assured me that probably less than 30 per cent preferred selling to other jobs listed. The irony is that, at least in many cases, respondents indicated selling because "they [the researchers] have ways of finding out who you are." They suspected that Rambeau managers had ordered the survey for some reason adverse to the interest of employees and would covertly act against those preferring other kinds of work. The researchers interpreted their findings to mean that high morale and job satisfaction were the factors in resistance to unionism. There *was* a job satisfaction among those participating in Rambeau's officially illegal but unofficially acceptable incentive system, but this was not what the researchers had in mind.

Reneging Informants

Obviously, withdrawal of this kind is a chronic threat to the researcher's every attempt to grapple with the unofficial. Moving in guarded areas, he expediently seizes promising words and incidents and comes to count on some informants more than others. But how does he, a student of selectors and selectees, go about selecting an informant?

My own theory was, first to look at issues and at who was involved. This was not easy, and I made mistakes, since I needed a changing and growing set of informants as I tried to deal with matters I had not foreseen. Most informants were not concerned with the whole organization nor as free to roam as I. Yet my freedom to move about was not freedom to probe unofficial matters but chiefly to deal with routine things. I was meeting more restricted but better-informed people in their area; they were closer to unofficial events but also variously dedicated to protecting departmental secrets. As nearly as I could, I next considered the state of my research: what consequences my closeness to some people would have later when I might wish to draw near to people they disliked or relative strangers to them. Here, I was influenced by established intimates. To speak grandly and to gloss over my mixed disappointments, irritations, and successes, intimates were my nucleuses for building a network of satellite informants who, in turn, aided me in expanding my universe of communications.

In assessing informants who were becoming relative intimates, I supplemented secondhand impressions with my own. Shortly, a small experience enabled me to judge them. I found that this must not be purely opportunistic, for naturally informants were also assessing me and resistive to being "used." To avoid this problem, I sought to build a relation of trust and exchange of favors. This was not always possible, for some persons gave me no choice of this kind; they selected me to report things to, to be their friendly ear, or to get aid of some kind for themselves. Some of these persons were malcontents who gave me qualified aid, as I have indicated above. Where they did not stand well with their group, I limited open interaction with them. Conversely, some of my supposedly committed informants reduced their interaction with me. For example, a currently cooperative person also in trouble might reveal more than I could absorb at the moment, but in a changed condition next week be cool, regret what he had said, and suggest that I was going in the wrong direction, had "misinterpreted" last week's remarks, and so forth. Since he hoped to conceal his withdrawal, this reneger offered no explanations. Possibly he thought me too obtuse to see his changed behavior, or he saw me as manageable or as unable to take—or above taking—reprisal. In dealing with such an informant, a second researcher might get a different initial response from mine, and I might be unable to duplicate my earlier results.

In such cases, I sought to hide my awareness of the change and my disappointment and continued friendly relations to avoid arousing his fears or to suggest that my resources were so limited that I was dependent on him. However, I used his new stance as a clue to his probably changed involvements, and without discussing him I checked with others for possible hints of factors in his change. Drawing on these, I followed, of course, consistent leads and utilized what had earlier seemed like contradictory data but were now consistent with the developing action of this person's group or clique. Renegers who openly withdrew as the inquiry deepened alarmed me. I thought that, like the other type, they feared labeling and punishment by their groups as persons who "talk out of turn." Apparently they feared me less than their groups or other personnel, for they excused the withdrawal, whether they explained or not. Sometimes I felt at fault for not having detected the informant's limits, for not more carefully preparing him, or for having moved too rapidly; but as with the other type of reneger, I continued on good terms. To reject the reneger is self-defeating and ignores the fact that he can feel friendship for the researcher and still fear the larger network of ties in which both are caught. Efforts to understand him often give a return. He may add new angles, reveal disturbances he shares with others, and expose misinformation they have given. Though possibly shifting from the role of confidant, he continues as a subject for analysis and a key to some areas of group feeling and action.

Objectivity in Covert Research

Getting at concealed matters obviously requires careful and personal contact. Studying such situations at a distance in the attempt to meet a false ideal of perpetual detachment, the researcher may be so "objective" that he misses his subject matter and cannot be sure just what he is objective about. My own concern was to fuse the personal and objective as nearly as possible, to have the subjectivity to catch and share elusive meanings; then, once I was sure of them, to assess and record them objectively. I thought first to immerse myself in the areas I must know, then to step out in the role of critic, reorient myself for another scrutiny, and re-enter. My greatest problem in these entries and retreats was to distinguish purely official behavior, written or understood, from the more elusive and exciting unofficial phases.

Part of the problem arose from my attempt to focus concurrently on the devotees of each: I wanted to cover the reactions by bureaucratic virtuosos to the irregularities they had to live with but at the same time catch the ingenious accords made by the masters of implicit procedure. Since the latter were more successful in reaching ends, escaping trouble, and maintaining poise, my impulse was to slight the importance of formal restraints and to forget their guiding influence on informal behavior. Repeatedly I caught myself moralizing in terms of my definition of formal codes which were to others, as I saw later, implicit or so vague as not to warrant fully the meaning I was reading into some informal activities. For a long time, as stated earlier, I saw all unofficial use of materials and services purely as theft. It was only after developing close relations with Merza and Berger of Milo and Nevers of Rambeau, and after long study of the web of incidents binding them to others, that I saw the oversimplification I was making and shifted from an interpretation of the too rigid and exclusive categories to a more realistic judgment about the whole.

Knowing Too Much

The inquirer who labors from the screen of his formal nonresearch role may forget important methodological facts as he strives to enrich his data. He may learn much of unofficial activities and get behind protective screens but forget that official roles have official images. That is, in his speculative prowling he is almost certain at times to forget that nonintimates see his formal function as embracing only a limited knowledge of unofficial events. Eager to learn more, he alarms some persons, even his fringe intimates, by accidentally disclosing bits of unofficial information they think it strange that he should have. Even formerly close work associates engage in confidential practices they would now withhold from him as one outside the group, and especially as one who already "has his nose in too many things." Variously committed persons misinterpret his slips, magnify what he knows, and fear that he will imprudently compromise them, or—in the changing scene —use his information for personal ends. In any case, they are likely to treat him as a red light and to alert others.

My own adjustment to these setbacks was to try to map the total role knowledge and defenses peculiar to the different levels and functions of prospective informants. Sensitizing myself to the role fronts of non-intimates and adapting my methods to them, I sought at the same time to remain close to their view of my role.

Obstructive Allegiances

Every experienced investigator knows this problem.[42] But in "live" situations in large organizations when he is under pressure to respond socially and is eager to start, maintain, or improve a flow of rewarding data, his theory of perfect detachment may languish. Escaping over-identification with any one group while maintaining some intimacy with all groups central to unofficial issues is a prime puzzle for the masquerading researcher.

For example, Randall and others at Milo and Jessup at Fruhling were Masons who gave me valuable aid. In our conversations, some of my actions were evidently interpreted as showing a wish to affiliate with the order. Accepting an invitation to attend installation exercises at a lodge is a tacit request to be accepted as a member. From both curiosity and to please Randall, who was a valuable informant at his level and had many good hunches to guide me to and among those things he could not prove, I attended exercises at his lodge. After days with no positive action from me, he said: "You talk a good fight. Maybe you'd find out a lot more if you'd get to know how we do things." Taking such remarks as hints for more proof of where I stood, I feared that failure to respond to cues for closer identification might be interpreted as deliberate and thus endanger the inquiry. On the other hand, I knew that entering the order would alienate helpful Catholics.

My "solution" was to follow the compromises I had long witnessed in these firms. I chose to hold my involvements with any group to the minimum level that would bring maximum workable findings. Undoubtedly this restricted some worthwhile details as the price for broader information. Knowing the deeper feelings of Masons who entered the order for career purposes—and of members who were formerly Catholics—might have given important side lights. The compromise assumes that this additional knowledge was not vital for the study as a whole and that the cost of acquiring it probably would have hurt the research in other ways.

• NOTES

1. P. W. Bridgman, "New Vistas for Intelligence," in E. P. Wigner, ed., *Physical Science and Human Values* (Princeton, N. J.: Princeton University Press, 1947), pp. 144-145.

2. Michael Polanyi, *Personal Knowledge* (Chicago, Ill.: The University of Chicago Press, 1958), p. 311.

3. *Ibid.*, p. 17. See also his rich speculation on problem-solving and steps in discovery, pp. 120-131. One must note, however, that objectivity may be less difficult for the natural scientist to achieve than for the social scientist, who studies entities from which he derives most of his attitudes (personality) and on which he relies greatly even to preserve his sanity.

4. *Ibid.*, p. 167.

5. *Ibid.*, p. 11. Polanyi's specific arguments for the unavoidable entry of unstated personal factors, which fill the gap between evidence and conclusion, into research method are well made on pp. 30-31, 49-65, 321-324. Other students, stressing cultural factors, report on both the play of personal concern and rivalry through the process (Robert K. Merton, "Priorities in Scientific Discovery: A Chapter in the Sociology of Science," *American Sociological Review*, XXII [1957], 635-659) and inability to define the precise course and time of discovery (T. S. Kuhn, "Historical Structure of Scientific Discovery," *Science*, MCCCLXI [1962], 760-764).

6. Albert Einstein, *Essays in Science* (New York: The Philosophical Library, 1933), pp. 4-5; R. Oppenheimer, "Analogy in Science." *American Psychologist*, XI (1956), 127-135; M. Johnson, *Art and Scientific Thought* (New York: Columbia University Press, 1949); Mario Bunge, *Intuition and Science* (Englewood Cliffs, N. J.: Prentice-Hall, 1962), especially pp. 67-111; E. E. Slosson, "A Number of Things," *The Independent* (January 26, 1918), p. 160, and *Creative Chemistry* (New York: The Century Company, 1919), pp. 72-73; D. L. Watson, *The Study of Human Nature* (Yellow Springs, O.: The Antioch Press, 1953), p. 122; *University Bulletin* (Berkeley, Calif.: University of California, November 13, 1961), p. 81; "The Night Prowler of Nobel's Laurels," *Saturday Review*, XLIII (1960), p. 68; W. B. Cannon, *The Way of an Investigator* (New York: W. W. Norton & Company, 1945), especially chaps. 5, 6, and 17; Anatol Rapoport, *Operational Philosophy* (New York: Harper & Brothers, 1954), p. 48; D. Cort, "World's Most Valuable Men," *The Nation*, CLXXXIII (1956), pp. 497-500; Bertrand Russell,

The Scientific Outlook (New York: W. W. Norton & Company, 1931), chaps. 2-3; Sir George Thompson, *The Inspiration of Science* (London, Eng.: Oxford University Press, 1961), chaps. 2, 10.

7. Polanyi, *op. cit.*, pp. 14-17. Also P. F. Schmidt, "Some Merits and Misinterpretations of Scientific Method," *The Scientific Monthly*, LXXXII (1956), 20-24.

8. B. F. Skinner, "A Case History of Scientific Method," in S. Koch, ed., *Psychology: A Study of a Science* (New York: McGraw-Hill Book Co., 1959), vol. 2, pp. 360-361.

9. C. P. Richter, "Free Research versus Design Research," *Science*, CXVIII (1953), 92. Also see parts of K. R. Popper, *The Logic of Scientific Discovery* (London, Eng.: Hutchinson, 1935).

10. Johnson, *op. cit.*, p. 153.

11. Polanyi, *op. cit.*, p. 30.

12. *Ibid.*, p. 94. Also see E. Nagel and J. R. Newman, *Gödel's Proof* (New York: New York University Press, 1958).

13. Oppenheimer, *op. cit.*, p. 127.

14. Bertrand Russell, *Philosophy* (New York: W. W. Norton & Company, 1927), p. 157.

15. Bertrand Russell, *Mysticism and Logic* (London: Longmans, 1925), p. 75.

16. Albert Einstein, "Geometry and Experience," in H. Feigl and M. Brodbeck, eds., *Readings in the Philosophy of Science* (New York: Appleton-Century-Crofts, 1953), p. 189. Among other physicists and mathematicians who speak of the "art," "beauty," and "human enterprise" in mathematics, see P. W. Bridgman, *The Way Things Are* (Cambridge, Mass.: Harvard University Press, 1959), p. 78; M. G. Mittag-Leffler and J. J. Sylvester in H. Ellis, *The Dance of Life* (New York: The Modern Library, 1923), p. 132; Bertrand Russell, *My Philosophical Development* (New York: Simon and Schuster, 1959), pp. 211-212; and L. Henkin, "Are Logic and Mathematics Identical?" *Science*, CXXXVIII (1962), 788-794.

17. A relative of the same root as *explicit*.

18. A negligible minority of psychologists and social scientists have spoken in some sense to this point. Among others, see Quincy Wright, *A Study of War* (2 vols.; Chicago, Ill.: The University of Chicago Press, 1942), vol. 2, Appendix 25, pp. 1355-1364; A. H. Maslow, "Problem-Centering versus Means-Centering in Science," *Philosophy of Science*, XIII (1946), 326-331; R. Redfield, "The Art of Social Science," *American Journal of Sociology*, LIV (1948), 181-190; A. M. Rose, *Theory and Method in the Social Sciences* (Minneapolis, Minn.: The University of Minnesota Press, 1954), chap. 14, pp. 245-255; H. Cantril and C. H. Bumstead, *Reflections on the Human Venture* (New York: New York University Press, 1960); B. F. Skinner, *op. cit.*; C. P. Richter, *op. cit.*; H. Blumer, "What Is Wrong with Social Theory?" *American Sociological Review*, XIX (1954), 3-10; C. Wright Mills, *The*

Sociological Imagination (New York: Oxford University Press, 1959), pp. 56-75, 119-131; Bernard Barber and Renée C. Fox, "The Case of the Floppy-Eared Rabbits: An Instance of Serendipity Gained and Serendipity Lost," *American Journal of Sociology*, LXIV (1958), 128-136. See also Barber's *Science and the Social Order* (Glencoe, Ill.: The Free Press, 1952), chap. 9.

19. Wigner, *op. cit.*, pp. 156-157. To deal explicitly with some of our concepts requires a forbidding, even though effective, elucidation. See Ralph Barton Perry's chapter devoted to defining "value" in *Realms of Value* (Cambridge, Mass.: Harvard University Press, 1954), pp. 1-14.

20. George Polya, *Mathematics and Plausible Reasoning* (2 vols.; Princeton, N. J.: Princeton University Press, 1954), vol. 2, p. 76. Also see the physicist, Bunge, *op. cit.*, pp. 67-90.

21. Usually, of course, these students are not participant-observers but specialists in the study of strictly visible behavior and the professed motivation of their subjects; frequently they deal with socioeconomically depressed and naïve subjects who, like primitives, have little to conceal and could not anyhow. Or—to reveal my own snobbery—their chief work is to interpret compilations. To the degree that science is successfully dispassionate, it becomes malodorous. Society implicitly assumes that science will function to aid society's moral purpose. Where the scientist avoids both the heat and risk of society's passions and derives equal theoretical delight from the victories and defeats of its interest groups, he is immoral in terms of the supporting society's interest. And the leaders of science, paid by society, may be supporting immorality when they urge perfect detachment in their protégés. See Perry, *op. cit.*, pp. 316-318.

22. In my opinion, science may influence ethics; that is, I do not think of ethics as always an independent variable.

23. Perry, *op. cit.*, p. 430.

24. "Realism" takes the unrealistic view that conduct is not moral unless it is *completely* unselfish, which means, in effect, that morality is impossible. As an ethicizing animal, man is paradoxical on this matter. In all his other doctrines and theories, he allows room for error and shortcoming, but on the topic of ethics only absolute conduct is acceptable. See the analysis by a philosophizing biologist, C. H. Waddington, *The Ethical Animal* (London: G. Allen, 1960).

25. Wayne A. R. Leys, *Ethics and Social Policy* (New York: Prentice-Hall, Inc., 1941), pp. 156-167.

26. This view is old but has been well stated in the setting of our geography and time. See Benjamin Cardozo, *The Nature of the Judicial Process* (New Haven, Conn.: Yale University Press, 1921).

27. Spinoza in the seventeenth century and William James in the nineteenth noted that name-calling may reveal more about the name-caller than the person called a name. Convinced of the point by observation, I came

to use name-calling as an emergency alert to new leads. See F. S. Cohen's focus on name-calling in another vein, "The Construction of Hidden Value Judgments: Word Choices as Value Indicators," in L. Bryson *et al.*, eds., *Symbols and Values: An Initial Study* (New York: Harper and Row, 1954), pp. 545-561.

28. This kind of behavior obviously parallels the strict versus liberal interpretation of state or federal constitutions, union–management contracts, legal precedents, religious and ethical principles, etc.

29. See instances below of the conduct of reneging informants, the formal study made of Rambeau by a group of students, the problem of getting the membership of personnel in the Masonic order, etc.

30. Immemorially, high figures in political, religious, and other bodies have won the aid of friends and relatives as agents and liaison persons to get information obtainable only by subtle indirectness or to carry out difficult missions. What is implied in *diplomacy*, confidential or open, is not confined to international relations. And its "exalted haggling" and "exquisite politeness" exist in many areas of private life and may spill over into social research with fruitful results.

31. To research ethicists, this may connote "sinister calculation" or even "Machiavellian simulation and dissimulation." (Simulation is now an accepted term in electronic data-processing, which may similarly, however, have suspect uses.) If so, the practice was widely shared with me and caused no disturbance, except among a few reneging informants (below) who, when the inquiry deepened, feared reprisals for revealing secrets but similarly practiced "injustice" when the risk was small. At the time, I silently thought their behavior supported Thrasymachus' observation that "mankind censure injustice, fearing that they may be victims of it, and not because they shrink from committing it."

32. They were the topic of focus. Union officers, production workers, and others would, of course, have defined ability in different ways.

33. For most of the pros and cons on the method of participant observation, see the numerous papers in Richard N. Adams and Jack J. Preiss, eds., *Human Organization Research* (Homewood, Ill.: The Dorsey Press, 1960); William F. Whyte, "Observational Field-Work Methods," in M. Jahoda, M. Deutsch, and S. W. Cook, eds., *Research Methods in Social Relations* (2 vols.; New York: The Dryden Press, 1951), vol. 2, pp. 493-513; "Communications," *American Sociological Review*, XXIV (1959), 397-400; R. W. Janes, "A Note on Phases of the Community Role of the Participant-Observer," *American Sociological Review*, XXVI (1961), 446-450; William F. Whyte, *Street Corner Society* (2d ed.; Chicago, Ill.: The University of Chicago Press, 1955), Appendix on Method.

34. Though returns for my effort were small, this experience with the priests gave me the vagrant notion that an individual researcher's effort, in projects like that at Milo, is often only a mask for a group venture that, theoretically

weak, still works well; if formalized, with egos given a basis for recognition and reward, the venture becomes less cooperative but theoretically sounder.

35. Early I established and followed the obviously essential policy that I must encourage informants to give me their views of others, but I should avoid making any comment about others—either in response to questions or spontaneously—that could complicate the research.

36. My appearance on these days inevitably raised questions among some of my acquaintances about why I was in the plant when I was known to be working part time on certain days. I adapted my answers to their known sensibilities or I gave the excuse of overtime, change of schedules with others, that I was out to pick up a check that I missed on payday, or to change my insurance.

37. The quickest means of getting close to the American-born was through discussion of sports, especially baseball. "Ultimate" rapport sometimes required the accessory topics of politics and sex. One student reports on the danger of too much rapport. See S. M. Miller, "The Participant Observer and 'Over-Rapport,'" *American Sociological Review,* XVII (1952), 97-99.

38. Several members of this group—in and out of my department—thought that I was insufficiently status-conscious about my job in terms of behavior and dress. One of these persons in the department believed I "spent too much time talking to workers." He suggested that I "must be trying to stir up trouble," on seeing me often with the grievance committeemen, Brady, Beemer, and Spencer. At this period, it was essential for the research that I spend considerable time with workers and shop stewards. I wore work shirts and old shoes similar to theirs with the thought that this would help build a bond. In my hearing, one of the first-generation Americans remarked, "We ought to make up a kitty [take up a collection] to buy Dalton a tie." The foreign-born defended me as "not stuck up" and as "a guy who don't show off." Privately and contemptuously they pointed out to me several of the American-born (first-generation and older) who had "been on relief" during an earlier business recession and who, alluding to their current material consumption, could again easily be in that condition.

39. Here, I was guided and greatly stimulated by H. L. Mencken's *The American Language* (New York: Alfred A. Knopf, 1936), chap. 10, and *Supplement Two* to the same work (1948), chap. 10.

40. In staying close to these data, I consoled myself with Santayana's observation: "A conception not reducible to the small change of daily experience is like a currency not exchangeable for articles of consumption; it is not a symbol, but a fraud." *The Life of Reason* (5 vols.; New York: Charles Scribner's Sons, 1906), vol. 2, *Reason in Society,* chap. 8.

41. Others have also discussed the problem in other terms: Herbert A. Shepard, "A Note on Cumshaw" (8 pp.; September 25, 1961), and Yvonne Treadwell, "The Relationship of 'Cumshaw Tolerance' to Authoritarianism and Other Variables" (22 pp.; January 1962). As confidential studies of a

reputable organization doing significant work, both papers are unpublished and restricted.

42. And at least from sometime before December 15, 1663, when young Pepys was shocked by the advice of an older bureaucrat, we have known that functioning members of bureaucratic structures must keep their personal enmities and friendships pliable.

4 The Biography of a Research Project:

*Union Democracy**

SEYMOUR MARTIN LIPSET

To return to the scene of a research study a decade after one has left it is a difficult task. It is not easy to recall one's feelings at the time. The fact of a printed volume, the reactions of readers, the work one has done since, all the other events which have occurred, mean that, when thinking back on what happened, one can only be engaged in the process of rewriting one's personal history. No matter how much intellectual change has taken place, there is a tendency to search for a strain of consistency, as well as of development. I know that I can argue the case that all of my work represents elaboration of similar themes. But such a judgment should not really be made by an author; it is for his critics to decide what has really gone on. I would like, therefore, not to concentrate this report on any methodological or theoretical rein-terpretation of the materials in *Union Democracy*, but rather to repro-duce as accurately as I can recall them some of the major events and decisions which led to the book's taking the shape which it finally did.[1]

INITIAL CONCERN

My first contact with the International Typographical Union came when I was quite young. My father was a lifelong member of the

* S. M. Lipset, Martin Trow, and James S. Coleman, *Union Democracy* (Glen-coe: The Free Press, 1956); reprinted in paperback (Garden City: Doubleday Anchor Books, 1962).

union. He had been a printer in tsarist Russia before he emigrated to the United States and had belonged to the printers union there. While in elementary school and high school, I frequently overheard discussions of union matters, and occasionally my father would take me to the monthly meetings of the New York local, which were held on Sunday afternoons at Stuyvesant High School—a set of experiences which was to play a role later in my conceiving of the "occupational community" as an important part of the environment of the union.

My special interest in the politics of the union stemmed, however, from certain political concerns which developed out of my experiences with and interest in socialist politics. I belonged to the Young People's Socialist League while in high school and as a freshman at City College in New York. At that time, the late 1930's and early 1940's, two of the main interests of socialists were events in the Soviet Union and Spain. Stalin's bloody purges and show trials were at their height, and in Spain the Communists used their increasing influence and power to destroy anti-Communist socialists and anarchists. The earlier triumph of Nazism in Germany and then of Franco in Spain pointed to the incapacity of the democratic socialist movement to act vigorously on its own as either an anti-Communist or an anti-Fascist force. And, in the eyes of a young radical, the trade-union movement, both in this country and in Europe, also seemed incapable of offering the type of effective leadership which might make significant social reform possible.

The experience of the left and labor movements in various countries indicated that the building of a large socialist or labor movement, or even its coming to power, was not sufficient to democratize a society. In the Soviet Union, a supposedly Marxist revolution had resulted in the creation of a new, even more exploitative, form of class rule than existed in the capitalist countries. The social-democratic parties and trade-unions in other countries seemed to exhibit more concern with organizational stability and survival than with the need to advance radical solutions to social problems. Though the situations and reactions differed greatly, it seemed that most movements which were dedicated to social reform, a reduction of class exploitation, and an increase in democracy did not act to further these objectives, once they held any significant share of power.

The question was why, and the best explanation I found was in Robert Michels' *Political Parties*,[2] a book which Philip Selznick called to my attention. Michels, though writing before World War I and the

Russian Revolution, anticipated the "betrayal" by the social-democrats of their internationalist commitment to oppose war among the European powers and the exploitative power orientation of any group of radicals which came to revolutionary power as the Bolsheviks were to do. He argued—and we summarized the argument in *Union Democracy*—that inherent in any large-scale social organization are the motivation and means to make the leaders of the bureaucratic apparatus of the organization place the retention of their superior position ahead of any commitment to democracy. Hence, any effort premised on the assumption that social reforms introduced as a result of political change would eliminate the power and privileges of the dominant strata was utopian, and ideologies supporting them only served the interests of new potential ruling groups. Inherent in complex organization was oligarchy—the rule of the few over the many.

In *Political Parties,* Michels specifically applied his analysis to the behavior of socialist parties and trade-unions. He argued that they were, and of necessity had to be, self-perpetuating oligarchies, that democracy within them was impossible. But here is where the ITU became relevant. The internal politics of almost all American trade-unions may be cited as examples of Michels' "iron law of oligarchy." In fact, two books illustrating Michels' thesis with American examples had been written which do just this.[3]

The record of the ITU, however, seemingly contradicted Michels. Here was a large trade-union which governed itself through an elaborate democratic political system. Members frequently defeated leadership policies in referendums. Perhaps of even greater significance was the turnover of union officers on a national and local level. International officers as well as the heads of large locals were defeated for re-election in contests conducted under rules that protected the rights of opposition and willingly gave up office to return to work in the print shop— phenomena which occurred rarely, if ever, in other trade-unions.

If there was an answer to the "iron law of oligarchy," it might lie in the Typographical Union. The very existence of the union's political system demonstrated that the "iron law" was not made of iron. And, when I decided to go on to graduate work in sociology, I thought of looking into the Typographical Union as a research topic for a thesis.

In large measure, my interests as a prospective graduate student flowed from my political concerns. In addition to wanting to find out whether there was a realistic "socialist" alternative to Stalinism—to

socialism simply developing a new, perhaps worse, form of class rule—
I wanted to know why all efforts to build any kind of socialist party
in the United States had failed miserably. Before beginning graduate
work, I had resolved on two topics which I wanted to study—the political system of the ITU and the emergence of a large socialist party,
the Cooperative Commonwealth Federation (CCF), in Canada. The
CCF, which attained large-scale support during the war years, was the
first mass socialist party in English-speaking North America. Its success
seemingly challenged the assumptions of those who argued that elements inherent in the history or current social structure of America
made an American socialism impossible. Most of the explanations of
"why no socialism in the United States" seemed to apply equally
strongly to Canada. Consequently, an analysis of the CCF might shed
light on the conditions under which a socialist movement could be
built in this country.

From an academic point of view, my interests in the CCF and the
ITU represented a concern with what I was to learn later from Paul
Lazarsfeld was "deviant-case analysis." I selected two major cases
which contradicted what appeared to be general laws and desired to
find out whether the general law was wrong or whether there were
some new or special factors present in the deviant cases which permitted the unexpected to occur. On the first day I registered as a graduate student at Columbia, I recall telling Prof. Robert Merton, when
he asked what interested me in sociology, that I wanted to do a doc-
toral dissertation on the CCF or the ITU.

In my first semester at Columbia in the fall of 1943, I wrote a paper
for Merton's course on social organization dealing with the ITU as an
exception to Michels' law. The paper described the system and indicated why various efforts in the literature to explain ITU democracy
did not seem valid (largely because all the factors cited were equally
present in other unions in printing trades, such as the Pressmen, which
was dictatorially oligarchic). On the positive side, my principal suggestion was that the presence of a high level of membership participation, opposition to leadership policies in referendums, and the turnover
of leaders were to be explained by the self-maintaining mechanisms of
a two-party system. Once institutionalized, such a system prevented
official salaries from becoming high, and the existence of a formal opposition meant that leaders who shifted from office to opposition retained
their status as important leaders—factors which reduced the pressure

on union officers to hold office at all costs, whereas the parties activated interest among the members and made them aware of issues which affected them.

Fortunately for me, Merton thought well of this paper; it served as the main basis of support for a fellowship which enabled me to finish at Columbia without outside employment. I intended to continue with a more thorough investigation which might result in a thesis on the subject, but Robert Lynd, with whom I also worked at Columbia, became extremely enthusiastic over the idea of studying the CCF. Looking at a successful socialist party some considerable distance from home seemed more attractive than further investigation of the printers in New York City. I applied for a fellowship from the Social Science Research Council to study the CCF in Saskatchewan, where it held office, with the mental reservation that, if I did not get the fellowship, I would go back to the ITU for my thesis. I received the fellowship and as a result spent the next three years (1945-1948) studying or writing about the CCF. The outcome of this was eventually published as *Agrarian Socialism*.[4]

THE DEVELOPMENT OF THE STUDY

In 1948, after three years in Canada (one in Saskatchewan and two teaching at the University of Toronto), I took a position at the University of California at Berkeley, which included a part-time research appointment in the Institute of Industrial Relations. On coming to Berkeley, I became involved in a study of social mobility, but I also revived my interest in the ITU. Both Clark Kerr, then the director of the institute, and the late Lloyd Fisher, its associate director, encouraged this concern. Fisher, in fact, himself began studying certain aspects of the ITU's legal system as part of his general interest in the emergence of a system of common law within industry, unions, and other private governments.

I elaborated on the paper which I had originally written as a graduate student and subsequently published it.[5] The elaboration essentially involved a much more detailed analysis of the history of the union and its party system and an examination of everything which I could find written on the party system. I discussed in detail the origins of the party system in the contest between the secret societies of printers formed in the latter half of the nineteenth century and their opponents. My essential argument was that a unique set of historical circumstances set in motion a reinforcing process in which two roughly equal factions battled each other for some years and, in the course of the struggle, institutionalized various democratic rights, much as has occurred in larger civil polities.

At this stage of the research, I thought that I understood the historical developments which had led to the competitive party system, but I felt dissatisfied with all efforts, including my own, to explain why this system had become institutionalized. The events which had given rise to the two-party system were relatively uncommon in the annals of American unionism, but clearly one could not accept an explanation that placed primary responsibility on a unique set of fortuitous circumstances, important as these might have been as precipitating factors. Many unions had histories of long-term, regularized factionalism which resembled events in the past of the ITU, yet none of these organizations had evolved a functioning, stable party system. The issue with respect to unions and other private governments, as with nations, was why comparable efforts succeeded in some places and not in others.

The conditions of the creation and maintenance of as complex a system as a competitive political democracy are never simple. Structural conditions which might facilitate it can do so only if the "right" pattern of historical events occurs; efforts to erect institutions for which the necessary structural supports are not present most often fail, even if the actors have the best intentions. It was necessary, therefore, to link the historical study to a detailed structural analysis. I planned to continue my investigations through systematic observation of various bodies of the union (I attended an international convention about this time) and through detailed interviews with informants who had been active at various union levels. I accepted a visiting appointment to Columbia University in 1950, in part because I thought that I could learn a great deal about the union while in the East.

During 1950-1951, I collected data from members and officers of the New York local. Sometime during this year, I became sensitive to the possible significance of what we were to eventually call the printers' "occupational community." That is, it struck me that the pattern of leisure activities followed by printers differed greatly from that of most other workers. There were a large number of organizations—social clubs, veterans groups, athletic associations, and the like—composed solely of printers. I do not now remember what reports or discussions made me most aware that this might be a major clue to the problem. Perhaps most important was my reading a number of weekly and bi-weekly newspapers published as private ventures which reported on the doings of printers and which had no formal relationship to the union. Fortunately, many of these papers had been preserved, either by individual printers who let me see their files or by the New York Public Library. As one read through these papers, some dating before the turn of the century, it became evident that this occupational community had been a major element in the lives of the workers. Clearly, many printers have spent much of their leisure time in activity with other printers. Many have held office in various of these organizations. At some periods, there were two papers reporting on printers' activities in New York, and these would refer to comparable organs in other cities around the nation. Interviews with union members indicated that the occupational community still persisted as a major pattern of printers' lives.

This focus on the occupational community was perhaps the decisive event of the entire study. Previous to this, I had been looking for mechanisms which would explain why printers and their leaders deviated from the Michels model. Michels had pointed to the "incompetence of the mass" to act politically and their reluctance to participate as factors contributing to oligarchy. My principal interpretation of the relatively high level of printers' participation had been that the party system itself operated to stimulate, recruit, and train men for union politics. The occupational community indicated the fact that men were tied to printing through their leisure activities which served to help recruit for the union's polity, that the occupational community developed leadership and organizational skills among men prior to any activity in union politics, and that it permitted politically relevant communications not under the control of union officialdom. For the first time, I had located some major possible explanatory variables which were not primarily aspects of the self-maintaining mechanisms of the political system.

Concern with the occupational community now shifted the focus of the study somewhat. A major question which had to be answered was why printers have an occupational community when other occupations do not have one. What was unique about the situation of the printer? As the discussion in *Union Democracy* indicates, a number of factors emerged as possibly relevant agents. The special hours of work of many printers was clearly one. Reading through the printers' newspapers indicated that much of their athletic activity—e.g., the once-numerous baseball leagues, fishing clubs, and golf tournaments—occurred during the day. Such activity was made possible by the fact that a large proportion of printers, particularly younger men working on newspapers, were employed nights and/or week ends. Such men had to find leisure companions at times when most workers in other occupations were at their jobs. Clearly, it seemed, they were pressed on one another; the most obvious and easiest group from which one could find friends was that of others in the same situation. The career cycle of printing meant that many men now working days had spent years on a night shift before earning the seniority to change over, so that their friendship patterns might reflect past experiences.

This was, of course, not the sole explanation. If the timing of their work suggested a push toward other printers, their attitudes toward their work suggested a pull. Interviews with printers as well as much documentary literature indicated that printers were extremely proud of their work and their status in the community. In trying to think of the factors which might affect participation in the occupational community, it struck me that these "subjective" elements were probably relevant. If men like their work, this should be a pull toward relating to others in the same occupation, to form intravocational friendships. But perhaps more significant was the question of status. Although printers are manual workers and very proud of their union, the self-image which came through in interviews was of a group, many of whose members saw themselves as very educated, highly skilled members of a semiprofession. In discussing the union's virtues, printers would use other occupations and their unions as negative points of reference. They would explain their own positive traits as a consequence of their being better educated than other workers. As I developed the sense that printers, in fact, thought of themselves as somehow better than other workers, it struck me that this, too, might be one of the elements sustaining the occupational community. The hypothesis became formulated in the assumption that printers would try to disassociate from lower-status manual workers, but not being quite middle-class or white-collar in point of education placed them in an intermediate position between the working class and the middle class, impelling the printers to interact with one another.

The high level of intravocational interaction and participation in the union also appeared linked to yet another characteristic; the effort to isolate unique elements inherent in printing somehow suggested the homogeneity of the occupation. If one compared the compositors, the constituency of the ITU, to the membership of the Pressmen's Union— a methodological procedure which I followed often when thinking of some new seemingly causal interpretation—it developed that the Pressmen had a large membership among pressmen's assistants, a lower-skilled, lower-paid, and lower-status group. Those conversant with affairs within the Pressmen's organization indicated that the union tended to function primarily for the skilled pressmen, that they supported the incumbent administrations, that, in effect, there were two classes in the union. A cursory look at other unions indicated that the phenomenon of multiskilled unions was quite common, not solely among industrial unions, where it was inherent in the structure, but even among many, if not most, so-called craft unions. Where such variations occurred, the more skilled often dominated.

Assuming that status homogeneity played a role in effecting general involvement in the occupational community and the union led to the general assumption that a high level of participation among printers— seemingly one of the causal factors which helped sustain a democratic polity—was a function of a relatively homogeneous, one-class community. In this community, various circumstances had resulted in the creation of a large number of community institutions—newspapers, clubs, and sports organizations—which served to link printers with the political institutions of the community—the union and its parties—and to train men in the skills of politics, creating a regular system of communications among members of the group.

At this point, I realized, much to my surprise, that I had come up with findings very similar to those which I advanced in *Agrarian Socialism* to account for the high level of political participation among wheat-belt farmers. In that book, I suggested that a primary cause of the many "class" political and economic organizations among them is the fact that wheat communities are essentially small communities of men in, roughly speaking, the same job and status who are forced to organize many community institutions themselves. I had estimated that about one out of eight Saskatchewan farmers actually held some community leadership post. As I recall the situation, I was never aware while I was focusing on the causes and consequences of the printers' occupational community that many of my findings and much of my analysis were similar to those in *Agrarian Socialism.* The coincidence only struck me much later; but, once realizing it, one cannot help wondering to what extent one set of findings determined the other. In essence, both studies advance a "theory" of political participation and democratic organizational government which suggests that conducive to these are situations in which a constituency is required by its social situation to take part in a large number of organizations, the membership and leadership of which are limited to people in similar occupations; in which there is a relatively small gap in status and background between leaders and followers; in which there are easy intragroup communication channels; in which the basic units are relatively small; and the like.

About this time, discussion of these hypotheses with others, particularly Daniel Bell, suggested the similarity of my analysis to de Tocqueville's discussion in *Democracy in America.* De Tocqueville had also been examining what in his time was a unique case, a democratic, postrevolutionary republic. From the experience of the failures of French democracy, he looked at the United States to see what made democracy work there, what was present in America that was missing in France. De Tocqueville emphasized the role of "secondary powers"—institutions independent of the central state authority which could act to lift the level of political involvement, could serve as sources of new ideas and means of communicating information independent of the state's interests, and could serve as sources of resistance to efforts to centralize state authority. De Tocqueville's view was that, in societies in which there was an unmediated relationship between the central state and the mass of the citizenry, the latter would be unable to resist the power or the blandishments of those who controlled state authority and the main media of communication. What distinguished the United States from France was precisely a plethora of voluntary organizations independent of the state, together with local autonomy of state and municipal authorities. These organizations and local governments served to prevent authoritarianism and helped to stabilize democracy.

Not only was the analytic approach similar to de Tocqueville's, but the explanation of ITU democracy which was emerging also corresponded closely to his. The organizations of the occupational community broke up the mass-and-leader relationship which exists in most trade-unions. As de Tocqueville suggested, not only the voluntary associations of the occupational community were relevant, but the autonomy of the formal subunits—in this case the shop and local union organizations—also appeared to contribute to the preservation of democracy.

As in many other craft unions, locals in the ITU have a considerable degree of autonomy. The historical record demonstrated that the opposition to a given dominant group which controlled the international union administration could usually operate through its control of certain large locals. Oppositions could maintain themselves, could train future international candidates, and could establish competing records through the fact that they always controlled some of the major locals. The autonomy of the locals in large part reflected conditions in the industry—printing is highly decentralized; there are almost no national concerns; prices and wages vary from one community to another. But fundamentally the locals served to maintain the political system because it already existed; their political independence is as much a consequence as a support of the democratic system.

The logic of the analysis also implied a look at the shop organization and its social environment. Each union print shop is organized as a chapel, with a chapel chairman and other officers and regular shop meetings. In a democratic organization with regular opposition, it is clear that the shop organization is, in effect, still another of the independent power centers and that shop or chapel officers are men who can wield independent power. A man may make a reputation as an effective leader while chapel chairman in a large shop and hence become a potential candidate for office, and the chapels provide an arena in which politics may occur.

At this stage, sometime in 1951, I had a fairly clear picture in mind of the factors which had created ITU democracy and those which sustained it. I also thought that I understood how it operated; how men were recruited to union politics; how parties gained and lost support over time; and how due process had developed and operated in the union to protect the rights of members, not only against employers, but against union officers. I now felt that I could write a book reporting on these matters in detail. I did write a long article, "The Political Process in Trade Unions: A Theoretical Statement," [6] which outlined in formal fashion much of what I had learned from combining Michels with the various hypotheses suggested by the ITU analysis.

THE SURVEY ANALYSIS

In thinking how to further the research, I began to consider the possibility of gathering survey data to actually test many of the hypotheses which I was advancing concerning the role of the occupational community, the self-image of printers, the effect of nightwork, and the conceptions which union and party leaders had of their occupation and their place in it. I was encouraged to do this by Charles Glock, Paul Lazarsfeld, and Robert Merton, then among the directors of the Bureau of Applied Social Research of Columbia University.[7] A proposal for funds for a survey study summarizing many of the interpretations up to the moment was submitted to the Rockefeller Foundation. The foundation official with whom I dealt told me that some of those who had been asked to comment on the proposal suggested that the only way to really test hypotheses such as these would be through a comparative study of many unions. My answer was that, though this was logically true, many of the hypotheses could be elaborated by internal analysis, by comparing those printers whose situations most resembled the assumed ideal-typical one with those whose environment was more like that of other workers. As the Introduction to the book makes clear, we did obtain a sum of about $10,000 from the Rockefeller Foundation.

Martin Trow, then a graduate student in the Columbia Sociology Department, had been working with me on the study in 1951, and he took over responsibility for administering the gathering of survey data. Trow had some previous survey experience at the bureau. The main task of the survey was to convert the hypotheses which had been developed earlier into questions for a schedule which could be administered to a sample of union members.

The final schedule is reprinted as an appendix to *Union Democracy*, and I will not report on it here. It is relevant to note, however, that many of the questions are not of the obvious, common-sense, or "shotgun" variety. Questions were asked concerning involvement in the occupational community; nightwork, both past and present; political experience in the shop; sources of information on union business; friendship patterns (occupations of friends, whether they visited other printers); occupational reference-images—what printers thought other people felt about various aspects of the printer's role; job satisfaction; their preferred career for themselves or children; the extent of their general political interest; participation in general (nonprinter) community organizations; general political opinions on items which might be classified on a left–right scale; and the like. In addition, of course, we asked many questions concerning the union and its party system.

In designing the sample, we made a number of decisions based on the need to test various system hypotheses. First, we decided to take three samples: (1) a sample of ordinary members, eventually 434; (2) a sample of chapel chairmen drawn from the same shops in which we would interview; and (3) the formal leaders of both political parties in the New York local. We could thereby compare attitudes and background factors among men at various levels in the system and see whether there was any major deviation among them. Second, we selected a "stratified sample of shops," rather than individuals. Since many of the propositions concerned diverse attributes of various shops (e.g., homogeneous for one party, two equal parties, large or small, etc.), it would be necessary to be able to talk about the attributes of shop environments as revealed by the interview data and to compare the behavior and attitudes of men with similar personal background features who were in widely differing social environments, e.g., large versus small shops. In so planning, I think that we were the first to design a sample survey seeking data on structures as well as individuals.

The methodological innovations evidenced in our sample design did not stem from any special concern with creative methodology. Rather, it and the questions reflected the fact that the survey was planned to test a highly complex, analytical model; it was a sophisticated survey design precisely because years of prior investigation of the attributes of a complex system had preceded it. Having the hypotheses formulated well in advance meant that the methodology almost created itself. Since we wanted to test hypotheses on the consequences of various structures, we had to have data relevant to structures, not only to individual members of the union. The need to have both was met by sampling individuals systematically in selected structural environments, as is described in the methodological appendix to *Union Democracy*.

I will skip over quickly the stages of data collection. Trow largely supervised this, together with staff furnished by the Bureau of Applied Social Research. All told, we were remarkably successful in securing the interviews. Most were actually done in the print shops. The union had agreed to the survey, leaders of both parties had been consulted, and I spoke to a meeting of the chapel chairmen about it. Our interviewers—largely sociology graduate students—made appointments with the men for an hour to an hour and a half. As we noted in the book, it attested to our unwitting assumption as to who controlled the pace of work in the shop that we never thought of asking permission from employers to interview their workers during work time, and we rarely heard from them. Occasionally an interviewer reported that a shop foreman had asked as a favor that the interview be done at another time for the convenience of the work.

While waiting for the survey data to come in, be coded, and be entered on IBM cards, Trow and I wrote a long article, "Democracy in the Printing Trade Unions," in which we summarized almost all our conclusions concerning the nature of the political system. It was our intention to have this in print so that subsequent readers of the book could see the extent to which the interview findings had led us to modify initial assumptions. Unfortunately this article, scheduled to be published in a book of original essays dealing with various trade-unions, never appeared in print for reasons related to the health of the book's editor. By the time the editor had released the article, we had finished a draft of the book including the survey data, and it did not seem proper to publish an article which had been superseded. This article was, therefore, never published, though much of it, naturally, was incorporated in *Union Democracy*.

After collecting the interview data, we decided to use the opportunity presented by a union election in May 1952 to send out a questionnaire to the men we had interviewed. Many of our hypotheses were "process hypotheses"; that is, we postulated effects of certain experiences over time. Since we had interviewed the men before an election campaign started, we assumed that the election campaign would operate on the members in ways which would enable us to test the hypotheses—e.g., which men would be most exposed to election materials, would be most likely to be concerned, would be most prone to shift their opinions, and so forth. The response to this questionnaire was very good—about 70 per cent of those interviewed returned it—and we could therefore engage in panel analysis, studying the characteristics of those who changed their attitudes or behavior during the campaign.

While our survey of a representative sample of the members and chapel chairmen was being processed, we conducted a number of long, exploratory interviews with leaders of the union and of its parties in order to fill out gaps in our knowledge of the occupational community, the party system, the local shop organization, and so forth. The respondents included substantially all the most active men in the local unions' political life. There were some thirty-five of these interviews, some lasting more than five hours, which focused on the processes through which these leaders had been drawn into the political life of the union and on the norms and behaviors of the most active men which sustained it. Trow and I were joined in this enterprise by James S. Coleman, then also a graduate student in sociology, who from that point onward was an active collaborator in the analysis of the data and the writing of the report.

By this time, no funds were available for other than IBM machine runs, and from September 1952 neither Trow nor Coleman received any pay. The only subsequent costs were for machine time and clerical and typing work. This fact in part explains why the study cost so much less than other surveys, since survey analysis requires a great deal of analytic time.

In dividing up the analysis, the three of us took separate sections. I was responsible for first drafts and primary analysis of the introductory, theoretical, and historical sections and of the analysis of the role of the occupational community—essentially what appeared in the final version as chapters 1 to 7; Trow dealt with the analysis of the shop organization and of the character of leadership (chapters 8-13); Coleman was basically responsible for the analysis of the electoral process, of political issues, and of the decisions of individuals, shops, and locals (chapters 14-17). These chapters constitute a summary of the material he submitted for his Ph.D. thesis, which he wrote as part of the study. I wrote the first draft of the final chapter, 18, and Coleman was the main person involved in drafting the methodological appendixes. Although initial authorship may be separated in this way, each of us spent considerable time rewriting the others' chapters and jointly planning further analysis of sections other than those under our prime responsibility. And, to get ahead of the story, Nathan Glazer ultimately helped edit the final product by cutting down a much lengthier manuscript and by improving our way of saying things.

I have been asked to specify what the survey study supplied for the book, other than to confirm points of which we were already convinced. Some friends who were conversant with the study as it developed and who do not share our commitment to testing social-science hypotheses by quantitative methods where possible have questioned whether the survey added much to our knowledge. They point to the fact that the basic concepts and interpretations were elaborated in articles written before any survey data were available. It has even been argued that the presence of many data tables in the text often gets in the way of the nonsociological reader and that the book would have more intellectual influence without them. A leading American historian strongly urged us to remove most of these tables from the book before it was published.

It is important to answer this argument, for it challenges the principal assumptions on which most sociologists' work rests. The best possible way to judge such assumptions is by the fruit they bear, by their effect on such specific intellectual products as *Union Democracy.*

This argument has two points: first, what does the research in question actually contribute to the problem at hand, and, second, should the quantitative statistics show up in the final product? The latter is a secondary question, and its answer depends on the audience and the nature of the problem. The former is the more fundamental, and it is this question of intellectual contribution that I will address.

Clearly, the value of the contribution of quantitative statistics, such as our survey and our statistical study of the vote in locals, will vary widely, depending on how much is initially known about the problem. In this case, years of study of the union had preceded the quantitative research, and thus a great deal was known about how the ITU functioned and about the structural supports to democracy in the union. Thus, the role of the quantitative analysis here was not one of discovering the major relationships, as it would have been if we had begun without such knowledge. Instead, it was a role of uncovering mechanisms, through looking at the fine structure of the occupation and the union. A list will indicate some examples of just how this worked.

(1) The strength of the occupational community was well recognized prior to the survey. But the survey showed just which elements in the occupation helped create the occupational community. We found that the substitute system, the size of print shops, and the great amount of nightwork constitute three important supports of the occupational community.

(2) The importance of the occupational community for support of the two-party system was evident before the research. But, again, we did not know the specific mechanisms. We found that formal organiza-

tions in the occupational community had a dual impact: by generating involvement of their members in union politics, but also by providing independent avenues of status outside the union's authority system for potential leaders. The best example of this was the Monotype Club, in New York, which performed a special job-agency function for mono-type operators. Many of the New York Independent party leaders had gained their initial prominence through this club.

We found informal associations to have a highly specific impact on union involvement. When carried out in the context of printing, they were effective; when carried out at homes and in the neighborhood, they were not.

(3) In previous publications, I had pointed to the fact that printers had many organizational and political skills—a background that made them able and eager to control their own union. But we were quite unprepared for the fact that sensitivity to political issues and an aware-ness of ideology were so important in maintenance of support for the "out" party. Our analysis of voting behavior showed that the core of support for the out party during its periods of drought was largely the "ideologically sensitive" printers, who in many cases had developed their ideological positions in civil politics. It indicated that a major reason for the greater strength of the Progressives was the larger num-ber of ideologically sensitive men among them.

(4) We were well aware of the importance of some large locals in providing opposition to the international incumbent. Yet the statistical analysis of voting behavior in locals of varying size showed us far more: the smaller the local, the more likely it would support the inter-national incumbent. If the union had been made up solely of small locals, the incumbent would almost never have been defeated. The small locals not only failed to maintain within themselves a two-party system but were unable to develop support for an opposition provided them by the political activity of the large locals.

This list could be extended greatly. For example, the panel covering the period of the local election showed us something quite new—the importance of the campaign for the out party. It suggested that in democratic politics a campaign has a peculiarly asymmetric function—drawing attention to the out party and allowing it to win back support that by default and disinterest had slipped to the incumbent.

In addition to uncovering important mechanisms in the functioning of a democratic system, the statistical analysis helped complete a factual description of the union. It showed, for example, the differential paths to leadership between the two parties in New York—in the Progressives, principally through directly political activity, as chapel chairman, or as active spokesmen in the union meeting; in the Independents, principally through nonpolitical organizations in the occupational community.

The survey showed us also that the New York union was composed almost equally of Catholics, Jews, and Protestants. This turned out to be extremely helpful for analysis, since we had enough members of each religious group to permit holding religion constant when testing the influence of other factors. The existence of these three distinct religious-ethnic groups allowed us to analyze the behavior of printers living in ethnic communities which presumably placed differing values on printing as an occupation. Thus, we assumed that Jews, a group preponderantly middle-class in New York, rated printing as lower in occupational status than did Catholics, most of whom were still in manual occupations which had less status than printing.

Our assumption that we could account for variations in the behavior of printers by asking them which occupational groups they compared printing to when they thought of its relative standing with respect to prestige or pay turned out to be invalid. The answers to these questions did not correlate with other behavior as predicted. However, we then learned that "objective" measures of the reference group of printers were much better predictors of behavior than were the subjective ones. That is, assumptions concerning occupational reference groups derived from our interviewees' religious backgrounds or from knowledge of the occupations of their neighbors helped us to understand who participated in the union and why. Those whose potential resources for nonprinter friends were men in manual jobs were much more likely to associate with other printers off the job and hence be involved in the union than were men whose friendship resources included many in middle-class occupations. This finding coincided with our assumption that marginal status between middle and working classes pressed printers to associate with other printers.

The analysis of the interviews of both members and leaders suggested a contribution of a democratic political system to union solidarity of which we had not been aware. In accounting for the prevalence of a one-party system in trade-unions, many have argued that such groups cannot afford internal cleavage because they are fundamentally conflict organizations. They must maintain strong solidarity in order to oppose the employer. It is assumed that sustained attacks by opposition factions on union policies will weaken the bargaining power and strike potential of unions.

In my early article, "The Political Process in Trade-Unions," I argued against this assumption by pointing to the fact that many unions had had open faction fights during their formative years, when they were presumably at their weakest, and yet they had survived and grown. The interview data suggested, however, that institutionalized competitive politics and the existence of regular opposition may actually contribute to the strength of unions. They showed that many men consciously distinguished between the incumbent officials of the moment and the union as an organization. Opponents of the administration would tell us: "The ITU is the greatest union in the world," and go on to describe various of its virtues, while placing the blame for what was wrong with it, not on the union as such, but rather on a given officeholder who should be defeated for re-election. It seemed clear that, in unions as in nations, democracy encourages loyalty to the larger collectivity in spite of serious misgivings about policies, whereas conversely, in one-party or dictatorial systems, it is presumably much more difficult for citizens or members to differentiate between the officeholder and the institution which he heads.

The survey contributed one important unanticipated substantive point relevant to conditions for a democratic polity. This came out in the analysis of the effect of the procedure for entering the occupation and of the hiring system in general on the existence of the occupational community. The interview schedule contained a question concerning unemployment experience. It was inserted because there had been considerable unemployment among printers in the 1930's, and some men were not able to secure regular full-time jobs even during the postwar prosperity. I do not recall which factors we expected to relate to unemployment background, but various studies of political attitudes and voting behavior have reported that a history of unemployment tends to predispose men to more left-wing sentiments. At some point in our analysis, however, it developed that those who reported unemployment were likely to participate in the union and to be involved in the occupational community; i.e., to be high in friendships with other printers and in memberships in printers' clubs.

In seeking to explain this finding, it struck us that an experience common to the background of most union members was having been a "substitute." In order to get a job in a union plant, a man files his union card with the foreman. He is then entered on the chapel list and has a "priority," or seniority, number. Men are then hired or laid off from regular employment according to their position on this list. In newspapers and many large job-printing companies, the need for labor varies considerably with the season and with the day of the week. Consequently, the industry requires that more men be available for work than are needed in slack periods. Such men are called "substitutes." Young men, on finishing their apprenticeship, or older members, who have lost their jobs, join the substitute ranks in some company which they have reason to think will have considerable need for "subs" through the year or which will have enough turnover or expansion so as to hire men from the substitute list as regular workers. The period required to move from the sub list to the regular list has varied greatly over the years, but, until quite recently, it was not unusual for it to take many years—during the Depression, ten or more. Substitutes usually "show up" every day. Sub hiring is not done by seniority; foremen may choose whom they want, men taking a day off may designate a replacement, or a lottery is sometimes used as a means of distributing the available work. In order to get work, many men show up every day of the week, including week ends and, if they do not get work on the day shift, often stay "downtown" until the work for the night shift is distributed.

This system of hiring means that the subs must find a way of spending their days when they are not working. The most obvious outlet is to go to a neighborhood bar near the print shop, visit a printers' club, or engage in some other activity with other subs. Hence, the hiring procedure, like the system of nightwork, presses printers to associate with other printers, quite often men about their own age. The system also encourages substitutes to try to become friendly with regulars, since the latter might assign a day's work to a sub friend. (Regulars must "work off" overtime.) There can be little doubt that this system of hiring has contributed greatly to the existence of the occupational community; yet we did not recognize its possible significance until the survey data suggested the need to find an explanation for the relationship between past unemployment and informal relations with other printers. If we had developed the hypothesis before designing the survey, our questions bearing on the effect of the sub system would undoubtedly have been much more illuminating. At it was, we had no direct information on whether individuals actually had been substitutes and how long this period had lasted. The proportion who reported periods of unemployment was clearly smaller than the actual number who must have been subs at some time. Our interpretation of this seeming discrepancy was that it was a result of inadequate questions.

One of the important gains of the survey analysis was that it enabled us to study the social-psychological processes through which the historical and structural characteristics of the union had their effects on the attitudes and behavior of its members. For example, before the survey, there was some evidence that the men who worked in the larger print shops and newspapers played, on the average, a more active role in union affairs than those in the smaller shops. But the precise extent to which shop size was related to union activity and the nature of the relationship could not be established impressionistically; moreover, our speculations on why size was related to activity had not gone far. The survey data allowed us to explore in some depth and detail the mechanisms linking shop size to political involvement. We were able to show how the size of the shop a man worked in affected his relations with his employer, with the union, and with his fellow printers—all in ways that tended to make men in the larger shops more involved and active in union affairs. It was even possible to find an "optimum" shop size from the point of view of political activity and to show how and why in shops larger as well as smaller than this lower rates of activity in union politics were found. The survey data indicated that the optimum size for involving workers in union activity was 200-300; in shops larger than that, those uninterested in union affairs were able to escape the influence of the ever-present minority of activists, whereas at the other end it became clear that shops with fewer than ten workers were too small to permit serious controversy or discussion about union matters without disturbing the harmony of the shop. Moreover, the survey data showed that the chapel officers in small shops, unlike those in large shops, did not really fulfill a leadership or communications role. Many of the small shops did not have a single man who was seriously interested or active in union politics. Consequently, many of the men in the small shops were not drawn into the political life of the union at all.

This analysis did not merely test hypotheses already held before the survey was conducted. Rather, the earlier hypotheses pointed to a fruitful line of inquiry, but many of the ideas and insights regarding the bearing of shop size on union politics emerged only in the course of the analysis of the survey data. The process of discovery involved (1) a sociological perspective on industrial life which dictated that questions about the printer's relations with his employers, his union, and his fellows be included in the interview; (2) knowledge of the structure and workings of this industry and union from other sources which led to guiding hypotheses regarding the relevance of shop size; and (3) intensive analysis of the survey data (supplemented by other kinds of data) which explored the nature of the bearing of shop size on union politics.

Our increasing interest in the effect of varying group sizes led us to a detailed statistical analysis of the voting returns of union locals the country over. This was done by classifying union voting returns for many decades by size of local. It demonstrated conclusively that turnover on the international level almost invariably stemmed from large locals turning against national administrations which they had helped to elect. The smaller locals—those with fewer than one hundred members—almost invariably voted with the incumbent administration, a pattern which may have reflected their dependency on them (small locals had no paid officers), and the fact that, like small print shops, such locals did not have much of an internal political life (there is no party organization in most of them). Hence, usually they are only well acquainted with the administration's side of the case. In a sense, we then argued, the situation in the small locals resembled the political environment of most one-party unions, in which members and locals get information from the incumbents only and hence "voluntarily" continue to support them. The last section of the book—the analysis of the vote decision—probably turned up the most "new" insights, since relatively little had been done with respect to this problem in the presurvey stage of the study. It revealed the impact of factors which affect liberalism or conservatism on these attitudes in a homogeneous occupational group.

Perhaps the most surprising and rewarding aspect of the survey analysis was to find the extent to which large and consistent differences appeared even though our basic sample only included 434 men. It suggests that, when social factors are really strong as causal agents and one controls for a variety of factors by dealing with a homogeneous group, the level of prediction can be quite high—much higher than has usually been obtained in studies of voting in a community or in the nation as a whole.

Methodologically, also, the survey demonstrated that it is possible to relate the characteristics of groups or units and even of formal leaders to the behavior of individuals. By interviewing enough people within the same unit so that one can specify the social environment of the unit, one may show how individuals with comparable personal traits behave differently in varying environments.

The interviews with the leaders also produced some new findings. We had not anticipated that the leaders of *both* parties were to the left of their supporters on general social as well as union-relevant attitudes. However, both membership and leadership data showed, as we anticipated, that the parties varied on a left–right scale, that the party which was more moderate on union issues was also supported and led by men who were more conservative on general political questions. The analysis of the issues also made us more sensitive to the impact of the need for issues not rooted in interest differences as a means of maintaining competitive politics within a private government.

In the civil polity, most of the major variations in attachments to parties are related to differences among groups with varying interests; e.g., the concerns of farmers, of businessmen, of Negroes, of unions, and so forth. The role of parties is to mediate among the various interests. This would be impossible in trade-unions, since interest issues there are clear-cut and unreconcilable by votes or shifting coalitions. Union members may differ between the skilled and less skilled as to what kind of wage demand to make, between those working for one type of employer or another, among those working in different regions. If an issue is clearly one which turns an immediate financial advantage to one group or another, it cannot be settled by majority vote; the larger group would always win, and the smaller group would look for means of seceding. In unions with diverse interest groups, such as most industrial unions, such conflicts are in effect arbitrated by the leaders of the union; they are rarely put before the membership for settlement by election. The need for such mediation procedure sharply lessens the possibility of competitive politics. Ideologically linked politics (left–right) or efficiency politics (ins versus outs), the types which characterize the contests in the ITU, are seemingly the only ones possible in unions on a long-term basis. The relevance of this point was not clear to us until we began to look for explanations of the fact that there had actually been few interest issues in the history of the union's political system.

At all times during the investigation, we were sensitive to the problem of comparative analysis. Were the factors which we concentrated on as characteristic of printers and the ITU actually specific to the union's system? What about other unions and occupations? If nightwork were conducive to an occupational community, did such groups exist among other groups of nightworkers? The evidence indicated that some of the others did have occupational communities.

Did printers, in fact, like their work better than those in other manual occupations? Here we looked for data from other studies of job satisfaction; these tended to verify our assumption that printers were in fact very high on the scale of job satisfaction. How about printers in other countries? Here we examined studies of labor unions around the world. These indicated that almost everywhere printers were among the first group of workers to organize into unions. The common explanation of why printers were unique in this way was congruent with our analysis.

Many writers pointed to the existence of an occupational community among printers elsewhere as facilitating trade-union organization; they often had vocationally linked organizations before they had unions. Others pointed to the effect of literacy, sense of status, and other factors which also concerned us.

Were printers unions more democratic in other countries? We could find no evidence of any precise parallel to the ITU's party system elsewhere, but in various ways printers unions showed interesting special modes of organizational behavior which seemed linked to the variables we pointed to as causal. Thus, in a number of countries in which almost all unions were highly centralized, the printers' organizations deviated from the national pattern; local autonomy and control remained strong. In some countries, the evidence pointed to a much higher rate of participation in internal union matters by printers than by unionists of other occupations.

Printers in many countries were reputed to possess a high degree of occupational and union solidarity. Their marginal position as the "aristocracy of the working class" has been noted by a number of observers abroad. Some have complained about the "snobbishness" of printers, the fact that they refrain from associating with other workers, that they prefer to associate with one another.

These efforts at securing comparative occupational and national data did not, of course, prove any point. They necessarily had to be cursory. Looking for them, however, did lend support to our belief that we had located significant factors and suggested means of testing these hypotheses further. In fact, since the original study was completed, we have had reports of studies in Sweden, India, Brazil, and Britain which have tested various hypotheses of the book, some specifically on printers in these nations.

• NOTES

1. I acknowledge with gratitude the help of Martin Trow and James Coleman in this paper. One could not ask for better friends or collaborators.

2. Glencoe, Ill.: The Free Press, 1949. This book was first published in Germany in 1911 and in the United States in 1915. I have written an "Introduction" to a new paperback edition in which I discuss Michels' importance to sociological and political inquiry, both past and present. See Collier Books edition (New York: 1962), pp. 15-39.

3. Sylvia Kopald, *Rebellion in Labor Unions* (New York: Boni and Liveright, 1924); William Z. Foster, *Misleaders of Labor* (Chicago, Ill.: Trade Union Educational League, 1927).

4. Berkeley, Calif.: University of California Press, 1950.

5. S. M. Lipset, "Democracy in Private Government," *British Journal of Sociology*, III (1952), 47-63.

6. In Morroe Berger, Charles Page, and Theodore Abel, eds., *Freedom and Control in Modern Society* (New York: Van Nostrand, 1954), pp. 82-124.

7. Merton served as the liaison director of the bureau for this study and contributed much to its development.

5 The Evaluators*

CHARLES R. WRIGHT AND HERBERT H. HYMAN

Rare is the student of social science today who, sooner or later, will not try his hand at collaborative research. Therefore, when invited to contribute to the current volume on chronicles of social research, we decided to set down some of our experiences in the role of codirectors in a project extending over five years in time (1955-1960), involving three directors at various stages, and culminating in a book about evaluation research which includes the findings from the study. The project was an evaluation of the Encampment for Citizenship, a summer institute for training young persons to be more effective democratic citizens.

Presenting the case history of a research project without reproducing much of the final report itself is extremely difficult. We will assume that the reader has access to the monograph, and we shall allude to it at times. For our narration, we have selected the literary device of a hypothetical interview about the project. Responses have been attributed to either of the authors at will; actually the views expressed are usually shared by both.

INTERVIEWER: How did you happen to get started on this piece of research? Can you remember its very beginning?
WRIGHT: My first awareness that there was such an institute as the Encampment for Citizenship came on a bleak day in February 1955, while having a late afternoon cup of coffee in a drugstore on Morningside Heights, New York. I was at that time on the faculty of

---

* Herbert H. Hyman, Charles R. Wright, and Terence K. Hopkins, *Applications of Methods of Evaluation: Four Studies of the Encampment for Citizenship*, University of California Publications in Culture and Society (Berkeley and Los Angeles, Calif.: University of California Press, 1962).

121

Columbia University. Another coffee-drinker was my previous thesis adviser, teacher, and friend, Dr. Herbert Hyman, who brightened the day by outlining a challenging research request that had come to the Bureau of Applied Social Research and inviting me to participate in the project with him. The request was from the Encampment for Citizenship for a scientific evaluation of the effectiveness of their program.

INTERVIEWER: Just what is the Encampment that was to be evaluated?

HYMAN: Each summer the Encampment for Citizenship brings together from throughout the United States and abroad approximately 125 young men and women, eighteen to twenty-four years old, of many races and diverse social backgrounds. These young persons live together on a school campus for six weeks, during which they are exposed to a program of lectures, workshops, discussions, and other didactic experiences and social activities designed to "prepare young Americans for responsible citizenship and citizen leadership, to educate them in the meaning of democracy . . . and to train and equip them in the techniques of democratic action." [1] The 1955 Encampment would mark the tenth anniversary of the program, and the sponsors felt that the time was appropriate for a scientific evaluation of its effectiveness. If the program proved effective, they hoped to expand their operations, perhaps sponsoring additional Encampments in other regions (the existing one was in Riverdale, New York). If it were ineffective, they hoped to learn how and why, perhaps obtaining guides for improvement or grounds for abandoning the enterprise. This, then, constituted the very practical problem that was posed to us.

INTERVIEWER: But what led you to do it?

WRIGHT: From our point of view, the Encampment's problem was very tempting. First, the task of evaluation presented a methodological challenge. Both Hyman and I had an interest in the methods by which one can evaluate the impact of such programs of social action as communication campaigns, educational institutes, training programs, and the like. At the time, there had been little codification of the procedures for applying the techniques of social research to such problems or of the principles of evaluation research. A study of the Encampment seemed to offer a ready-made opportunity to develop or apply evaluational techniques and procedures and to grapple with questions of short-range and long-term effects, ceiling effects, and other methodological problems.

HYMAN: There were substantive issues, too, that appeared promising from the perspectives of social-psychological and sociological theory. Here was an opportunity to gather new data on factors affecting changes in the attitudes, opinions, and other dimensions of character of young persons. In addition to evaluating the effectiveness of the specific program, we could collect data on the dynamics of change, hopefully relating the phenomena to previous theoretical and empirical work on opinion formation, reference groups, and the like. In short, the Encampment project struck us as an exciting opportunity to merge the practical needs of a sponsor with certain of our intellectual interests as social scientists.

WRIGHT: At the same time, we had certain apprehensions. Did the sponsors realize the consequences of commissioning a scientific evaluation? Would they be willing to cooperate with the researchers and yet permit us to have complete freedom of inquiry? What would be their reactions to possible negative findings? Would they agree to the necessary steps to preserve the anonymity of the responses of the campers and staff and allow complete control over the data and analysis by the researchers? In short, did the Encampment officials really want a proper evaluation badly enough to stick with it to the end?

HYMAN: There is reason for the researcher to be apprehensive about such matters. The staff of such institutions as the Encampment often have invested considerable time, effort, and sentiment in their programs. They may be ego-involved in their activities. They may be sensitive to the cold-blooded, objective probings of the scientific researcher. Even under favorable circumstances, it is common to find that action-oriented and dedicated persons are unreceptive to social science and efforts to "measure" such seemingly complex and qualitative phenomena as human values, attitudes, and "spirit." How much more likely a hostile reaction may be if such measurements threaten to reveal unfavorable information!

WRIGHT: There were more mundane grounds for apprehension, too. Research is expensive in time and money. A complex evaluation such as would be required here might be impractical; we needed to estimate how much could be done within the limits of a budget acceptable to the client.

INTERVIEWER: How did you resolve such doubts?

WRIGHT: We attacked the latter problem first. Although it was impossible to anticipate fully the amount of time and money required for the evaluation until a detailed study design was completed, certain minimum research costs seemed certain. For example, a proper evaluation would involve at the least a before-and-after research design, preferably with some controls and also measures of longer-range effects. To execute such a project and have a completed report within a year—the time specified by the client—would require a commitment of a substantial part of the working time of each of the directors for the full year, plus additional time during the summer of the Encampment under study. Other administrative and research costs could be estimated, so as to have a realistic budget.

HYMAN: The problem of whether the sponsors were fully aware of what an evaluation entails was more delicate. Preliminary conferences were arranged between the educational and executive directors of the Encampment and ourselves. In these meetings, we discussed the nature of evaluation, the issues of freedom of inquiry, anonymity of responses, and the like. The Encampment officials described their motives for desiring the research and quizzed us about what an evaluation involves. Each party became acquainted with the other—the researchers with the prospective client and the client with the "evaluators"—before either was committed to the project. Here was the opportunity to clear the air and to back off or forge ahead with the research, as either party chose. Both decided to go ahead. We had our mandate, then, to get down to specifics of research design and execution.

INTERVIEWER: How did you go about that?

HYMAN: The plan of the research evolved during six meetings between the project directors in March 1955. From the outset, it was apparent that we were faced with a problem that required not one study but two: an analysis of the impact of the forthcoming Encampment and an assessment of the long-range effects of the institution. The project's notes on the six planning meetings reflect this dual concern.

INTERVIEWER: You have notes on these early meetings? What do they show?

WRIGHT: Without entering into details, I can give you a brief review of what research decisions were made at each meeting.

At the first meeting, on March 1, we produced a tentative outline of a study design for determining the effects of the camp on the next class, 1955, and another design for determining long-range effects, primarily through a survey of former graduates. On March 8, another conference was held with the Encampment's directors in order to obtain a background on the Encampment and its objectives. Later the same day, Hyman and I held a second research-planning meeting, devoted to spelling out some possible effects of the Encampment which would become the major dependent variables in the study. A revised study design was developed during a third meeting, on March 16. Two meetings were held on March 18. One produced an elaboration of the study design for the survey of Encampment alumni; the other was devoted to a fuller conceptualization of independent, dependent, and intervening variables to be included in the basic study of the current Encampment. At the sixth meeting, March 22, a time schedule was made for the entire project—from its design in March 1955, through the anticipated report in March 1956.

HYMAN: The month of planning also involved several conferences with officials and staff of the Encampment, an intensive study of the institution's records and correspondence from former campers, a search for books or articles reporting evaluations of similar educational enterprises, and considerable but unmeasurable mulling over of the problem. The results of this initial phase of work were recorded in a memorandum (April 6) from us to the Directors of the Bureau of Applied Social Research, accompanied by a final detailed study design, proposed work and time schedule, and budget.

INTERVIEWER: Do you have a copy of that memo? May I see it?

WRIGHT: Be my guest.

EXHIBIT A

April 6, 1955

To: Drs. C. Y. Glock and E. DeS. Brunner[2]
From: H. Hyman and C. Wright
In re: Design of the Evaluation Study of the *Encampment for Citizenship*

In accordance with our original conversations, we have been engaged for about the last month in planning this study. We have had about six meetings to work out our own views of the design. In addition, two conferences have been held with Dr. Black and two with Mr. Shannon on the clarification of the goals of the Encampment and on factual details needed for planning. In addition, we attended one meeting of the staff of the 1955 Encampment which provided opportunity for informal observation, for gaining rapport with staff members, and for presentation of the general outline of the project design herein described.

Detailed study of a large volume of the total correspondence had already been accomplished. Our study of these files confirms our initial judgment that the correspondence is *not* worth systematic treatment as a source of scientific evidence on effects. However, insights into the process of change and into independent variables which relate to change have been obtained. Further study of the correspondence will not be repaid in much additional insight since it is rapidly running thin as concerns any new findings.

Incidental to such examination of the files, however, we have unearthed a file of past evaluation forms, which were systematically administered to all campers at the end of the summer for years 1946-49. These data, which had not been mentioned to us before and which have never been processed thoroughly by the Encampment, would constitute a source of valuable evidence on effects and the programmatic factors related to effects. We will use them to supplement the old application forms as a base line for measuring change among alumni. It is our recommendation that the Encampment process these, and if time and money permit we may advise on a method for coding and tabulating the findings.

On the basis of the planning sessions, conferences with staff, and study of the files, a detailed design, work schedule, and budget have been worked out. While the design departs from the original proposal to the client, it represents, in our judgment, a desirable approach and has met with the approval of the client, as informally expressed to us in recent meetings. After you have had time to study the enclosed materials will you let us know, and we'll arrange a conference.

INTERVIEWER: What was the study design?

WRIGHT: The initial design for the main study called for basic measurements of the 1955 campers at the beginning and end of the Encampment and two mail questionnaires sent to them at home, one six weeks prior to the Encampment and the other six weeks after its completion. Comparisons of responses to the first mail questionnaire and those given at the start of the Encampment would provide a measure of the stability of the young persons' attitudes prior to training. Changes in attitude during the summer would appear between the second and third measurements. Stability of response and possible longer-term effects would be detected in the follow-up mail questionnaire.

H Y M A N : The nature of the program under study precluded a classic controlled experiment, with persons assigned at random to an experimental and control group. Therefore, special efforts were made to obtain data that might approximate answers to some of the problems ideally solved by the classic control group. We describe these procedures in detail in Part I of the monograph. We planned to conduct a two-wave inquiry with applicants who were unable to attend the Encampment in 1955. Changes in their attitudes during the summer might indicate how much the campers would have changed had they not come to the program. We also planned to use data from previous or pending national surveys of American young adults in order to determine the effect of nonexperimental events on the stability of the characteristics under study and in order to provide some norms for these characteristics against which to compare the campers. This decision required introducing into our questionnaires items similar to those used in national surveys. Finally, we decided to design certain parts of the study in such a way as to permit contrasts with the results achieved in evaluations of similar programs of citizen-training. In particular, we obtained permission from Dr. Henry Riecken to use certain scales that he had introduced in an evaluation of volunteer work camps sponsored by the American Friends Service Committee.[3]

I N T E R V I E W E R : Did you rely solely on questionnaires?

W R I G H T : At one point in our planning, we considered the collection of data through participant observation, but we decided against this for several reasons. First, the limitations imposed by the available funds prohibited hiring a full-time observer. Consequently one of us —probably I—would have had to assume that role. We feared that a full-time immersion into the Encampment culture would have cut seriously into my time available for such other essential research tasks as the design of questionnaires and would also have entailed the risk of reducing my objectivity as an evaluator as I, perhaps, became socialized by the program. Second, we feared that constant observation of the campers during forty-two days of the program would make the research too obtrusive and interfere with the natural atmosphere of the Encampment, thereby even spoiling the major evaluation itself.

H Y M A N : We thought also that the major burden of proof of the impact of the program would more properly come from measurement of the campers than from impressions of the observer. The most useful function of observations appeared to be to provide direct accounts of the educational program in action (in contrast to the formal description of the program that might be obtained from its officials), an opportunity to witness the informal social life, unanticipated incidents, and selected events and to obtain informal impressions and information through conversations with the campers and staff and insights into factors in the process of change. Therefore, we decided that one of the directors would assume the major responsibility for making planned periodic observations of the Encampment, using a time-sampling system, and the other would join him occasionally. The periods for observation were "selected on the basis of (1) consideration of budget and allocation of research time and (2) conferences with the Director of the Encampment and his staff concerning probable significant points in the six weeks experience." [4] We decided to make observations on the following days: June 27 through July 1, the opening phase of the Encampment; July 15-16 and 24-26, intermediate points; and August 3-6, the closing period. In our plan, we anticipated that such timing would capture "the early formation of spontaneous patterns of groupings, later a period of stabilization and possible satiation, and ultimately a phase of nostalgia, reflection, and possible disruption as the campers face departure." [5] We also made it a deliberate point to observe the various types of activities during these time periods—lectures, discussion groups, workshops, meals, town meetings, and evening educational programs—in addition to making casual observations of the program and informal social life.

INTERVIEWER: Did you plan to use any other sources of information in addition to the questionnaire and observations?

HYMAN: We conducted a quasi-experiment similar to classic experiments on "prestige suggestion" in order to measure the effect of the training in reducing susceptibility to prestigious political symbols. The experimental design, however, was secretly smuggled into the regular questionnaires by varying the wording of several questions in different sets of questionnaires and controlling the assignment of these different forms to subgroups of campers during two waves of the study. We regarded the different forms of the questions as "stimuli" and interpreted campers' answers to them as responses to these different stimuli rather than as evidence on the distribution of opinions on each issue itself. This procedure is not novel, being similar to the use of split-ballot techniques in measuring the reliability of items and to tests of the effects of question-wording, but the use of the questionnaire in this manner as a quasi-experimental method has hardly been exploited in evaluation research. Therefore, we wanted to try it out on a small scale without diverting resources from our major efforts at evaluation.

One of the Encampment's criteria for effective citizenship is the ability to judge social issues on their own merits, unswayed by emotional reaction to persons or groups associated with the issue or by propagandist terms in which the issue is phrased. We decided to limit the experiment to this realm and to test whether campers developed a resistance to having their opinions influenced by the "prestige symbols" of traditional political-party labels. At the start of the Encampment, campers were asked whether they favored or opposed each of six hypothetical issues presented as political-party planks in a presidential campaign; for example, setting a deadline for desegregation in the schools of the South. Half of the campers received forms of the question that attributed each policy to the Democratic party and the others received forms attributing each policy to the Republicans. To measure resistance to these political-party symbols, we compared the proportion of favorable responses to all issues among the answers given by campers who received a form of the questionnaire attributing the policies to the political party which they preferred to the responses given by campers who received questionnaires attributing the same policies to the party which they did not prefer. If the first group gave a higher proportion of favorable responses than the second, it would indicate the campers' relative susceptibility to the party labels, that is, that their opinions are a reaction to the party label as well as to the issues themselves. Then each camper was asked the same questions, in identical forms, at the end of the Encampment. If the campers became less susceptible to the symbols during the Encampment, then the differences in responses between the two groups should have been smaller at the end of the summer than at the beginning (assuming an initial random assignment of the two forms of the questionnaire).

INTERVIEWER: And were there any other sources of information that you planned to use in the over-all evaluation of the Encampment?

WRIGHT: Yes. We also planned to use the Encampment staff for supplementary data collection. For example, they could provide accounts of attendance at the various educational activities which would permit us to identify campers who, because of such defections, failed to receive the complete program. Also we thought that illustrative cases and biographical accounts of selected campers might be obtained through the cooperation of the staff.

INTERVIEWER: Did you use them?

WRIGHT: Only to a limited extent. The accounts of absences were useful, but we abandoned the idea of biographical accounts, primarily because we were fully occupied with the questionnaire and observational procedures. Since these illustrative cases were supposed to provide qualitative flavor rather than quantitative evidence of impact, they were "luxury" materials for the evaluation and were dropped under the pressure of the research.

INTERVIEWER: You mentioned a detailed conceptualization of the problem, especially the dependent variables. Could you elaborate on that?

HYMAN: All together, there were more than forty dependent variables used to measure the objectives of the Encampment and other unanticipated consequences. Many of these measurements involved batteries of items forming scales or indexes; some were open-ended questions, others check lists and fixed responses. There were far too many items to repeat here, although all are spelled out in the monograph. Briefly, we conceptualized the problem as involving measurements of different regions of possible change in the campers' character, including areas hypothesized as least subject to change as well as those most likely to be influenced by the program. There were seven such regions: basic values, action orientation, cognition of social problems, salient social attitudes and opinions, perceived relationship with the rest of society, skills or capacities, and conduct.

Our independent and intervening variables included such things as campers' predispositions, expectations about the Encampment, satisfactions and dissatisfactions with the program, absences from didactic and communal events during the summer, traumatic experiences, difficulties in following the program, support from prior friends and family, new friendships formed during the Encampment, use of reference groups, ethnic, educational, and other statuses of the campers, function of the staff as role models, the nature of the home environment, persistence of Encampment reference contacts at home, use of ex-campers as reference groups, and many others.

INTERVIEWER: Were you able to include items on all the planned variables in your questionnaires? And were they actually used in the analysis?

HYMAN: Yes, to both questions. Of course, not every variable was included in each wave of questionnaires. For example, the realm of conduct was explored more extensively in the questionnaires sent to campers after their return home than it was during the Encampment. We felt that many forms of democratic conduct relevant to the Encampment's goals had to be measured in the community setting in which ex-campers now had to function as young citizens.

INTERVIEWER: While we are on the subject of the questionnaires, were there any special problems there? Did you pretest them, and how did they work?

WRIGHT: No, there were no special difficulties. We were able to pretest the Wave 1 and alumni mail questionnaires and made some minor adjustments in question-wording as a result. The rest of the waves of questionnaires could not be pretested, because we could not intrude on the campers any further without risking spoiling our major measurements. So we had to rely on our experiences with Wave 1 in forming the second questionnaire, and on the latter in forming the third, and so on. Of course, some of our items had been used in other studies and therefore were proven effective.

We had good rates of return on the 1955 campers' questionnaires, including the mail as well as the self-administered waves. The returns from alumni were not as high. Waves 2 and 3 were administered to the campers at one "sitting" by Hyman and me. All campers who were there cooperated fully. We missed a few persons who arrived late for the Encampment and a few others who left before the Encampment was completed, and we had to consider these sources of possible bias in the analysis, in addition to estimating bias in the alumni returns.

INTERVIEWER: Let us go back to the memorandum that you drafted at the start of the research. You mentioned that a time schedule for the project had been worked out in March 1955. Do you have a record of that now?

WRIGHT: Yes, here it is:

EXHIBIT B

Time Schedule for Evaluation of EFC

STUDY I—CAMPERS AND CONTROLS		STUDY II—ALUMNI
3/15/55–4/4/55	*Design*	*Design*
3/25	Establish basic design, concepts and variables.	
3/31	Scrutinize testimonials and correspondence for ideas.	
4/4	Meet with E.F.C. Executive Director and his secretary re: needs of project.	
4/4 4/5	Conferences with E.F.C. Executive and Educational Directors on translations of objectives into variables, and on incidental facts relevant to project.	
4/5–5/8	*Instrument: Wave No. 1*	
4/5–4/22	First draft of Wave No. 1 for campers and control rejects.	
4/23–5/1	Revise.	
5/2–5/6	Preparation.	
5/8	Mailing.	
5/8–6/1		Draft and prepare alumni questionnaire. First mailing 6/1
6/2–6/27	*Instrument: Wave No. 2.*	Second mailing 6/18
6/2–19	Trial tabs on Wave No. 1. Consult questions from other comparative studies. Check on national surveys.	
6/20–26	Draft and prepare final version.	
6/27	Administer Wave No. 2.	
6/27–7/1	*Observations* (*Monday–Thursday*). Process during opening phase of Encampment.	

7/1–8/3	*Instrument: Wave No. 3.*	
7/1–14	Trial tabs on Wave No. 2. Check national control.	
7/15–16	Observations (Friday, Saturday).	
7/17–23	Draft questionnaire.	
7/24–26	Observations (Sunday, Tuesday).	
7/27–8/2	Prepare Wave No. 3.	
8/3	Administer Wave No. 3.	
8/3–8/6	*Observations (Wednesday–Saturday).*	
8/7–9/16	*Instrument: Wave No. 4.*	
8/7–14	Trial tabs Wave No. 3.	Trial tabs on mail questionnaire.
VACATION		
9/5–12	Draft questionnaires for panel and rejects.	
9/12–15	Prepare Wave No. 4.	
9/16	Mail Wave No. 4 (second mailing later).	
10/1–1/31/56	*Processing and Analysis*	*Processing and Analysis*
10/1–15	Code Building.	Code Building.
10/15–22	Coding.	Coding.
10/22–11/1	Machine-punching and cleaning.	Machine-punching and cleaning.
11/1–1/31/56	Analysis.	Analysis.
2/1/56–3/1/56	*Report Writing.*	*Report Writing.*

INTERVIEWER: Were you able to do all the things called for in the initial study design?

HYMAN: Yes, with a few exceptions. The original plan to use rejected applicants as a control group had to be abandoned because of insufficient cases. We discuss alternative solutions to this problem in the monograph. And, as mentioned, we did not make any attempt to get biographical case studies of individual campers. Also the old "evaluations" that we had discovered were more useful as sources of insight than as evidence.

INTERVIEWER: How realistic was this time schedule?

HYMAN: We were extremely fortunate. I would say that we were able to keep to it within a few days of each date set. The report was completed on schedule.

INTERVIEWER: How did it come out?

WRIGHT: Mimeographed, initially, and a brief oral summary was presented by Hyman at an EFC anniversary dinner in March 1956. Now it is integrated into our published monograph, *Applications of Methods of Evaluation.*

INTERVIEWER: I mean, rather, what were the findings, and how were they received?

HYMAN: The findings were quite favorable to the Encampment, and that made the presentation of our report less of a crisis than it might have been. We had no idea in 1955 how the data might turn out, of course, but we had feared the worst. Frankly, we dreaded the moment when we might have to report that the venture did not achieve its purposes, for the history of so many endeavors in character education is that their good intentions are shattered on the granitelike resistance of people to change. We were sympathetic to the Encampment's desire for scientific evaluation; this took courage on their part, for many agencies prefer to go about their activities with little knowledge of their efficiency. But we were bound to be objective and rigorous; as a matter of fact, we did everything in our power to be conservative in our measurements of change produced by the Encampment in order to feel certain that whatever measure of success was demonstrated would be compelling proof.

WRIGHT: Our report detailed the many ways in which the youth of the Encampment changed in directions favorable to the program's goals: they became more appreciative of our traditional civil liberties, more tolerant of unpopular views, and stronger in their defense of civil rights for minorities. It also considered possible unanticipated effects: they did not become more "radical" in their political ideology. And we were able to make reasonable inferences that the changes were produced by the Encampment rather than by other causes. At the same time, we had evidence on aspects of character that were not changed: basic values of the young persons as indicated by such measures as their vocational goals, the goals they considered most worthy of great personal sacrifice, and other responses to items in the questionnaire. But, on the whole, the evaluation demonstrated that the institution was successful.

INTERVIEWER: Your report was the end of the study, then?

WRIGHT: We thought so at the time. Actually it was not, and the subsequent developments were as exciting as they were surprising to us. In 1957 and again in 1958 the Encampment requested additional research on their work. By that time, I was no longer with the Bureau of Applied Social Research, having joined the Department of Anthropology and Sociology at the University of California, Los Angeles, in 1956. Therefore a new codirector, Dr. Terence Hopkins, joined Dr. Hyman in executing the new studies. My role in this phase of the research—until I rejoined the project in 1959—was a minor one, involving informal consultation and administering the questionnaires to a newly formed West Coast Encampment in 1958. I shall leave the history of this phase to Hyman.

H Y M A N : These new developments provided us with an unparalleled opportunity for replications of the original research. When one is engaged in evaluation of a program such as the Encampment where the classic controlled experiment is impossible, regardless of the safeguards employed there is always the lurking possibility that the results are spurious—the compounded product of a variety of circumstances peculiar to one instance of the program being evaluated. How much more confidence could be placed in measured changes apparently caused by the program if they should recur in other, independent evaluations of the program in different years! We were delighted at the prospects.

Furthermore, the new research would provide evidence on many of the interpretations about the dynamics of change set forth in our 1955 report. Besides providing an opportunity to retest some of these initial hypotheses and findings, the new studies would permit us to gather new kinds of data relevant to the dynamics of change. This became especially evident in the third study (1958), after the second findings (1957) had supported the pattern of changes in campers discovered in 1955. That is, once a replication had supported our initial conclusions that the Encampment was indeed effective in achieving many of its stated goals, we could afford to divert some of our research resources and energies into exploring such potential causal factors as the development of friendship groups and reference-group phenomena.

In addition, the 1957 and 1958 studies introduced measurements of the impact of the program on college students who attended—a problem that had become of particular concern to Encampment officials.

All told, there were three replications of the 1955 evaluation: on the original Encampment site (New York) in 1957 and 1958 and at a new Encampment in California in 1958. The design of these three replications was essentially the same as the 1955 study, except for the new concern with college students and dynamics of change. The Wave 1 and Wave 4 mail questionnaires to campers in their home settings were restricted to college campers in order to measure the Encampment's impact in this context. Additional evidence, of a sociometric nature, was gathered on each camper's friendships, mealtime companions, and the like. (This latter phase of the research is undergoing separate and intensive analysis by Hopkins.)

Then, in the spring of 1959 we obtained additional support for a follow-up study of the original 1955 group. We sent them a mail questionnaire. The returns permitted us to determine directly the long-range results of the Encampment's work, a problem that had been treated previously through the inferential method of studies of alumni groups from the years before 1955. The role of contacts with ex-campers and use of them as reference groups as factors affecting the stability or loss of Encampment gains in social attitudes and other phenomena were explored. Perhaps these two charts will be helpful in outlining the chronology and design of the several studies and illustrating some of the comparisons used in the evaluation. (See charts I and II.)

In the summer of 1959, Wright returned to Columbia University to teach in the summer session and participate in the analysis of the new studies.

WRIGHT: I find it difficult to describe my reactions on returning to the project. But a few reminiscences may prove useful to the student who may someday be involved in "removal and return" to a long-term project. I have no idea how common this experience may be.

When I returned to the project, all of the 1957 data and two waves of the 1958 data had been processed. A small roomful of coders were busily at work on the balance of the 1958 materials and the 1959 follow-up. My first reaction was a sense of being lost in my own house. The scene was familiar—it was the right house—but there had been changes. The original project, although extensive in coverage, had been executed primarily by the codirectors, a part-time research assistant, and the help of various staff members of the Bureau of Applied Social Research, such as their machine-tabulation group. As a codirector, I had had my fingers on the data at all stages; for

CHART I

Chronology and Design of Studies Evaluating the Encampment for Citizenship[6]

GROUP UNDER STUDY	STAGES OF MEASUREMENT				
	1 Pre-enrollment "self-control." Six weeks prior to Encampment (By mail)	2 Start of Encampment (In person)	3 End of Encampment (In person)	4 Short-term follow-up. Six weeks after Encampment (By mail)	5 Long-term Follow-up (By mail)
Initial group (1955 Encampment)	A_1	A_2	A_3	A_4	A_5 (1959)
First replication (1957 Encampment)	B_1	B_2	B_3	B_4*	
Second replication (1958 New York Encampment)	C_1*	C_2	C_3	C_4*	
Third replication (1958 California Encampment)	D_1*	D_2	D_3	D_4*	
Alumni mail survey (1946–1954 Encampments)					E (1955)
Control groups (college)	B'_1 (1957) C'_1 (1958)			B'_4 C'_4	

* Measurements B_4, C_1, C_4, D_1, and D_4 include college students only, among the campers.

Comparative "control" groups:

(1) H. Riecken's evaluation of volunteer work camps (1952);

(2) National Opinion Surveys (1955);

(3) R. Dentler's evaluation of summer interns (1958).

example, Hyman and I had done much of the coding ourselves, with one assistant. But now I had no sense of day-to-day familiarity with the new data. I would need time to become immersed again. Furthermore, the new developments in the project, resulting in three replications and a new follow-up, made it patently impossible for the directors to handle so many routine tasks themselves, nor was it nec-

CHART II

Examples of Evaluation Equations and Comparisons[7]

EXAMPLE	OPERATION	MEANING
(1)	$A_3 - A_2$	Immediate impact of 1955 Encampment Program, plus changes caused by growth and extraneous events during forty-two days, plus practice and sensitivity from initial testing, plus instability of instruments.
(2)	$A_2 - A_1$	Changes caused by growth and extraneous events during a "normal" period of forty-two days, plus instability of instruments and practice.
(3)	$(A_3 - A_2) - (A_2 - A_1)$	Change during 1955 Encampment less changes probably attributable to six weeks of growth, extraneous events, instability, and practice. Or, immediate impact of Encampment plus changes caused by sensitization to it from prior testing.
(4)	$(A_3 - A_2)$ vs. Riecken's or Dentler's findings on effects	Relative effectiveness of Encampment program (1955).
(5)	A_4 vs. A_3	Short-term stability or change (impact of return to home or college community) and conduct.
(6)	A_5 vs. A_3	Long-term stability or change; new conduct.
(7)	A_5 vs. A_4 vs. A_3 vs. A_2	Patterns of stability and change (e.g., steady erosion; immediate deterioration; constant stability; fluctuations).
(8)	$C'_4 - C'_1$	Changes between spring and fall semesters for college students with various summer experiences.

cont. on next page

(9)	$C_4 - C_1$	Changes between spring and fall semesters for college students among 1958 Encampment Group—New York.
(10)	$(C_4 - C_1) - (C'_4 - C'_1)$	Short-term effectiveness of Encampment for college students, less "normal" changes during that time, plus changes caused by sensitization to program from initial testing.
(11)	$(A_3 - A_2)$ vs. $(B_3 - B_2)$ vs. $(C_3 - C_2)$ vs. $(D_3 - D_2)$	Replications for verification of 1955 findings; comparative effectiveness of different Encampments; increased generalizations; and specification of dynamics of change.
(12)	E vs. A_5	Comparison of cross-section vs. longitudinal evidence on long-term effects, stability and changes, and conduct.

essary, because many of the 1955 research decisions could serve as directives to the staff working on the 1957 and 1958 replications. As an example, many coding decisions made in 1955 had to be kept in the three replications, to preserve comparability between studies. These could be transformed into formal instructions and conventions to be followed by a staff of coders under the ultimate supervision of one of the directors. In addition, we were fortunate that the original research assistant on the project was available to help with supervising the coding and providing other services during the replications. So there was continuity not only among the directors of the research but also in the staff. This continuity made my reorientation to the project much easier, as both Hyman and Hopkins as well as the research assistant could instruct me on what had been taking place. They informed me also of the new decisions with which I needed to become familiar, such as the treatment of the college campers.

Looking back, it seems that the summer's work divided itself into three overlapping phases: (1) About three weeks were spent getting reoriented and directing the final processing of the new data. (2) About two weeks were spent in planning and ordering the new machine tabulations needed for replication and for analysis of the follow-up. (3) About a month was devoted to analyzing and beginning to draft the final report integrating the original 1955 study, the replications, and the follow-up. Then it was time to return to California for the fall semester.

INTERVIEWER: But the report was not finished. How did it get done?

HYMAN: During the fall of 1959, the three of us worked on various chapters and exchanged drafts by mail. Then we revised our work in the light of one another's criticisms and suggestions. Wright had a complete set of the cards, code books, and tabulations with him in California; and Hopkins and I had the original sets in New York, of course. Consequently, it was possible for any one of us to obtain whatever special tabulations we needed to supplement the basic analyses completed during the summer. Then, in January 1960, the three of us got together to thrash out our next-to-final draft of the monograph. We literally locked ourselves in an isolated house in the country, well stocked with food and liquid refreshment, and worked on nothing except the project for about a week. Then, in the spring we each put the finishing touches on the work, collaborating again by mail. The final report appeared in May 1960.

INTERVIEWER: You seem to have done a lot of collaboration at long distance.

HYMAN: Very long distances; at one time the three directors were scattered in Turkey, New York, and California, but it was manageable.

INTERVIEWER: And now the report is published?

WRIGHT: Amen.

• NOTES

1. Encampment for Citizenship, official brochure (privately printed, 1954).
2. Then directors of the Bureau of Applied Social Research.
3. H. Riecken, *The Volunteer Work Camp* (Cambridge, Mass.: Addison-Wesley, 1952).
4. H. Hyman and C. Wright, "Study Design for Evaluation of Encampment for Citizenship 1946-1955" (Typed memorandum, Columbia University Bureau of Applied Social Research, April 5, 1955).
5. *Loc. cit.*
6. Adapted from Herbert H. Hyman, Charles R. Wright, and Terence K.

Hopkins, *Applications of Methods of Evaluation: Four Studies of the Encampment for Citizenship,* University of California Publications in Culture and Society (Berkeley and Los Angeles, Calif.: University of California Press, 1962), for presentation at the annual conference of the American Sociological Association, Washington, D.C., 1962.

7. *Loc. cit.*

6 Research Chronicle:

*Tokugawa Religion**

ROBERT N. BELLAH

The search for beginnings leads ever deeper into a bottomless well, but it is perhaps not amiss to trace the beginnings of this particular research back into my undergraduate years. I had concentrated as a Harvard undergraduate in the field of social anthropology (one of the four subfields in the Department of Social Relations) because I was interested in exotic cultures and also for the more sober reason that I believed one might achieve greater insight into the understanding of human action through the study of simpler societies than through studying more complex ones. Gradually it became evident to me, however, that my real desire was to understand the advanced societies of the West, particularly my own, and that the study of primitive societies presented too circuitous and delayed an approach to that end. Believing, nonetheless, that a comparative approach would ultimately yield the greatest rewards (and retaining an interest in the exotic), I turned to the study of the major non-Western civilizations as giving a more immediately useful framework of comparison for the understanding of Western society.

Here the somewhat fortuitous fact that in my senior year (1949-1950) I took Social Sciences 111, "The History of Far Eastern Civilization" (locally known as Ricepaddies), undoubtedly precipitated the decision to concentrate on East Asia in my graduate studies. It was Japan, in particular, which fascinated me because of the many analogies to Western history, especially in the development from feudalism to capitalism. At the time, I thought largely in Marxist categories about the problems

* Robert N. Bellah, *Tokugawa Religion* (Glencoe, Ill.: The Free Press, 1957).

of historical development, and much of the version of the Japanese development then current fitted neatly into my preconceptions. If both Japan and the West, with quite clearly different historical antecedents, developed in a parallel fashion, might not the study of the parallels lead to a deeper understanding of the general laws governing these developments? That was how I felt at the time.

However, in the same senior year in which I was introduced to East Asian civilization I underwent another important intellectual experience when I took Talcott Parsons' Social Relations 130, "Comparative Institutional Structure." Up to this point in my undergraduate career, I had sedulously avoided taking any courses with Parsons because I had found some of his lectures, which I had heard as a sophomore, frustratingly incomprehensible and also because what little of his position I had understood I felt to be incompatible with my then Marxist assumptions. In fact, I believed the still current notions that the theory of action was devoted to the defense of the *status quo*, that it postulated a static, equilibrated society, that it had no means to understand social change, and so forth. David Aberle, my honors-thesis adviser at the time, told me quite bluntly that I was being foolish and that I ought not to leave Harvard without hearing what Parsons had to say. With misgivings, then, I took the course, and while I underwent no immediate conversion, I did find my intellectual framework greatly expanded as a result. Not only did I discover Parsons; I also discovered Weber, who figured prominently in the reading list. Already, then, in my senior year thoughts about Japan and its parallels with the West, on the one hand, and the ideas of Weber and Parsons, on the other, were in my head. Though not yet connected at all clearly, these were the germ plasms out of which *Tokugawa Religion* would eventually grow.

I entered the Harvard graduate school as a candidate for the Ph.D. in sociology and Far Eastern languages in the fall of 1950. The following spring, I took Parsons' seminar, Social Relations 206, "The Theory of Social Systems." The seminar that year was organized around a critical reading of the manuscript of *The Social System*,[1] then in dittoed form. The students' papers were to take off from theoretical issues raised in the manuscript. It is perhaps worth recording that I found the manuscript extremely difficult—my undergraduate course with Parsons notwithstanding—and it took weeks of reading, rereading, and careful outlining before I began to have a glimmering of what was going on.

For my paper, I decided to apply to the case of Japan the sugges-

tions at the end of Chapter 5 of *The Social System* concerning the use
of pattern variables to classify the value patterns of whole societies. In
so doing, I utilized much data and many ideas which had come out of
John Pelzel's course, Social Relations 135, "Social Structure of China
and Japan," which I had attended in the fall of 1950, as well as subsequent reading (all of it in English, as I was just beginning to study
Japanese at that time). The title of the paper indicates the position I
took at the time: "Japan: An Example of the Particularistic-Ascriptive
Value Pattern." Looked at now, it almost seems to be a draft of Chapter
2 of *Tokugawa Religion*, "An Outline of Japanese Social Structure in
the Tokugawa Period," though the material was drawn mainly from the
modern period. The paper had the same purpose as the later chapter,
namely, to define the basic value pattern in pattern-variable terms and
show how this value pattern is related to the major aspects of the
structure of the system. The main difference was in the structural categories used for organizing the data. In the 1951 paper, these were
drawn from *The Social System,* as should be evident from the following
outline of topic headings.

Japan: An Example of the Particularistic-Ascriptive Value Pattern

I.   The Problem of the Japanese Pattern
II. Role Complexes
   A. Family and Community
   B. Instrumentally Oriented Relationships
      1. Small Enterprises
      2. Large-scale Enterprises
      3. Professions
      4. Government
         a. The Parties
         b. Civil Bureaucracy
         c. The Military
      5. Government and the Army at the Local Level
      6. Miscellaneous
III. Expressive Aspects of the System
IV. Integrative Structures
V.   Sources of Strain
VI. Some Personality Considerations
   A. Shame-Guilt
   B. A Socially Preferred Pattern of Deviance
   C. Sincerity and Expertness
VII. The Particularistic-Ascriptive Type of Value Orientation

Later, when I recount the writing of Chapter 2 of *Tokugawa Religion,* I shall return to the problems of this outline and the subsequent changes in it.

The second difference between the paper and the later chapter is even more important: in 1951 I characterized the Japanese value system as "particularistic-ascriptive," whereas by 1955 I was calling it "particularistic-achievement." In order to indicate the kinds of difficulties of which I was already aware in 1951, I quote the opening section of the seminar paper in its entirety.

THE PROBLEM OF THE JAPANESE PATTERN

A superficial glance at Japanese society suffices to place it with some confidence in the particularistic-ascriptive category. The enormous importance of the family, the extension of family-type relations into many other spheres (such as small business and the army), the importance of personal and status loyalties, the importance of inherited status, and the emphasis on the forms of status all lead to the same conclusion.

If, however, we were to consider certain other factors first, we might expect anything but the particularistic-ascribed pattern to apply to Japan. Japan is certainly one of the great modern industrial nation-states. Considering only this, we might expect that universalism and achievement orientations would be primary, as they tend to be in other industrial nations. Furthermore, on the basis of most known particularistic-ascriptive societies (primitive societies, Latin America) we might expect a relatively static, stable, traditionalistic society. Japan, however, is a society in a state of most rapid change and one shot through with tensions. Even though these seeming contradictions might be avoided by assuming that Japan is in a state of transition away from particularistic-ascribed values to another pattern (and this would certainly seem to be the case), we are still faced with the problem of understanding how this society functioned during the last fifty years or so. Since there is no reason to believe that the analysis of Western industrial societies which pointed up universalism and achievement as related to an industrial economy is wrong, we will be especially interested in seeing the place these value patterns have in Japanese social structure and, especially where they are absent, in the problem of how functional equivalents fulfill their functions.

Why I subsequently changed my ideas about the Japanese value pattern will become evident below.

It is interesting that in the seminar paper completed in May 1951 there was none but the most passing mention of religion. Some major concerns of *Tokugawa Religion* are already discernible—for example, the relation of values and modernization—but the specific concern with religious orientations is almost totally lacking. The first reasonably clear statement of the later thesis which I can discover is in a notebook under the date August 20, 1951.

> T H E S I S T O P I C : Nothing less than an Essay on the Economic Ethic of Japan to be a companion to Weber's studies of China, India, and Judaism: The Economic Ethic of the World Religions.
>
> It would be necessary to clarify and reformulate Weber's thesis in order to "test" it or at least to contribute toward making it "testable." Focus of course would be on the relation of value pattern and social structure in Tokugawa and modern Japan. Problems would have to be specific and limited—no general history would be attempted— since time span is several centuries. Field work in Japan on the actual economic ethic practiced by persons in various situations, with, if possible, controlled matched samples from the U.S. (questionnaires, interviews, etc.). This topic would involve much work on economic history in the West, on capitalism per se, on the corpus of Weber's works, etc.
>
> Central, theoretically, would be a reworking of the concept "class" so as to make it usable in the analysis. Perhaps considerable theoretical contribution could be made on subject of collectivity structure and value pattern, etc.

As is usually the case with Ph.D. theses, the original conception was considerably more grandiose than the ultimate product. A number of the aspects of the initial notation—for example, carrying the study up to the present time, utilizing field work in both Japan and the United States, extensive work on Western economic history, and so on—did not become parts of the final project. Still, the basic idea is clearly present. I am not sure how I arrived at this idea, when it seems totally lacking in the paper of a few months before, but it may well have developed out of the reading I was doing in the summer of 1951. It is my recollection that that summer I first worked through Parsons' *The Structure of Social Action,*[2] in which Weber's comparative studies are given considerable attention, and began an extensive reading of Weber himself.

Lest it be thought that the subsequent road to the thesis was entirely smooth, let me indicate that there were a number of other thesis ideas which appear in my notebooks for the next year and a half or so, though the entry quoted above is the first of these and recurs in muted form in a number of the later ones. On the same day as the entry and just following it, I noted the idea of comparing the breakdown of Ashikaga feudalism in Japan and the recrystallization in the Tokugawa period with the breakdown of high feudalism in the West and the subsequent "baroque crystallization" there. On November 19, 1951, there is a notation for a thesis on the primary group in Japan, its structure and relationship with Japanese culture and society. This was apparently to concentrate on the contemporary scene and be carried out through field investigation.

On February 14, 1952, there is a notation amplifying the one of November 19. It again envisions a field study of the relationship of collectivity structure to "the belief and expressive symbols with the patterns of value orientations serving as the crucial link." But it was to include also a history of collectivity structure from Tokugawa times about which there is the following comment: "The whole study would undoubtedly uncover a development similar to that which would be revealed by a study of the rise of Protestant Calvinist communities from their heyday to the period of advanced industrialization." Finally, in the fall of 1952, when I was in the Parsons-Stouffer-Florence Kluckhohn seminar on social mobility, there is a notation for a thesis project involving the study of occupational aspiration in Japan. This was to be based mainly on field research, but here, too, there was to be a historical study reaching into the Tokugawa period.

In spite of the meanderings with respect to thesis topics indicated in my notebooks, the next major research paper which I wrote was on Japanese religion, and it contains the germs of a great many ideas later developed in *Tokugawa Religion*. The paper was written in January 1952 for Parsons' course, Social Relations 164, "Sociology of Religion." In it there appears the notion of a general synthesis of religious traditions in the Tokugawa period and an analysis of this synthesis which foreshadows that contained in Chapter 3 of *Tokugawa Religion* with differences which will be noted later. Shingaku is mentioned as an example of a popular syncretic sect embodying this Tokugawa synthesis, though at the time of this paper I had read no Shingaku texts and was going on remarks of Serge Elisséeff in his lectures on Japanese literature. Following the general discussion of religion in the Tokugawa period, there is a section raising the question of the parallels to Weber's argument with respect to the Protestant Ethic. Since this discussion is the first full statement of the theme of *Tokugawa Religion*, it is perhaps worth quoting extensively.

In the first place the Japanese ethic is certainly this-worldly. Further, the concept of the "calling" is present and is developed at least to the point to which Luther brought it in the Reformation. The concept of man as an instrument of divine will is at least latently present. Furthermore, a motivation analogous to anxiety over salvation is perhaps latently present in the implicit threat of rejection discussed in relation to the problem of evil.

However, there are difficulties. The Japanese concept of the calling was not necessarily as dynamic as the Calvinist. The Shin sect even went so far as to uphold the view that even if we lie we are showing our gratitude to Amida because our whole life is an expression of gratitude. The necessary fulfillment of obligations to one's work could be interpreted as merely fulfilling the traditional prerequisites. The anxiety over rejection might strongly sanction conformity but would probably not per se lead to systematic rational achievement.

Further, the incentive to rid the world of magic, though present, was not as strong as in Calvinism. The Shin sect of Buddhism opposed all magic and even prayer on the grounds that salvation came only through faith and no human action was of any value. Zen Buddhism also tended to be iconoclastic, and Confucianism, as in China, was hostile to magic and ritual. The Konkō sect of Shintō was also opposed to magic and ritual on grounds of their inefficacy. In spite of these important tendencies there was no concerted drive to "rid the world of magic." Many sects of Shintō and some Buddhist sects employed magic, divination, and similar techniques.

Lastly, though asceticism was latent in many features of the Japanese ethic it was by no means expressed by all groups.

Having stated these difficulties I would now like to show how certain emphases and a later restructuring change the picture.

First, we have seen that certain features of the Japanese ethic, though not necessarily producing effects comparable to those of the Protestant Ethic, can do so if sufficiently stressed. Under what circumstances will they be "sufficiently stressed"? Before attempting to answer this question, let us see where this stress turns up in actuality. I think we may safely say that socially it is most prominent in the Samurai class and is reflected ideologically most closely in Bushidō and in independent Confucianism, both of which were largely confined to the Samurai class. Here I think we find a real development of this-worldly asceticism, at least equaling anything found in Europe. Further, in this class the idea of duty in occupation involved achievement without traditionalistic limits, but to the limits of one's capacities, whether in the role of bureaucrat, doctor, teacher, scholar, or other role open to the Samurai. How can these differences be explained in relation to the general system which the Samurai class shared with other groups?

Looking at that system, we see that the type of demand made by a superior is of crucial importance in determining the response. If what is demanded is merely the performance of traditional tasks, then such performance will be adequate. But if expectations are higher, performance must follow unless the system as a whole is to break down. It must certainly be said that performance expected from the Samurai class was considerably higher than that expected from other classes. The type of loyalty and service demanded by the Daimyo or the Shogun could be extremely exacting and could in fact override every other obligation quite drastically. The calm dispatch with which suicide was committed is but one example of this. We may safely say that not only were the normative demands higher, but also the level of anxiety was higher and the fear of rejection more developed. In fact all of this did lead to a personality type admirably suited to rational, systematic labor in a calling. The calling, however, was not in the economic sphere. Thus, we have the anomalous situation of a personality type amenable to the capitalist spirit but which is debarred from the economic sphere (this, of course, before Meiji). One further point on the Samurai: their main orientation was to their lord rather than to Amida or Amaterasu. This is why Bushidō is called the Religion of Loyalty. However a perusal of Section III will show that the system outlined there holds for any referent in the sacred world of superiors and not just for those referents which we easily identify as gods. Thus we have here a fundamental difference from our conception of deity. In Japan deity is continuous (with man), whereas with us it is radically discontinuous.

If, however, the Samurai class was the most "pious," to use a Western term, we have already seen at the end of Section II that this piety was seeping into the lower classes through various movements of popular piety. The Shingaku movement among the merchant class has been mentioned in this connection, but the relative importance of any such ideological features in the rise of the merchant class and the beginnings of capitalist-type undertakings during the pre-Meiji period is a problem which I cannot now answer. Ninomiya apparently had rather considerable success in raising the level of peasant production and efficiency in the areas where he worked.

Let us turn to a restructuring which occurred at the end of the Tokugawa period and had, I believe, profound effect on the rise of industrialism during the Meiji period and later. This is the re-emergence of the emperor cult as the centerpiece of Japanese religion. . . .

I believe the new emperor cult admirably filled all the deficiencies discovered in the preceding synthesis with regard to functional equivalence to the Protestant Ethic and in a way perfectly consistent with Weber's argument.

I believe the newly entrenched position of the emperor gives a leverage to the system quite analogous in function (though drastically different in origin) to that of the absolute God in the Protestant Ethic. Carrying out the Imperial Will and Divine Mission of Japan became a limitless obligation on all Japanese. The overwhelming position of the emperor lent an authority to his commands which quite overpowered traditionalist opposition and succeeded in ridding the world of magic, if not as cleanly as Protestantism, at least sufficiently so that it did not interfere with the growth of industrialism. The obligation to selfless devotion led to a high regard for and practice of inner-worldly asceticism in all walks of Japanese life.

My point about the emperor cult is not that it alone created all this, the elements were almost all present before, but that it galvanized the system into dynamic purposeful action. Of course those groups who were most ready were the first to respond. There is a fair amount of evidence that leadership in the Meiji Period in all spheres of life was drawn first from the Samurai class. But this was not solely the case. And as Weber has pointed out leadership alone is not enough. A disciplined, work-oriented labor force is also essential for capitalism. And this Japan had practically from the start, an enormous advantage over almost all other "backward" areas. Lockwood has pointed out [3] that, especially in the early years before the First World War, it was small business and industry which was the backbone of Japanese economic success. The rapidity with which small producers sprang up, took over new production techniques, adjusted their output to new types of market demands, and steadily expanded their output can indeed be looked upon as the very foundation of the growth of modern Japan. It is my opinion that is it impossible to understand this development without considering the religious system which I have been describing in this paper.

It is not necessary here to point out all the statements in the quotation above which need correction. Suffice it to say that the emphasis on the emperor cult is far too great, a mistake which is only partly corrected in *Tokugawa Religion.* One corrective to what might seem to be an extreme argument as to the basic parallelism between the Reformation and Japanese developments was included in the final paragraphs of the paper under discussion. There I pointed out that the Japanese religious pattern had favorable consequences for industrialization only in the situation of Western stimulus and example. I argued that in the absence of that example, even a renewal of the emperor cult would have been very unlikely to produce industrialization.

It should be noted that in the paper on religion there was no conscious attempt to utilize the pattern-variable scheme or to relate the analysis directly to the earlier analysis in the paper on Japanese social structure. These two early papers have been quoted and discussed at length, as they give an objective base line of the state of my thinking before the actual research on *Tokugawa Religion* got seriously under way. In their absence, I would have had to rely much more on the always hazardous resources of memory in order to trace the development of the ideas which emerged in the book, for the research notes and brief reports of the intervening period seldom give an over-all interpretation comparable to these papers.

Sometime during the fall of 1952, it became evident that I would be unable to go to Japan as I had earlier hoped, so the decision to do a thesis based on historical sources was forced upon me by objective necessity. It is, of course, possible that this would have been my ultimate decision in any case, although, as indicated above, my notebooks of 1951 and 1952 contain a number of thesis ideas involving field work. Any research project is constantly shaped by what is practically feasible as well as by what is scientifically interesting, and my inability to go abroad for reasons having nothing to do with the research is only a particularly clear example of a general fact.

On January 1, 1953, in an application to the Social Science Research Council for a research fellowship, there appears the first definite statement about my thesis research plans after it had become clear that I would not be going to Japan.

> The program of study and research which I plan to undertake will involve theoretical work of a rather general nature as well as analysis of empirical data. The problem which interests me is set up by Weber in his "Economic Ethic of the World Religions" and especially in his *Religion of China*.[4] I would like to narrow my empirical focus to the early modern period in China and Japan, that is late nineteenth and early twentieth centuries, in an effort to apply the Weberian hypothesis (on the relation of values and other aspects of social structure to economic motivation) to the particular problem of the response to the West in China and Japan during that period. It is my hope that specific hypotheses can be developed which will yield a better understanding of the differential nature and rate of industrialization in the two countries.

This is a fairly clear statement of the main theme of *Tokugawa Religion,* though the focus is slightly different from what finally emerged. The comparative reference to China (later in the SSRC application it was made clear that the work on China would be merely a supplement to Weber and that the main concentration would be on Japan) was greatly reduced and is now evident mainly in the Conclusion and parts of Chapter 2 of *Tokugawa Religion.* No systematic research on China was in fact carried out, the job on Japan proving to be more than large enough. Also, the time period mentioned in the application was somewhat later than that finally chosen.

In the spring of 1953, I was preparing to take the special examinations for the Ph.D. and in the summer and fall was involved in carrying out and writing up a field project (on a Mormon community in New Mexico), so that it was not until the end of the year that I began work in earnest on the thesis project. Now, for the first time, the problem arose of choosing a suitable body of data against which to test the ideas which had been developing for several years, as outlined above. Previously, I had simply read as widely as I could and then tried to make sense of what I had come across in terms of some kind of conceptual scheme. It now seemed necessary to work with a more clearly controlled body of data and to consider especially the kind of evidence which might negate my analysis. Here, there were a number of problems, both theoretical and practical.

I was well aware that the logic I was employing was somewhat different from that of Weber. He was arguing for the special significance of the Reformation and the importance of ascetic Protestant groups, compared to Catholics and others, in the development of the new economic order. I knew that there was no Reformation in Japan in the Tokugawa period, nor was the population divided into analogues of "Protestants" and "Catholics." My argument was rather that there were elements in the Tokugawa religious system and value pattern which, *given the stimulus and example of the West,* would be conducive to rapid industrialization.

For my argument, then, I had to show the existence of such favorable ideal factors *before* the major impingement of the West and indicate that they were significant in the industrialization process. The actual degree of significance would be difficult to prove from the available data, and I did not intend to attempt to show *exactly how much* of the later developments could be attributed to religious and value factors. I certainly did not intend to argue that such factors were exclusively or even mainly responsible; I was aware that other factors were also very important, but I did want to argue that such factors were present and may even have been of strategic importance when Japan is compared with other countries. Since the structure of my argument was different from Weber's, I did not need to show differences in the religious orientation of different groups, but my argument would certainly be strengthened if I could discover the presence of the values which I held to be important among groups that proved to be significant in Japan's modernization. I had already argued in the paper quoted above that the Samurai, from whom most of the Meiji leaders came, were especially imbued with the values I was holding important. But the question arose at this point, how should I select the data for the thesis research itself?

This involved a number of practical problems. First of all, there was really no one at Harvard who was particularly qualified to help me, who had specialized in or was signally interested in Japanese religion, or more specifically Tokugawa religion. There proved actually to be quite a bit of material, including a number of translations concerning Tokugawa religious and ethical thought and movements in Western languages, but the material was scattered and did not provide very much depth at any point. The relevant Japanese material in the Harvard-Yenching Library, on the other hand, was enormous and would have taken a lifetime to assimilate. Moreover, at that time I still read even modern Japanese with painful slowness, being unable to skim, and Tokugawa Japanese I had to take apart sentence by sentence.

I felt it essential, however, to get some firsthand acquaintance with primary sources because I knew how misleading translation can be. I felt that some control over actual material from the Tokugawa period would greatly strengthen my ability to interpret secondary work, both in Western languages and in Japanese. As a result of the theoretical and practical considerations mentioned above, I decided to choose a single religioethical movement for intensive analysis as a case study. The reasons for the choice I made should become clear from the following paragraphs, quoted from my thesis prospectus of March 1954.

> The Tokugawa period saw a series of ethical and religious movements in the various classes, partly shaping and partly giving expression to the status ethic of those classes. My thesis will be primarily concerned with one of these movements called Shingaku ("Mind Learning"). This movement originated in the merchant class in the early eighteenth century and was largely restricted to it in the course of its development in the eighteenth and nineteenth centuries. It was perhaps the most popular such movement in the merchant class and is probably the most conscious expression of the status ethic of that class. . . .
>
> My interest in Shingaku is then primarily in (1) its effects in encouraging the development of an attitude toward work comparable to that of the "calling" in the West and (2) its effects in helping to legiti-

mize the economic sphere of activity in terms of the dominant values of the culture. Both of these effects have great social implications. The development of such an attitude toward work in the merchants and other city classes certainly would have a considerable importance for the ease of adopting Western capitalistic and industrial institutions and techniques. That the merchant class did indeed furnish important cadres in this development (though the leaders came from the lower Samurai class) is, I think, in part attributable to such an ethic. The legitimizing and validating of the economic sphere of activities in terms of the major values of the culture would have an importance for the ease with which members of the Samurai class could move into the economic sphere, an event which, it has already been noted, happened with considerable frequency. . . .

There is an abundance of material on Shingaku in the Harvard-Yenching Library, and it is this which I plan to use as the data for the thesis. Four large collections of the works of Shingaku teachers are available. They contain a total of fifty-four works by twenty-four men. Some of these works are extensive collections of sermons, and almost all of them are intended for popular "mass" audiences. All of the chief teachers are represented as are works from the whole course of the movement's existence. In addition there are extensive secondary works, some of which give statistics on the growth and spread of the movement, case studies of its development in various cities, etc. All this material is in Japanese, there being only one article in any Western language, German. I will use the Shingaku tracts to sample the teachings of the school in the various periods of its development and the material on the organization and spread of the school to measure its impact on the merchant class. For this latter purpose, a study of material contained in the house codes of some of the merchant families may prove of considerable interest. . . .

In order to make the study of Shingaku more meaningful, I plan to include a rough descriptive analysis of the Tokugawa institutional structure, especially occupational structure, and a comparison of Shingaku with other religious and ethical movements of the period. The burden of the thesis will, however, be concerned with interrelations between values and the social system in the effect of the Shingaku movement on the merchant class. I hope that this thesis will contribute a case study which can be meaningfully compared with the large amount of work which has been and is being done on the relation between religious and other values and the rise of industrial society in the West.

In addition to these reasons for my choice of a particular sect, it should be noted also that I had already begun to study Shingaku texts with Prof. Elisséeff in the fall of 1952.

The thesis did not go quite as the prospectus envisioned, for a number of reasons. The Shingaku material proved to be extremely difficult, and work on it progressed slowly. When I learned, for reasons having nothing to do with my research, that it was imperative that I complete my requirements for the Ph.D. in the spring of 1955, it became clear that anything like a complete study of Shingaku was out of the question. Actually, a careful study was made only of the life and writings of the founder, Ishida Baigan, and later developments were summarized from secondary sources. Partly because the research on Shingaku had to be truncated (the research and writing of the thesis took about fifteen months, from December 1953 to February 1955), the treatment of this sect was not as central in the final study as first imagined.

To compensate for the decreased importance of Shingaku in the study, more attention was given to other religious and ethical movements as they were reported in Western languages. Here the work could proceed rapidly, and a virtually complete coverage of everything available at the time was possible. This actually had the advantage of giving a kind of random sample of the total material. One could argue that there would be a bias in what was selected for notice by Western scholars. However, the fact that there was material from Western scholars with varied interests from a number of countries and over a period of nearly a century reduced the probability of bias. On the basis of considerably greater knowledge today, I feel that no major aspect of Tokugawa religion or values was, in fact, overlooked, though a number were only tangentially mentioned. The data were, then, I have some reason to believe, a fairly representative sample of all available material on Tokugawa religion and values. What difference did the data make to the ideas with which I had started the research?

On the whole, the data confirmed my expectations; in fact, overwhelmingly so. I found over and over again the themes which I had already discovered in the paper on Japanese religion. A really different pattern of religious and value orientation was hardly to be found. Of course, there were many nuances of formulation and expression. It was only in the research for the thesis, for example, that I discovered the two main types of religious action reported in Chapter 3 of *Tokugawa Religion* (pp. 70-77), but both types turned out to support the kind of ethical orientations in which I was interested. Similarly, the structure of many of the religious movements reported in chapters 4 and 5 and their implications for the status ethic of various social groups became clear to me only in the course of the research. But though this considerably filled out the analysis, it did not cause me to alter my basic conceptual framework. Yet my basic framework did change radically in the fall of 1954.

It would probably be wrong to say that the shift did not arise at all from the data. There were certain assumptions in my thinking which the data must have been gradually undermining. These concerned the crucial link in my argument which connected religious and value orientations and the process of industrialization. In the paper on Japanese religion, great causal weight was given to the emperor cult in linking these two elements, as appeared in the long quotation above. But in the thesis prospectus there was not a word about the emperor cult, and the emphasis was on a direct relation between values and the economic or occupational sphere. To quote again from the prospectus:

> One factor of great significance is the existence of an occupational structure with a considerable degree of specialization and of ethical developments which linked up the major value commitments of the culture to motivation in the occupational sphere. It is this aspect of the problem which especially interests me.

Neither the earlier nor the later assumption accounts satisfactorily for the data, and this must have become evident to some degree. The issue would have been forced at the time of writing the thesis when I came to the "rough descriptive analysis of the Tokugawa institutional structure" which was projected in the prospectus. Though continuing to operate on many of the assumptions of the analysis made in my first paper on Japanese institutional structure, I had never reworked my analysis of social structure after gaining my much more extensive knowledge of the religious and value systems. Such a reworking was not necessitated during the course of the actual research itself, which was confined largely to religious and value materials. But before I came to interpret the results of my research for the larger question of the relation of values and social structure, I was, as must often be the case, exposed to a new set of conceptual tools which, when applied to the materials I had in hand, rendered substantial changes in my analysis.

The first awareness of a possible new approach to my problem came at a conference in New Mexico in September of 1954, when Talcott Parsons, fresh from a year at Cambridge University and with a draft in hand of what became *Economy and Society*,[5] presented a new scheme for categorizing social institutions. In this new scheme, he categorized the major subsystems of a social system in terms of the four dimensions of action (as derived from a combination of the pattern variables and Bales's four system problems and reported in *Working Papers in the Theory of Action*).[6] These subsystems were called economy (adaptive dimension), polity (goal-attainment dimension), institutional subsystem (integrative dimension), and pattern-maintenance subsystem (latency dimension).

Actually, this categorization had been foreshadowed in his essay of 1953, entitled "A Revised Analytical Approach to the Theory of Social Stratification,"[7] but I had not caught the implication or even very clearly understood the paper on first reading. What was new to me was the linking up of values as defined in pattern-variable terms, dimensions of action, and particular social subsystems; for example, the link between particularism and performance, the goal-attainment dimension, and the political subsystem, or polity. It seems to me now that even during those days in New Mexico it had already occurred to me that it might be very illuminating to think of Japan as a society stressing "political values," goal-attainment, and the polity.

During the fall, I read a copy of Parsons' Marshall lectures,[8] pored over a set of dittoed diagrams showing the major subsystems of a social system and their input–output relations (never published and since much revised), and went back to the earlier article, "A Revised Analytical Approach to the Theory of Social Stratification." Out of much cogitation on this material, I came to a new conceptualization of the link between values and the industrialization process in Japan and in December wrote Chapter 2 of *Tokugawa Religion*, which spells out the analysis of Japanese social structure and the major processes I wished to analyze.

It is probably unnecessary to go into any detail as to what finally emerged, since that is all contained in the book, but it might be useful just to indicate some of the changes from earlier positions. The new emphasis on the polity forced a change in the pattern-variable categorization of Japanese values from particularism-quality to particularism-performance (in recent years Parsons has tended to speak of performance-quality instead of achievement-ascription). There are still a number of problems here, but the new categorization did, I think, prove useful.

Another important change will become evident in comparing the major headings of Chapter 2 of *Tokugawa Religion* with the outline of the early paper given above. The categories of *The Social System* have been scrapped, and a three-level analysis, as indicated in the revised paper on stratification, has replaced them. The three levels are (1) the value system, (2) functional subsystems of the society, and (3) concrete structural units. The heart of the analysis was contained in the discussion of the functional subsystems, for here it was that I worked out the link between values and economic action.

Of course, the new set of categories did not affect just the analysis of social structure in Chapter 2 but the whole organization of the book. Chapter 4, "Religion and the Polity," and Chapter 5, "Religion and the Economy," are specifications of some of the input–output relations suggested in Chapter 2. The new understanding of the relation between values and economic development which use of the new categories had generated is summarized in Chapter 7, the conclusion. There I argued that the most important way in which religiously backed values affected economic action was not directly, though there were important direct consequences, but indirectly through the polity. This was a shift away from the emphasis in the thesis prospectus, which had been exclusively on direct influences of religion on the economy.

In a sense, it was a shift toward the considerably earlier analysis of the paper on Japanese religion, which had emphasized the emperor cult. But the new position was much more highly generalized. Though the emperor in *Tokugawa Religion* probably is still overemphasized, the argument no longer rests on the particular development of emperor worship but is a generalized argument about the relation between religion and political structure in Japan. A few readers have been thrown off by my rather special use of the term "political." I do not mean to equate polity with "government." Even the most nongovernmental collectivities have a "political" aspect, namely, that which is concerned with the attainment of group goals. It is this aspect which I found to be central in the Japanese case.

It is my impression, in which most of my Japanese readers concur, that the main contribution of *Tokugawa Religion* was in ways of conceptualizing and not in uncovering new facts. Though some things, particularly about Shingaku, were not generally known in the West, there is almost nothing in the way of factual information which is not only known but also, I would say, embarrassingly obvious to Japanese scholars in the field. Yet the book has received considerable notice in Japan, and in 1962 it appeared in Japanese translation.[9]

The usual textbook notion of social research is that one forms a hypothesis and then proceeds to gather data to confirm or negate it. In many instances, this may be a tolerably accurate description; but, at least in the field of comparative and historical sociology, the researcher often finds himself with an abundance of data, and the problem is how to make sense of it. This situation does not seem likely to change in the near future, since the historians and philologists continue to unearth many more facts than they can make sense of.

Making sense of other people's data would seem, then, to be the main job of the sociologist who ventures into this field. Though this conception of the historical sociologist's role may not fit with everyone's definition of empiricism, it nevertheless does seem to fit rather broadly into the framework of empirical science. For the new interpretations offered by the sociologist give a stimulus to new research. Often, as in the case of Weber's famous Protestant Ethic hypothesis, those undertaking the new research have failed so utterly to grasp the structure of the original argument that their research casts little light on it. But eventually a body of further research does build up which renders the sociologist's hypothesis more-or-less tenable.[10]

• NOTES

1. Talcott Parsons, *The Social System* (Glencoe, Ill.: The Free Press, 1951).

2. Talcott Parsons, *The Structure of Social Action* (New York: McGraw-Hill, 1937).

3. W. W. Lockwood, "The Economic Development of Japan: Growth and Structural Change, 1869-1938." Ph.D. dissertation, Harvard, 1950.

4. Max Weber, "The Social Psychology of the World Religions," in H. H. Gerth and C. W. Mills, trans., *From Max Weber: Essays in Sociology* (New York: Oxford University Press, 1946); H. H. Gerth, trans., *The Religion of China* (Glencoe, Ill.: The Free Press, 1951).

5. Talcott Parsons and Neil Smelser, *Economy and Society* (Glencoe, Ill.: The Free Press, 1956).

6. Talcott Parsons, R. F. Bales, and E. A. Shils, *Working Papers in the Theory of Action* (Glencoe, Ill.: The Free Press, 1953).

7. In Talcott Parsons, *Essays in Sociological Theory* (rev. ed.; Glencoe, Ill.: The Free Press, 1954).

8. An earlier draft of Parsons and Smelser, *loc. cit.*

9. R. Bellah, *Nihon Kindaika to Shūkyō Rinri* (Tokyo: Miraisha, 1962).

10. I had originally hoped to carry the story of *Tokugawa Religion* up to the present, but considerations of space prevent this. I would have especially liked to discuss the Japanese criticisms of the book, notably those of Maruyama Masao, many of which I have come to accept. A brief discussion of some of the issues involved can be found in Robert N. Bellah, "Reflections on the Protestant Ethic Analogy in Asia," *Journal of Social Issues,* XIX (1963), 56.

7 Cross-Cultural Analysis:

A Case Study*

STANLEY H. UDY, JR.

A thesis topic, like greatness, is something which some students are born with, others achieve, and still others have thrust upon them. Since I was not so fortunate as to have been born with one, nor so unfortunate as to have had a topic thrust upon me, I was obliged to "achieve" one.

My approach to this perennial problem of the graduate student was to try consciously and systematically to choose that topic which combined the greatest possible number of my interests. After some exploration, the amalgam on which I settled was comparative social analysis, organization theory, industrial sociology, pleasure derived from reading ethnography, and curiosity about what one could do with the Human Relations Area Files. This combination (actually the result of putting together some work of Profs. Marion Levy, Wilbert Moore, and Melvin Tumin, under all of whom I had studied) suggested some kind of cross-cultural study of work organization in nonindustrial societies, assuming that such a study was possible and could be reasonably expected to contribute anything to organization theory.

I had read enough ethnography to be persuaded that some study along these lines was possible, but I did not yet know specifically what kind of study it should be or what particular questions it should attempt to answer. Furthermore, as yet I had no real justification—beyond the notion that it might be fun—for studying work organization

* Stanley H. Udy, Jr., *Organization of Work: A Comparative Analysis of Production among Nonindustrial Peoples* (New Haven, Conn.: Human Relations Area Files Press, 1959).

161

cross-culturally. Why should that approach be used instead of some other one? In other words, is there some distinctive set of important problems in organization theory which not only can be studied cross-culturally but cannot readily be approached in any other way?

Let us begin to answer this question with a brief overview of the general relevance of cross-cultural analysis. Broadly speaking, cross-cultural analysis is appropriate to any research problem involving variables whose values differ from one society to another but remain more-or-less constant in any given society. Since most highly general institutional variables tend to be of this variety, one of the great advantages of the cross-cultural method is that it allows maximum variations in basic institutional structure to be introduced into the research design. Cross-cultural research is thus obviously relevant to the development of theories about basic institutions and entire societies. But cross-cultural analysis is also relevant at certain points to the development of theories about organizations that always appear in the context of some society. Inasmuch as such organizations are not self-contained but are dependent to some extent on the society within which they exist, the question eventually arises as to the degree to which their structural variations are accounted for by influences of the social setting, as opposed to pressures arising from internal consistencies. In order to answer this question, one must use a design that allows the setting to vary. The most extreme variations can be achieved by comparing organizations in different societies, particularly in nonindustrial societies, since there is considerable reason to believe that industrial societies are substantially alike in many basic essentials.

But why do variations in the setting need to be so "extreme" as those afforded by a comparative study of nonindustrial situations? One can certainly introduce contextual variations in a comparative study of, say, work groups in American industry, or even work groups in the same plant. The value of studies which do this is undeniable. However, it is equally undeniable that organization theory has been carried to its greatest level of sophistication in its treatment of purely internal aspects of administrative operations with influences from the setting held constant. Part of the problem is that we do not really know what the major relevant dimensions of influence of setting on organizations are. Investigation of contextual influences on organizations is thus still pretty much at the exploratory stage; it is not a question of systematically introducing variations to test hypotheses but rather a question of discovering what the most strategic variables are and how they seem to operate. In such circumstances, there is great value in using data which may be expected to exhibit extreme variations, simply because extreme variations are more visible and hence more likely to be noticed in an exploratory study than are minute ones.

I thus became persuaded (1) that exploratory studies of relationships between organizations and their technical and social settings can contribute to organization theory and (2) that cross-cultural analysis, by reason of the relatively extreme variations which it permits one to introduce into one's design, would be one of the best ways to conduct such a study.

My purpose thus became the exploration of work organizations in different societies in order to discover the major relevant dimensions of influence of the setting on the organization. I took as my unit of investigation the *production organization,* defined as any social group manifestly engaged in carrying on one or more technological processes. My main working hypothesis—that the structure of any production organization is partly a function of the characteristics of the technological process which it is carrying on and partly a function of the nature of the social setting within which it exists—followed from the purpose of the study. The aim was to explore a number of different production organizations carrying on widely different kinds of technological processes in widely different social settings and to do so in such a way as to give specific content to the working hypothesis. In other words, my objective was to discover, insofar as possible, what particular characteristics of technology and what particular aspects of the social setting influence which specific elements of organization structure. And I proposed to do this by studying a number of different work organizations in a number of different societies.[1]

PRELIMINARY SURVEY

According to the logic of research procedure, after one has defined the problem, one next draws a sample. But that is not what I did next, nor is it probably what any exploratory cross-cultural researcher would be likely to do next. I did not yet know how many societies I wanted to use. I had decided to attempt generalizations rather than purely descriptive "comparing and contrasting." But I did not know whether it would be better to study a few work organizations—say, eight or ten

—"in depth" or to study specific features of the largest possible number of organizations. Also, I had not decided whether to do the whole thing in words or to use numbers and statistical techniques, though this issue was not a very important one at this stage. (In fact, a decision was not made on this score until rather late in the research.) But, above all, I was not yet sure just how much descriptive information was really available on nonindustrial work organizations.

Actually, many of my problems at this point resulted from my own background. I had been trained as a sociologist, rather than as an anthropologist, and thus was not systematically familiar with ethnological literature. I had read a considerable number of anthropological monographs—partly because I did, and still do, find them enjoyable as well as instructive. In particular, I was familiar with the various works of Firth and Malinowski, especially as they deal with primitive economic systems, and with much of the literature on the Australian aborigines and the Plains and Northeastern American Indians. I also knew most of the standard works on primitive economics, such as those of Thurnwald and Herskovits. I had read Murdock's *Social Structure* in some detail.[2] I had also skimmed through, at one time or another, quite a few nineteenth-century works in the early "classical" comparative evolutionist tradition, and I must admit that I was greatly impressed by some of them, such as Julius Lippert's *Kulturgeschichte.*[3]

But that was it. I had read just about enough to be persuaded that this study was possible. So I tackled a list of representative ethnographic monographs, reading them, perforce, from a sociologist's point of view. I found out what anthropological journals and monograph series existed that I had not known about and began looking at them. At this point, I felt ready to survey the literature to find out if anyone had ever done what I had planned to do. My survey yielded Nieboer's *Slavery as an Industrial System* and Buxton's *Primitive Labour*, published in 1900 and 1924, respectively, as the only general works I could find which dealt directly with my topic.[4] And, although there was enough information in monographs to convince me that a study of nonindustrial work organization was possible, it nevertheless appeared impossible to study eight or ten organizations "in depth," largely because variations even on the most general level seemed so great that I had no way of deciding which eight or ten organizations to choose for purposes of generalization. I did not know what controls to make. It appeared that part of the purpose of the study would have to be to discover what types of work organizations would be strategic for a depth analysis. So I decided to use a large number of cases, to keep the analysis on a very general level, and to try to discover "what is going on here."

Having decided in a general way what I wanted to do and having gained at least a nodding acquaintance with the literature, I now needed to conceptualize my problem so that I would know what data to look for, how to classify it, and what to do with it after I had it, assuming I could get any. Professors of sociology seem particularly fond of telling students that one cannot just "go out and study something" without having some kind of conceptual framework. It is interesting that this is probably one of the most "okay" statements to make in our field, despite its palpable question-begging character where exploratory studies are concerned. It appears that one must know what variables are relevant in order to be able to explore data to find out what variables are relevant. This is, of course, not so, despite the fact that much of our methodological literature leads to this paradox. The problem is that such literature does not have much to say about the exploratory, "what is going on here" type of study, where the problem is more one of starting on a general level and working toward more detailed information through successive rounds of exploration." And one can begin on a very general level, indeed.

My main working hypothesis suggested three very broad categories: technology, organization, and social setting (i.e., the physical nature of the process being carried out, the structure of the organization doing the work, and external cultural influences on the organization). It also suggested a broad method of analysis, namely, to explore relationships between technology and organization, holding social setting constant, and to explore relationships between organization and social setting, holding technology constant. The exploratory problem emerges as a trial and error of various ways of subclassifying these categories in the context of this procedure.

Further reflection, aided by common sense, plus the literature on industrial sociology, suggested the desirability of a fourth broad category: reward systems. For one presumes that all motivation to work ultimately derives from goals held by organization members in the social setting, that rewards attached to work constitute one way of making organizational activities means to such goals, and that probably, therefore, certain things about reward systems will vary, depending on conditions in the social setting and the organization.[5]

A fifth category suggested itself as a result of preliminary reading; namely, one which would include all activities other than production carried on by the organizations studied, including magical and ritual activities connected with work. Particularly in view of the work of Malinowski, it seemed advisable to keep track of such things separately from the outset, though this category is probably a special case of the "social setting" and, indeed, eventually proved to be analyzable in this manner. I felt that these five categories would be adequate for preliminary study and would also serve as a framework within which systematic exploration could take place later.

The next problem was to develop a system for searching the literature and recording data. I needed to be able to peruse ethnographic works, locate production organizations, and record data describing their internal structure, the technological nature of the tasks in which they were engaged, and those aspects of the social setting influencing their structure. I also wanted to be able to do all this rather quickly, since I was planning to use a large number of societies (I still did not know how many, or what kind of sample I would employ) and at the same time use uniform methods for all societies. Further, the data had to be compiled in such a way that they could be subsequently explored, not only for the purpose of discovering interrelationships among variables but also in many cases to discover what variables should be studied in the first place.

The cross-cultural researcher is fortunate in these regards, because a considerable part of this work has already been done for him, thanks to the Human Relations Area Files.[6] This facility represents an extremely ambitious effort to collect and classify all extant descriptive data relevant to social science on all known societies. Needless to say, this objective has not been reached and probably never will be; but fairly complete files exist at present on a surprisingly large number of societies, enough to make the files an indispensable research tool in any cross-cultural analysis involving a large number of societies. The societies in the files are classified in six major culture areas, with each such area subdivided into ten subareas. At the time of my work, the files contained too many gaps to enable them to be used directly in sampling. But, once a sample had been drawn on other grounds, I found that the files could be used wherever they were complete and that doing so saved innumerable weeks of work.

The file on each society is divided into two parts. The first part consists of complete reproductions of all important source materials on the society in question, with a bibliography of materials not contained in the files. The second part consists of the same material classified in categories in such a way that material in the categories is comparable across societies—at least as comparable as possible, given the present state of ethnography and coding operations. It is thus apparent that the Human Relations Files can be of inestimable aid. To the extent that he can rely on the files, the researcher finds that the bibliographical work on each society has already been done and that the data have been collected and classified for him. All he needs is to go through the relevant categories in the files and run their contents against one another (whether verbally or statistically), and he will have a cross-cultural analysis.

Unfortunately, it is not that simple. The Human Relations Area Files are not like the *World Almanac,* despite the fact that one may be tempted to use them as such. One cannot just "look things up" in them, unless the categories relevant to his research happen to be the same as the categories used in the files. This is unlikely to be the case. I found that in order to use the files, I had to design a procedure for searching them and to do so in such a way as not to be bound in advance by the HRAF categories. This latter point was important for my purposes, as I was not yet sure what detailed categories I ultimately wanted to use.

I thus posed for myself the following problem. Assume I were to read an anthropological monograph on a particular society all the way through in detail and carefully retrieve any information from it dealing with work organization. What method of searching the files would result in my retrieving the same information with less time and effort? Accordingly, I procured from the library various ethnographic works on the Tikopia, Trobriand Islanders, Chukchee, and Crow. (I had no particular reason for choosing these societies, except that I wanted to include Firth, Malinowski, and one older work—in this case, Bogoras on the Chukchee—and to take societies from roughly different parts of the world.) [7]

All the works I chose are also excerpted and classified in the Human Relations Area Files.

I then read these books carefully, recording all the production organizations reported in them and classifying the data on each such organization in the following broad categories: internal organization structure, technological characteristics of each production activity reported for each organization, other activities performed by the organization (including ritual and magic in connection with production), the system of rewards for work, and mutual influences between the organization and its setting. I then went to the Human Relations Area Files and tried various systematic search patterns to see if I could find one which would yield the same data with less time and effort. I did all this by trial and error, tempered with what I conceived to be a common-sense emphasis on those file categories which appeared obviously to be more relevant than others. As a result, I was able to devise a standard procedure which involved routinely searching the following categories:

22. Food Quest
23. Animal Husbandry
24. Agriculture
31. Exploitative Activities
32. Processing of Basic Materials
33. Building and Construction
342. Dwellings
62. Community
46. Labor
47. Business and Industrial Organization

Categories 22 through 342 represent those areas in which most of the organized work done in nonindustrial societies is carried on. Search of these categories therefore usually resulted in locating most of the production organizations and in finding any available information about their technological and internal structural characteristics. But the data contained in these categories proved only to allude in varying degrees to the other matters on which I desired information. Such information, I found, could usually be obtained by searching Category 62 and "lining up" the information in it with the data previously obtained from the other categories. It was thus not ordinarily necessary to search all the institutional categories, since Category 62 proved to include most of the available information about interrelations of subgroups in the locality group. And due to the limited ambiences of nonindustrial peoples, this information was often sufficient. Where it was not, any other category that was clearly relevant in view of information obtained up to this point was consulted. The most frequently consulted ones were:

59. Kinship
60. Family
64. Government

These categories, however, were consulted only when clearly necessary.

Categories 46 and 47 were routinely searched as a check on the entire procedure, since in principle all production organizations are at least alluded to there. Any additional leads provided were followed up *ad hoc.* It had been my original intention to use these categories initially to locate production organizations and then follow them up as appropriate. But categories 46 and 47 proved to be too incomplete for such a procedure.[8]

In the four societies comprising my preliminary survey, I found that this search routine yielded the same information for my purposes as a complete reading of the monograph material and only consumed about 10 per cent as much time. I estimated that by this procedure I could "do" about four or five societies per day instead of one society every two days, assuming the societies in question to be included in the Human Relations Area Files.

SAMPLING

I was now satisfied that a study of the kind I had in mind could be made and that it ought to proceed by exploring general relationships among a large number of cases. I had devised a broad conceptualization which, in view of my preliminary work, seemed to be a realistic and probably fruitful point of departure for an exploratory study. I had established the fact to my satisfaction that data could be collected and put into that conceptualization from existing ethnographic sources, and I had designed a routine for use with the Human Relations Area Files wherever use of the files proved possible. I could now confront the problem of collecting a sample of cases to compare.

Any research study of any type whatsover which seeks to make generalizations beyond the material studied involves problems of sampling. It makes no difference what field the study is in, how many cases are analyzed, whether they are "in depth" or in general terms, whether the analysis is done with words or numbers, whether or not "statistical techniques" are employed, or whether or not the research is exploratory. In any of these instances, whenever the researcher proposes a generalization, he is implicitly identifying a larger population, of which his cases purport to be a representative sample, and contending that certain relationships observed in his sample could not have occurred there by chance. I say these things only because in some quarters impressions seem to exist to the contrary, particularly in connection with cross-cultural studies. It is simply not true that one can avoid sampling problems by proceeding in words instead of numbers or by avoiding the use of statistical techniques, though it is unfortunately true that by avoiding such methods one can often keep sampling problems from becoming explicit. To be sure, it is also unfortunate that the exploratory researcher is at something of a loss to be able to say definitively whether his relationships could or could not have occurred by chance. But this is not because he may be using numbers or statistical methods; it is because his research is exploratory, and he cannot solve this problem by being purposely vague about it and contending that because the research is exploratory, "anything goes."

Thus, it did not occur to me not to pay some attention to problems of representativeness, independence of cases, and randomization of choice where controls cannot be made. But it is admittedly frustrating to pay too much attention to such problems in a cross-cultural analysis, since one cannot really solve them. There is no reason, however, why one cannot endeavor to make the best of the situation and randomize sampling decisions whenever it is possible to do so. This is what I tried to do.

I had already decided to generalize to a population of all nonindustrial production organizations. Also, as explained earlier, I had decided to assume that production organizations in the same society are never independent events but that production organizations in different societies always are. These considerations suggested that I could draw a sample of nonindustrial societies and then draw one production organization from each society. The assumption that production organizations in the same society are not independent leads one into problems with larger societies which have discontinuous structures and hence, perhaps, afford the possibility of independent production organizations. But I could not devise any means of solving this problem which did not get me into other problems that seemed to be worse. Thus, I did not change this assumption, and as a result my sample was probably biased against production organizations in large societies.

The second assumption, that production organizations in different societies are always independent events, is probably one which, in retrospect, I would be inclined to change. If I were to do it over again, I think I would limit myself to a smaller number of cases drawn from different culture areas and thus be in a better position to argue their independence. My reason for not doing so initially was to maximize sample size for exploratory purposes, but in doing so I probably ran considerable risk of interaction.

It is obviously impossible to draw a random sample of all nonindustrial societies. One can, however, select a sample by design to represent the major culture areas of the world. On this basis, Murdock provides a guide for sample selection in cross-cultural research.[9] He divides the world's population into six basic ethnographic regions: Africa, Circum-Mediterranean, East Eurasia, the Insular Pacific, North America, and South America. He then subdivides each region into ten smaller areas which "insofar as possible" observe boundaries between recognized culture areas. Within each of these smaller areas, he lists from five to fifteen societies, chosen so as to include those societies which (1) are the most populous, (2) are the best described, (3) exemplify each basic type of economy, (4) exemplify each linguistic stock or subfamily, (5) are relatively distinctive. He avoids selecting two societies from any area which are either geographically contiguous or characterized by mutually intelligible languages. The result is a list of 565 societies, more-or-less evenly distributed over the world.

Unquestionably, Murdock's selection, like everything else, is open to criticism. One might, for example, have preferred a grouping within regions which followed recognized culture areas more closely, rather than an insistence on dividing every region into ten parts. However, after tinkering for a while with various possible alternative procedures, I found that even when I thought I might have a better approach on purely technical grounds, I could not really evaluate it relative to Murdock's simply because I did not then, do not now, and am very unlikely ever to know as much about ethnography as he does. I thus decided to accept Murdock's sample as a base from which to work. For reasons already explained, I had decided to use the largest possible sample. But I could not use all 565 societies, because the particular material I desired was not available for all of them. In the interests of representativeness, however, I thought I had better end up with about the same number of societies in each of the six ethnographic regions and also space the societies in each region as evenly as possible over the ten areas. The size of my sample thus came to depend on the amount of data available for the most sparsely covered area.

At about this point, I joined the faculty of Yale University and met Prof. Murdock himself, who became extremely helpful. He advised me that material on South America was the most sparse, so I started there. I found that the largest minimum number of societies for which I could count on being able to get data consistently was two from each of the ten areas there. So I proceeded on this minimum basis, drawing two societies at random from those listed by Murdock under each area throughout his entire sample. I then looked at the ethnographic litera-ture on each society to see whether the material I required was avail-able. If it was not, I kept drawing at random until I found a society in the same area where it was. If I exhausted Murdock's list at any point, I substituted other societies from the same area, often making use of advice from Prof. Murdock, who was very helpful on this score, as in other ways. In only a very few instances was I unable to obtain data on at least two societies from the same region. The result was a sample of 120 societies.

The next step was to collect data on every production organization clearly reported in each of these societies. Using the Human Relations Area Files wherever possible, I did so, classifying the material accord-ing to the provisional scheme that I had decided on earlier, taking notes summarizing what the ethnographer said, and attempting no further coding at this stage. I took notes on all the organizations I could find because I thought my later explorations might require strati-fication of the sample in various ways which I could not at that time foresee exactly, and I wanted to make sure that I had enough material for such a purpose. At this point, I realized that as a minimum I wanted to be able to introduce gross variations in both technology and social setting, so I thought I had better check my sample to see if that was going to be possible. So I classified my organizations by type of techno-logical process, using a traditional scheme of seven categories: tillage, hunting, fishing, collection, construction, animal husbandry, and manu-facturing.

It turned out that I had relatively few construction and manufac-turing organizations. I thus tried to add a few more societies offering information on construction and manufacturing (I succeeded with construction, but not with manufacturing), as well as some more soci-eties which had centralized governments (another characteristic on which I thought I might wish to stratify). The net result was to throw the sample out of balance regionally, so I drew a few more societies at random in other regions to equalize the distribution as much as possible. Quite arbitrarily, I did not allow more than four societies to be included from any single area. The result was a final sample of 150 societies, from each of which I drew one production organization at random, except that I stratified the sample to yield approximately equal numbers of cases of each technological type. I felt that this sample would enable me to explore variations adequately in both technology and social setting, that in drawing it I had randomized my procedure as much as possible in the circumstances, and that where such was impossible I had tried to make my selection criteria explicit.

EXPLORATION, PROCESSING, AND ANALYSIS

I now had a sample of 150 production organizations from 150 non-industrial societies, for which I had assembled descriptions classified according to the provisional five-way breakdown already described. I now asked myself what the ultimate results would look like and how they would be presented, and I decided on something akin to Murdock's *Social Structure*. This implied systematic subclassification of my major categories such that I could count the frequencies in each category and cross-tabulate to show the various specific aspects of the relationships described by my original working hypothesis plus the further working hypothesis about reward systems. I thus faced the following steps: (1) exploring my existing data plus any relevant theoretical literature to suggest classification schemes and hypotheses; (2) analyzing and recoding all my data in terms of the resulting scheme; and (3) mechanically tabulating it. I decided not to "run everything against everything else" but to limit my runs to hypotheses either generated from the theoretical literature or previous studies or deduced from other hypotheses whose validity in my own sample had already been demonstrated. This did not make my procedure especially simon-pure, however, since the process of exploration automatically excluded inapplicable theoretical perspectives. The entire design of the study did not permit me to propose hypotheses and then accept or reject them on the basis of statistical tests; it simply permitted me to describe what I found.

Conceptually speaking, the final taxonomy involved seven concrete types of technological process crosscut by four aspects of any technological process; four concrete types of production organization crosscut by five aspects of organization structure, with each such aspect possessing anywhere from two to five subdimensions; at least three major relevant institutional areas in the social setting with each organizational type related to one or more of these three areas in up to seven different ways; and four kinds of reward systems. Ultimately, all these items were related to one another by 64 propositions, 61 of which described cross tabulations, one of which was an "existence theorem," and two of which were true by definition but did not seem obvious unless explicitly stated. It may seem a bit ridiculous to say that I frankly have trouble describing exactly how, by combining an exploration of my data with a study of the theoretical literature, I reached this elaborate conceptualization. But, however absurd that statement may seem, it is so, despite the fact that this process received my full-time attention for about seven months. I can, however, indicate in a general way what my approach was and describe some specific experiences I had in the course of the analysis.

Basically, it seems to me that there are three different ways by which one can arrive at relevant typologies from a mass of relatively undifferentiated verbal descriptive data and a virtually endless procession of theoretical works, all of varying relevance to one's problem. First, one can ransack the theoretical literature and try out all the existing typologies that one finds there. Graduate students, in particular, often seem to have a tendency to do this, and it is all right, as far as it goes. The trouble is that usually one's own research problem is not exactly like the problem discussed in the source one is using. At one point or another, the researcher will have to stop fishing and start cutting bait.

A second approach, particularly suited to "cutting bait" and starting from one's own data, has been termed "analytic induction." [10] Essentially, analytic induction is a piecemeal approach whereby one selects any two units that he is comparing and, on a purely descriptive level, makes all the general statements he can which apply to both units. He then takes a third unit and similarly compares it with the previous general statements, iterating this procedure until he finds a stable set of general statements, from which he then extracts a typology. Such an exploratory procedure can be extremely helpful, but it, too, suffers from various defects. First, unless the researcher is extremely cautious, he is quite likely to find himself straying from his original working hypotheses, since he is obliged to move "wherever the data take him." Second, analytic induction focuses on characteristics which are present rather than those which are absent and thus tends to yield a list of concrete attributes which everything has in common rather than a set of universally applicable variables or analytical categories. And finally, in a related vein, analytic induction says nothing about the process of generalization itself; it implies that one begins with a *tabula rasa,* which is impossible.[11] At best, analytic induction simply describes a way of going through one's material in conjunction with other methods.

The third type of exploratory procedure may seem to be a peculiar one to mention, but it is nevertheless an important one. That procedure is simply to sit and think about it. Before doing so, one should digest as much of his descriptive data as possible. Sometimes it is also a good idea to bear in mind a great deal of theoretical literature while doing so, because under those conditions one can usually better describe later where his ideas, if any, ultimately came from. At the risk of appearing either brutally frank or ridiculous, however, I must say that there were times in this study when I found it desirable to try to forget everything I knew about sociology and simply to ask myself how "any reasonable man" would go about attacking the problem. A case in point is the typology of reward systems which I employed. After repeated unsuccessful attempts to make any sense of the material in a way pertinent to my working hypotheses, I finally tried the following "common-sense" assumptions: (1) nobody will work unless he (*a*) is paid for it, (*b*) is forced to, or (*c*) feels very strongly that he ought to; (2) people who swap things around among themselves feel closer to one another than people who do not.

The first assumption suggested some hypotheses about relationships between the ways in which people are recruited for work and the strength of the reward system. The second assumption suggested in a vague way that the strength of the reward system might be estimated by variations in patterns of exchange of raw materials, produce, and the rewards themselves within the organization. It turned out that Max Weber discusses an analogous problem in connection with taxation, differentiating movements of goods and payments up and down the status hierarchy. I tried applying the same thing to production organizations, and it turned out, indeed, to provide at least a rough way of measuring the relative strength of different arrangements, although the over-all results were very crude.[12]

As one may infer from this example, I found it desirable to use a combination of the three exploratory methods described. I would usually start with one or the other in particular and then shift methods after I had carried the first one as far as I was able. After reaching a provisional classification via one method, I would then try retrospectively to see if I could also reach it starting with the two ways I did not use initially. This usually resulted in some revisions. Then, if I had not already done so, I would search the theoretical literature to see if I could, from that, justify what I was doing and also to find hypotheses about connections with other schemes in other relevant parts of the study.

In the case of technological processes, I used a standard concrete scheme directly from the anthropological literature (tillage, hunting, fishing, etc.), since it seemed that such a scheme involved types with widely different task problems. The next question was: along what principal, organizationally relevant dimensions do these types vary? At this point, I resorted to analytic induction to describe the major attributes of each of the seven types of processes. At the same time, I found a standard typology of division of labor used by many European economic historians of the early twentieth century and discovered that with some modifications in both I could generate that typology from a standard set of process dimensions derived from the contemporary industrial-engineering literature. I then—again with some mutual modifications—combined the result with the results of my earlier inductive analysis of my own data. This yielded four categories—work load, complexity, outlay, and uncertainty; indicated some subdimensions of complexity; and suggested some ways of measuring them in standard engineering terms.[13]

How did I know that this classification would be better than some other scheme, that is, that I could predict more about organization structure by using it than some alternative taxonomy? I did not. One never does, for in principle there exists an infinite number of ways of classifying anything. However, I did know that two of its sources— the economic-historical and industrial-engineering schemes—had been found by other researchers to be related to organization structure, and by "working back" to these sources from my own I found that I could arrive at some hypotheses—as, for example, how complexity might be related to authority structure—which could be "translated" into my own framework and which also suggested some leads about relevant organizational categories. It also appeared to me that an argument could be made to the effect that the four categories which I had reached were special cases of Parsons' four functional problems of any social system, in that they seemed to represent four different ways in which technology posed adaptive problems for administrative structure (work load could be assigned to the goal-attainment cell, complexity to the integration cell, uncertainty to the latency cell, and outlay to the adaptation cell).[14] I simply took this possibility as evidence that these categories might be peculiarly relevant to administrative structure and made no further effort to carry out its implications systematically at that time.

The five aspects of organization structure—division of labor, authority, solidarity, proprietorship, and recruitment—were devised largely by starting with Levy's *Structure of Society* and sitting and thinking about how its contents might be applied to organization structure. Whatever one may think of the "functionalist" posture assumed by Levy, it does provide an orientation which I found quite useful from a practical standpoint in thinking up taxonomies. I would contend that once one has devised a taxonomy by this means, it must be justified on other grounds, but that does not alter the usefulness of "structural-functional requisite analysis" as an exploratory tool. While doing this, I also studied the industrial sociological literature and other theoretical writers, trying to "plug" any organization variables I could find into my nascent taxonomy and checking my data to see if I had enough information about them to use them. During this phase of the work, I "theorized" in the aforementioned fashion during the afternoon, read my data and checked it against the afternoon's results during the evening, slept on the outcome, and tried proposing hypotheses in the morning. I found it necessary purposefully to plan my time in this way to avoid being carried away by theory or getting bogged down in data.

As one may imagine, the ultimate result was wildly eclectic and contained some strange theoretical pedigrees. My dimensions of authority were for the most part straight from orthodox organization theory, whereas my five forms of recruitment represented the results of thinking about a combination roughly consisting of various works of Sir Henry Maine, Melvin Tumin, R. M. MacIver, Wilbert Moore, Melville J. Herskovits, and Richard Thurnwald in the afternoon, combined with analytic induction from my own data during the evening, plus efforts to see whether it worked the next morning.[15]

The "permanent versus temporary" aspect of solidarity I conceived to be obvious and a result of common sense, but my failure immediately to think of the equally obvious common-sense distinction between organizations which are autonomous as opposed to those which add and then drop auxiliary members from time to time resulted in a work standstill for about two months. I needed a typology of organizations based on recruitment to use as a bridge to considering influences of the social setting, but I simply could not seem to get my organizations to fit into any set of recruitment categories very well. The trouble was that in many cases different parts of the organization seemed to recruit in different ways. Finally, an almost chance passage from Weber's *Theory of Social and Economic Organization* suggested what was retrospectively obvious: divide the organizations into those with and without auxiliary members and classify the way the basic core of membership is recruited, separately from the mode of recruitment of auxiliary members.[16]

This procedure enabled me to move via recruitment into the realm of the social setting, with innumerable bouts like the ones I have described. I think there is no point in going into further details, except to note that I have said very little here about how I devised hypotheses. This process is, I think, adequately described in the book itself, so I shall not treat it here.

This entire taxonomy-building process thus eventually resulted in a set of detailed categories—and even with a few honest-to-goodness variables, such as technological complexity. And inasmuch as analytic induction had been a part of the procedure at all points, all my categories were operational relative to my data, though, as we shall see presently, gaps in information existed in quite a few particular instances. Having completed the exploration, I could now formalize my procedure by coding my cases, counting frequencies, and cross-tabulating.

The coding operation proved to be very tedious "dog work," in the worst sense of the term. I had to do it all myself, and herein lies a basic defect of the study. Clearly, it would have been advisable to have other people do some coding so as to be able to check reliability. But this was impossible at the time. I also found that although I could code a great deal of the material from the notes I had taken—particularly where very general distinctions were involved—very frequently I had to have recourse to the original source materials. This took a great deal of time and was not very interesting, since I had been through all the material before and was now attempting to resist, rather than encourage, flights of imagination. I was also forced to make some decisions which were intellectually rather unsatisfying.

First, I had to assume that the ethnographer was always right. Researchers who use secondary sources are always open to the charge that they are cavalier and uncritical in their use of source materials, and cross-cultural analysis—particularly where large numbers of societies are used with information taken out of context—is particularly vulnerable to such criticism. I felt, however, that I had to face the fact that I did not know enough to be able to assess the validity of factual reporting by any ethnographer and that therefore I had to believe what he said if I were going to make this study. To decide otherwise would be to contend that ethnographers are inclined to be knaves or fools and that anthropology as a general discipline is hence useless. Needless to say, I cannot accept such a position.

Secondly, I had to accept the fact that there were gaps in the data about which I could do nothing. I would, for example, have liked to know more about the ideology of work and about prestige as a possible reward for labor. But I could not include them in my study. Also, I found that I could not use the entire sample of 150 for all my variables, for the same reason. Lastly, I had to decide how to handle situations where I wanted to classify something as being either present or absent under conditions where the ethnographer simply said nothing about it. Did this mean that the characteristic in question was really absent, or did it mean that the ethnographer had simply not investigated the matter? This is clearly a tricky question to decide, and I adopted the following convention. In cases of this sort, where my source was as a whole rather sketchy, I reported "don't know." This caused a great many South American societies to drop out of the analysis at various points, but there was nothing that I could do about them. In cases where two or more thorough and detailed sources were available, with neither reporting the characteristic in question, I reported the characteristic as "absent." Where only one source was available but was detailed and thorough, I read it carefully to see whether the ethnographer had discussed other closely related matters. If he had, I reported the unmentioned characteristic as "absent"; if not, I reported "don't know." [17]

The actual process of coding may be illustrated by the following example.

The *total complexity* of any process is defined as the number of *tasks* plus the maximum number of *specialized operations* ever present in any one task plus one or zero, depending on whether *combined effort* is ever present at any time in the process, or always absent, respectively. A *task* comprises all work performed during one period of organization assembly; an *operation* is any physical action which leaves raw material in such a condition that it can remain untended without further changes taking place during the time required by the task at hand; operations carried on simultaneously are *specialized; combined effort* is present whenever several persons work in unison according to some established rhythm. The coding problem is thus to discover the total complexity of any process ethnographically described, using these definitions. "Wet" ricefield cultivation among the Betsileo of Madagascar provides a good example.

Tasks. First, rice seed is planted and tended in a nursery. This appears to be a rather simple activity, and is carried out by the women of a residential family group. It is not clear whether one or two clearly differentiated tasks are involved (i.e., the extent to which planting shades over into tending). I coded this as a single task, as the differentiation over time did not appear to be clear. It is also not entirely clear whether there is any specialization or not in this task. I judged that there was not, on the ground that specialization seemed to be discussed in some detail elsewhere in the monograph. In any case, it seemed probable that there is less specialization in this task than in subsequent ones; if such is the case we do not need to know exactly how much occurs at this point for purposes of index construction.

Second, the ricefield proper is prepared and the rice shoots are transplanted from the nursery into the ricefield. This is done over a period of one very long day at maximum effort, with considerable specialization, as will be shown presently. This task involves getting an even flow of water into the field and controlling it by dikes; getting lumps out of mud in the field; removing plants from the nursery; transplanting them to the main field, and planting them.

Third (there may be some intervening tasks involving care of the crop), the rice is harvested. This too is done at maximum effort with considerable specialization, but not so much as is the case with preparation and planting.

Conservatively, I thus reported three tasks.

Specialization of Operations. The greatest degree of specialization appears to occur in the second task, that of preparing the ricefield, and transplanting shoots from the nursery into the ricefield. Five different operations are performed at once. The young men start across the field, driving cattle around in the mud to reduce the soil to an even consistency. Following immediately behind them are the middle-aged men with hoes who level the field and even it up. During this time the old women are digging up the plants from the nursery for transplanting and are transporting them to the younger women in the field. And the old men are repairing the dikes and regulating the water level around the side of the field. I thus reported five specialized operations.

Combined Effort. Combined effort clearly occurs in the third task, where the men manifestly reap in rhythm.

Total Complexity is thus coded as nine.[18]

The last step—apart from the actual writing—was the analysis of data. Since I had reached my "hypotheses" by a combination of verbal reasoning and ex post facto induction, the appropriate method in most cases appeared to me to be to array frequencies in fourfold tables, with controls as appropriate. This I did, proceeding from one hypothesis to another through a verbal chain of reasoning. I am now inclined to think that a more sophisticated approach might have been preferable. However, the nature of the study was such that this phase actually constituted a description rather than an analysis, properly so called. It seemed to me that any statistical devices used should be confined to those which were as close as possible to the verbal reasoning process which I actually employed. I did, however, at this point decide to present statistical tests, since it seemed to me that they accurately described the way in which I had actually been thinking about the material.[19] When anyone asserts that two things are related, he is implicitly saying that he has somehow been led to believe that the association which he has observed between them did not occur by chance. In other words, he is implicitly comparing this association with a random model, regardless of whether he is using numbers or words. I took advantage of the fact that I had tabulated numerical frequencies, which fact enabled me to report explicitly what I had been doing.[20]

Such is the story of one cross-cultural analysis, which was subsequently written up and published. If I were to do it again, there are many things that I would undoubtedly do differently. I have indicated some of these during the course of the discussion. I have, however, tried to describe as accurately as possible exactly what I did, even to the point of appearing rather farfetched at times and at the risk of sometimes providing a description of what not to do. I feel that there is much to be gained from cross-cultural research and that efforts to perfect its methodology are therefore well spent. I think I learned quite a bit from doing this study. I also enjoyed it.

• NOTES

1. Stanley H. Udy, Jr., *Organization at Work: A Comparative Analysis of Production among Nonindustrial Peoples* (New Haven, Conn.: Human Relations Area Files Press, 1959), pp. 1-4.

2. George P. Murdock, *Social Structure* (New York: Macmillan, 1949).

3. Julius Lippert, *Kulturgeschichte der Menschheit* (Stuttgart, Ger.: F. Enke, 1886-1887).

4. H. J. Nieboer, *Slavery as an Industrial System* (The Hague, Neth.: Martinus Nijhoff, 1900); L. H. D. Buxton, *Primitive Labour* (London, Eng.: Methuen, 1924).

5. Wilbert E. Moore, *Industrial Relations and the Social Order* (2d ed., New York: Macmillan, 1951), pp. 253-270.

6. See George P. Murdock, C. S. Ford, *et al.*, *Outline of Cultural Materials* (New Haven, Conn.: Human Relations Area Files Press, 1950).

7. Specifically, Raymond Firth, *We the Tikopia* (London, Eng.: George Allen and Unwin, 1936); Bronislaw Malinowski, *Coral Gardens and Their Magic* (2 vols.; London: George Allen and Unwin, 1935); Waldemar Bogoras, *The Chukchee* (3 vols.; New York: G. E. Stechert, 1904, 1907, 1909); Robert H. Lowie, *The Crow Indians* (New York: Farrar and Rinehart, 1935).

8. Udy, *op. cit.*, pp. 4-9.

9. George P. Murdock, "World Ethnographic Sample," *American Anthropologist*, LIX (1957), 664-687 (included in Frank W. Moore, ed., *Readings in Cross-Cultural Methodology* [New Haven: Human Relations Area Files Press, 1961], pp. 193-216).

10. Florian Znaniecki, *The Method of Sociology* (New York: Farrar and Rinehart, 1934), pp. 235-331. Also see Donald R. Cressey, *Other People's Money* (Glencoe, Ill.: The Free Press, 1953), pp. 16-17.

11. See Marion J. Levy, "Some Basic Methodological Difficulties in Social Science," *Philosophy of Science*, XVII (1950), 294.

12. Udy, *op. cit.*, pp. 97-116; Max Weber, *General Economic History*, trans. by F. H. Knight (Glencoe, Ill.: The Free Press, 1950), pp. 95-336.

13. Udy, *op. cit.*, pp. 10-35; Carl Bücher, *Industrial Evolution*, trans. S. M. Wickett (New York: Henry Holt, 1912), pp. 83-149; Max Weber, *The Theory of Social and Economic Organization*, trans. A. M. Henderson and T. Parsons (New York: Oxford University Press, 1947), pp. 225-226; William G. Ireson and E. L. Grant, eds., *Handbook of Industrial Engineering and Management* (Englewood Cliffs, N. J.: Prentice-Hall, Inc., 1955), pp. 291-292.

14. Talcott Parsons and Neil J. Smelser, *Economy and Society* (Glencoe, Ill.: The Free Press, 1956), p. 199.

15. Sir Henry Maine, *Lectures on the Early History of Institutions* (New York: Henry Holt, 1889); Melvin M. Tumin, *Caste in a Peasant Society* (Princeton, N. J.: Princeton University Press, 1952); R. M. MacIver and Charles H. Page, *Society* (New York: Rinehart, 1949); W. E. Moore, *op. cit.*; Melville J. Herskovits, *Economic Anthropology* (New York: Alfred A. Knopf, 1952); Richard Thurnwald, *Economics in Primitive Communities* (Oxford, Eng.: Oxford University Press, 1932).

16. Weber, *The Theory of Social and Economic Organization, op. cit.*, p. 148.

17. Swanson, with more resources available, followed a superior procedure. See Guy E. Swanson, *The Birth of the Gods* (Ann Arbor, Mich.: The University of Michigan Press, 1960), pp. 37-54.

18. H. M. Dubois, S. J., *Monographie des Betsileo* (Paris, France: Institut d'Ethnologie, 1938), pp. 434-440.

19. Andre J. Köbben, "New Ways of Presenting an Old Idea: The Statistical Method in Social Anthropology," *Journal of the Royal Anthropological Institute*, LXXXII (1952), 129-146 (included in F. W. Moore, *op. cit.*, pp. 175-192). I would agree with Köbben that there is nothing distinctive about "statistical methods." But it is important to recognize that there is nothing distinctive about other methods, either. Neither are the problems which he describes the result of using statistical methods nor can they be avoided by avoiding statistical methods. The advantages of statistical methods are twofold: (1) they tend to make problems visible which might otherwise be obscured by verbal discussion; (2) they make it easier to manipulate large amounts of data systematically.

20. For a critical discussion, see Harold E. Driver, "Introduction to Statistics for Comparative Research," in F. W. Moore, *op. cit.*, pp. 303-329.

8

Research Chronicle:

*The Adolescent Society**

JAMES S. COLEMAN

I shall describe at some length the intellectual developments that led to my research on the adolescent society and its impact on my subsequent intellectual development and further research. I do this, not because I feel that the mechanics of carrying out research once designed are unimportant, but because I feel that the most difficult matters to communicate in science and scholarship, and yet the most crucial, are the linkages that lead from ideas to research, to new ideas, and on to further research. If a student can obtain a feeling for the movement from ideas to research—a realization that his deepest concerns about society can become the basis for his research—then he need see no dichotomy between these two. If he can come to see, in turn, that the research, grounded in these deep concerns, is his principal means of refining and developing them, then he will hardly settle for their separation.

My major interest, from the time I entered graduate school, has been in the relation of the individual to society. The dilemma that confronts each society, on the one hand, to maintain social order and, on the other, not to restrict the freedom of the individuals within it, is the center of my concern. This concern led further toward an interest in cultural and political pluralism and a general concern with the future of such pluralism. I felt, in a hazy way, political pluralism to be one of the few answers to the dilemma and felt it to be dependent on cul-

* James S. Coleman, *The Adolescent Society* (Glencoe, Ill.: The Free Press, 1961).

tural pluralism—stable sources of cultural diversity within the social system. Yet the old sources of cultural diversity, stemming from the autonomous development of various cultures, were vanishing, as communication destroyed such autonomous development. Thus, my interests centered on the general questions of how a social system generates diversity within itself and whether such diversity can support political pluralism. I saw the critical political question of the future as one to which socialism is irrelevant: not in *whose* hands the means of production will rest but in how many *independent* hands.

At this time—the spring of 1952, my first year in graduate school— I was interviewing for the International Typographical Union study (which later came to be published as *Union Democracy*[1]) and helped in the distribution of a mail follow-up questionnaire. Fortunately for me, Lipset was off to teach in Germany that summer, Trow was off to help C. Wright Mills build a cabin on an island in Michigan, and I had a chance to start some analysis and try my hand at writing. A major problem of the ITU study—the sources of political cleavage in the union—centered precisely on my focus of fascination. For one of the fundamental problems in a homogeneous trade like that of the printers union was the springs of political cleavage which maintain a two-party system. Lipset had established this as a principal aim of the research, and I was able to examine a portion of that problem, to trace the history of issues, to show the role of various issues, and to examine the part played by locals in the maintenance of a loyal opposition. For me, the first provisional and partial answer to the question of political pluralism lay in the economic structure of the union and the differing interests generated by that structure.

The research on the ITU further centered my interest on the question of how pluralism can be generated within a social system. Also this research narrowed the question somewhat. For it was evident that in order for such political pluralism to exist, a crucial prerequisite was the existence of independent avenues of status—paths of status and visibility that were not dependent on the central power. In the ITU, these avenues depended in part on independent means of communication, in part on independence in the control of economic resources. For certain activities in the printers occupational community, such as sports leagues and benevolent associations, it was important that the leaders, who had gained status within a small segment of the membership, have avenues of communication so that they could become more widely known and so they could disseminate dissident political views.

A smaller set of institutions, however, played a more crucial role, for their power was directly linked to the economic interests of the members. Nationally, the most important of these was the local, for large locals in this union have a great deal of autonomy. The large locals, whose size relative to the national union approximates that of states relative to the United States, have power over local wage negotiations and over most of the matters that are important to a printer. The status of the local's leader, in turn, depends largely on the printers who elect him, not on the international officers. This source of independent power meant that the local could afford political dissent without fear of retribution. This element was missing in most trade-unions and present in the ITU, with its strong locals, its clubs and organizations, and its informal occupational community. But the more fundamental question was the theoretical one: how such independent paths are generated by a social system and the conditions under which a system does generate them.

As a kind of motivation behind this (along with the motivations described above), I had a strong interest in knowing the social conditions that promoted psychological freedom. In a social system that rewards many avenues to success, the members have a kind of freedom from normative constraint that does not exist in a more monolithic system.[2] In part, this was evident from the ITU study, in which the pluralism of authority (neither the employer, the international union officials, the local officials, nor the shop chairman was omnipotent) gave printers an exhilarating sense of their own autonomy. As one printer said in an interview: "How free can a man be? I own my job. I'm as free as free can be." In part, of course, it had its roots in my basic interest in pluralism and how pluralism can be generated in a society.

Unfortunately, these interests were interests in the macrostructure of society, and they were not easily treated with methodological rigor. My methodological training and interests (largely from Paul Lazarsfeld) were in quantitative statistical analysis of the type one usually does with survey research; yet survey methods were hardly appropriate to the study of the sources of status pluralism in social systems.

There was, however, a concurrent methodological development which centered on a kind of modification of survey techniques in sociological directions. A number of us were beginning to engage in "contextual analysis," which took explicitly into account (ordinarily as independent variables) characteristics of the social unit of which an individual was a member, and "relational analysis," which took social relationships rather than individuals as the units of analysis. This had its proximate origins in Paul Lazarsfeld's classification of group characteristics,[3] in Philip Ennis' examination of contextual effects on voting,[4] in ITU analyses of chapel and chapel-chairman effects in printers' voting, and in McPhee's examination of contextual effects in the Elmira voting study.[5] (That this was no major intellectual revolution, but merely an application of common-sense notions to solve sociological problems, is not important. The point is that heretofore survey research had restricted itself to individualistic problems—problems of aggregate psychology, if you will. This new development opened up many new sociological problems that had been closed to survey research.)

The conjunction of these methodological developments, my interest in sources of status pluralism sharpened by the ITU study, and the conclusion of formal graduate training led me, in the fall of 1954, in the general direction of the research that ended as *The Adolescent Society*. The requirements that the research had to fulfill were (1) that it substantively deal with the problem of sources of status pluralism in social systems; (2) that it use quantitative methods in doing so, which seemed to imply a comparative analysis of several social systems; and (3) that it be of such a nature that financial support could be found for the project and for me during the life of the project. These requirements did not, of course, dictate high schools as the locus of study. Primitive tribes that differed in the pluralism of their status systems would have been better for some parts of the problem (though poorer for others).

But my research skills lay elsewhere, and I felt it was possible to isolate parts of American society that were nearly enough self-contained social systems to sustain full-fledged status systems. Yet an important attribute of adult life in Western society is its role segmentation. This means that there is an interlinking of many subsystems except in the smallest communities, and it is hard to define the boundaries of an adult's relevant status-giving environment. It means also that if an adult does not find his self-esteem in one of his roles, he may focus energies on another, so that any social system as delimited for research purposes is likely to have been psychically abandoned by those of its members who were not rewarded by it. Any effect of the social system's rewards on the psychological states of its members was thus likely to be attenuated. This is far less true among children in school, whose relevant social environment is still small and whose freedom to choose their own activities is far more limited.

The specific choice of high schools, however, came as a result of casual discussion one evening among Martin Trow, his wife, my wife, and me, comparing our high-school experiences. Trow's high school was Townsend Harris in New York City, which had been an academically elite, selective school. His wife's was a private girls' school in Georgia; my wife's, a high school in a small, basketball-crazy town in Indiana. My own high-school experience had been split in half, starting in a small-town school in Ohio and ending in a large public boys' technical school in Louisville, Kentucky.

It was difficult for each of us to understand the others' high schools, for the systems of status, the frames of reference they provided, were so different from our own. It suddenly struck us that the four of us had spent those years in four rather different worlds, and we were like Bushmen, aborigines, Moslems, and Mongols, trying to comprehend the others' customs and status structure and how it would feel to be in such a society.

This discussion focused my attention on high schools. Once the subject of high schools had arisen as a possibility, a number of other advantages were evident: they existed in abundance; they had a rapid and intense impact (the "life cycle" was only three or four years in length, and a new generation arrived each year); their members were old enough to look to each other for a large proportion of daily rewards, yet young enough to have some plasticity to the environment; the institution provided a boundary to the system, cutting its members off, in part, from the outside. In retrospect, this is one element of the research that I would not change. For my purpose and, indeed, for numerous problems in the functioning of social systems, high schools can constitute the sociologist's "fruit fly" for study.

Working at the Bureau of Applied Social Research in the winter of 1954-1955, I wrote the proposal for a study of high schools. I hoped to use schools associated with Columbia University Teachers College in a self-study group, in which case the number of schools would have been larger and each school less intensively studied. Other than this, the proposal at this time was not greatly different from that finally accepted by the United States Office of Education in the fall of 1956.

Because this study included for me—and, I think, for survey research generally—new developments in methodology, I shall dwell at some length on the roots of its methodology. Because it constituted also a focusing of my substantive interests, a statement of the substantive focus of the study as proposed will precede this somewhat lengthy methodological discussion.

THE PROBLEM OF THE STUDY AS PROPOSED

As outlined in the 1955 and 1956 proposals, the study was to measure three classes of variables: the "social climate" of the school, the determinants of the social climate, and its consequences for adolescents subject to it. The term "social climate," rather than more precise terminology, was a compromise to give the proposal's readers a heuristic sense of what the study was about. Throughout the research, it turned out to be a particularly appropriate term for communicating the general intent of the study to nonsociologists, most of whom remembered that the "social climate" that existed in their high schools differed from that which they faced in later life. In more precise terms, this means two attributes of the adolescent status system: the amount of pluralism in the status structure (that is, how many different paths to informal status there were) and the degree to which the paths to status were in accord with educational goals, independent of them, or antagonistic to them. The pluralism derived from my basic interests; the pro–anti school dimension was included because of its importance for education.

The rather simple view of the structure of status systems as pluralistic or monolithic was elaborated in my ideas at that time, but even more as the study progressed. For example, pluralism can arise through the existence of separate subcultures, each holding one value as important, or without subcultures, where all members of the system hold several values to be equally important. I shall return to this point in discussing the substantive ideas that developed from the study.

The focus on status systems instead of a total "social climate" was justified on the grounds that status systems are patterns of reward established in a social system, and it is such reward patterns that shape the activities and aspirations of the system's members. This implies a lack of interest in such other aspects of the social climate as the customs (the particular styles of clothes or the special linguistic terms) or the mores (the levels of sexual morality). I was not concerned with how free the sexual restraints were, though I was concerned with how much time and attention were devoted to sociosexual pursuits and their relation to the patterns of reward.

The determinants of the "climate" that I felt such a study could examine consisted of two: attributes of the community and attributes of the school and staff. The consequences included, particularly, the questions of deepest concern to me. Would a pluralistic system allow more persons to have high self-esteem? Would the level of over-all achievement be higher? Would adolescents in such a system find it easier to hold to their personal tastes when they differed from others? Would they be less constrained in occupational choice? In addition, there were questions concerning the consequence of value-conflict between adolescents and staff: Would such conflict reduce the correlation between ability and scholastic achievement by pulling adolescents' attention in other directions? Would it carry over and generate negative attitudes toward the community as well as toward the occupational role of teacher?

THE METHODOLOGY OF THE STUDY AS PROPOSED

In essence, what I wanted was to use survey techniques of data collection and analysis to do something new—to study the functioning of a social system. We had bordered on this in the ITU study, and some of the research mentioned earlier had begun to use group characteristics to analyze the effects of social environments on individuals, but I wanted to do more than any of these studies did. I wanted to take the social system rather than the individual as the unit of analysis.

Parenthetically, this has led to comments on the study, as it turned out, from a number of colleagues to the effect that I had no dependent variable. I think this will be a characteristic of such studies; in contrast to most social research, they will have no single dependent variable. The difference is analogous to the difference between (1) finding the coefficients of a regression equation to account for the variance in a dependent variable and (2) finding the coefficients of a system of simultaneous equations which link together a number of variables. It is quite obvious that the latter research will appear less focused than the former, for it attempts to lay out the structure of relations in a system rather than to explain the variance in a single variable.

I was far from completely successful in this aim to study ten social systems by survey methods, but it should be clear that this was my aim. I did not plan to relate one characteristic of an individual to another characteristic of that individual, and the study did not do so. The statistical analysis at nearly all points classified persons according to their position within the social system of the high school (as members of the leading crowd, as persons who were outside but wanted in, as persons whom other adolescents wanted to be like or be friends with, as best students, or best athletes) and related *these* characteristics to attributes of their background, to their activities, their aspirations, their achievement, and their attitudes. As a consequence, the units of analysis were roles and statuses in the social system, not individuals— and the statements that could be made from the analysis were statements about the system, inferred from the status accorded to particular roles, the kinds of persons who were found in these roles, the self-esteem held by people in particular roles, their attitudes and activities.

The methodological difficulties in testing our ideas in the ITU study with a single case and my more ambitious aim here, to use survey techniques to study the functioning of a social system, led me to look toward a design in which many social systems would be studied comparatively or a few would be studied through a combination of internal analysis of each system and comparative analysis of the several systems. The study developed into the latter, though it could easily have been the former if chance had ruled thus. As the first proposal was written, in the early spring of 1955, the research involved the use of a large number of schools associated with a Columbia University Teachers College self-study group of schools.[6]

This methodological question (see *Union Democracy*, pp. 478-480, for a discussion) has yet to be solved satisfactorily. There are many advantages to an intensive single-case analysis which in effect examines the working parts of the system. Yet many of our data-collection and statistical-analysis methods are designed to show a static cross section in time, and so long as they are thus a comparative analysis of several systems (in my case, schools) is more easily carried out. A caveat to this may be made where it is possible, as in the present research, to study roles and statuses in the system statistically. By sampling roles rather than individuals and using roles as the units of analysis, as was done in this research, it becomes possible to infer something about the functioning of the system. If that were not possible, the present research would have failed completely, for the number of schools was too small for a statistical analysis based wholly on the school as the unit.

It is painfully evident to anyone who attempts to study a social system that our quantitative research techniques are in their infancy. For, by sensitive observation and description (as exemplified, say, by William Foote Whyte's *Street Corner Society*),[7] we can trace the functioning of a social system. Yet, when we attempt to carry out quantitative research in such a system, we find ourselves stymied. We shift from a sensitive examination of events, in which intimate sequence in time suggests causal relations between events, to a crude measurement of "characteristics" and a comparative cross-sectional analysis that relates one characteristic to another. That is, when we shift from qualitative reporting to quantitative analysis, we change our very mode of inference.

In qualitative description, we report a stream of action in which the interlinking of events suggests how the system functions. An example from *The Adolescent Society* illustrates this, though such reporting was rarely done.

> [How do you get in the top clique?]
> Well, I'll tell you, like when I came over here, I had played football over at. . . . I was pretty well known by all the kids before I came over. And when I came there was . . . always picking on kids. He hit this little kid one day, and I told him that if I ever saw him do anything to another little kid that I'd bust him. So one day down in the locker he slammed this kid against the locker, so I went over and hit him a couple of times, knocked him down. And a lot of the kids liked me for doing that, and I got on the good side of two or three teachers.

In order to carry out quantitative analysis, we must collect a number of such systems (in this case, high-school student bodies) and then carry out a comparison to show the interrelation of characteristics. The stream of action is gone, and the whole character of the analysis changes.

Is it not possible to combine quantitative research with the study of a single system? I will offer some suggestions toward an answer near the end of this paper, but for the present it is sufficient to say we cannot do it now.[8]

The design of the present study can be seen as stemming directly from the methodological statement in the Appendix to *Union Democracy*. The discussion of characteristics at different levels of a social system shaped the plans for data analysis, and the discussion there, stating the pros and cons of single-case and comparative analysis, led to the compromise—using ten schools rather than one or many and using a two-wave panel (to capture some of the system's dynamics) instead of either continuous observation or a single cross section.

The methodological design, then, was to consist of questionnaires at two points in time (initially to be a year apart, but, as it turned out, at the beginning and end of the school year). Because the research was to depend on internal analysis, and since it was impossible to sample roles and structures in a loosely structured system, the total student body was included. Because determinants of the social systems were to be looked for in the community and the staff of the school, questionnaires were to be mailed to the parents, and teachers were to be interviewed. (The design showed a lack of realism at this point, for the number of teachers could be estimated even at that point to be around five hundred, and to interview so many persons is no small research project in itself. Instead, the research depended on questionnaires from teachers in the schools.)

THE INITIAL STAGES OF THE RESEARCH: SELECTION OF SCHOOLS, DESIGN OF QUESTIONNAIRE

After the initial failure to find financing, the research was set aside, first for six months while I worked on other projects at the Bureau of Applied Social Research, and then for a year while I was at the Center for Advanced Study at Palo Alto, California. During the summer of 1956, before I left the center, I learned from Ralph Tyler that the United States Office of Education had just initiated a cooperative research grant program under which this study might be financed. I submitted a proposal through the University of Chicago, and in January 1957 the grant was made. (Simultaneously, a proposal was made to the National Institute of Mental Health, but was rejected.) The final proposal is reproduced in the Appendix of *The Adolescent Society*.

The selection of schools was made in several stages. I wanted to select schools (within a one-hundred-mile radius of Chicago) with as wide a variation in adolescent status system as possible according to variations in the social climate. In such an exploratory study, it is probably always a good idea to sample on the basis of the variable of principal concern (ordinarily the dependent variable) to ensure that enough variation exists to make the study worthwhile. However, I wanted also to select within the framework of other variations: size of school; rural, urban, suburban community; and some variation in social-class composition of the neighborhood or community.

The first step was to select seventy-five schools from Chicago and the eight counties adjoining Chicago, using these school-size and community criteria, the latter obtained from census data. Fortunately for the research, this step covered a wide range of school size (from about fifty to about five thousand), wealth, and urbanism; and fortunately, Illinois has local school districts rather than county-wide districts, thus allowing for more variation.

At this initial stage, Kurt Jonassohn and, shortly thereafter, John Johnstone, both graduate students in sociology, joined the project. We sent for the yearbooks of these schools with the aim of using criteria from the yearbooks (the attention devoted to different activities, the activities of the senior-class president, and so forth) which would suggest the importance of different avenues of status in the school. These criteria were coded and used to aid the selection. In addition, we went to persons with special knowledge of schools: in particular, assistant state superintendents of public instruction who visit each school periodically for accreditation and state aid. These men gave us a valuable perspective on high schools in northern Illinois, and I visited several schools with them—my first contact with a high school since leaving my own thirteen years before.

On the basis of initial selections with these two criteria, I visited the state superintendent's office and copied information concerning approximately twenty-five schools from public records filed by the schools each year. This included information about course offerings, expenditure per pupil, degrees of teachers, and proportion of seniors who planned to attend college.

These criteria allowed us to select twelve schools, including two Chicago public schools. The selection, as implied above, was not done according to precise criteria, for we wanted to include wide variation in social climates and wide variation in size of school and type of community, yet not have these two completely correlated.

We wrote letters to each school superintendent, and I visited each to explain the study. Our data-collection needs were one class hour from each student in the fall, one in the spring, questionnaires to be filled out by teachers, some interviewing of selected students, and copying of school records for each student. In some cases, the school superintendent could make the decision himself; in some cases he went to his school board; and in one case, he let his department heads vote on the study. (In the last case, they voted against it, following the objections of the football coach, and this was one of three schools outside Chicago for which we had to substitute.) In general, the cooperation of the schools was gratifyingly high, especially since they themselves would realize little gain from the research. Altogether, three of ten schools outside Chicago declined to be studied, and Chicago's Board of Education declined to have any of its schools studied. Two schools were added outside Chicago, to make nine, and a Catholic school in the center of Chicago was contacted and agreed to become part of the study. The sponsorship of the research by the United States Office of Education was an important aid in obtaining this high degree of cooperation.

Still in the spring of 1957, we designed a pilot questionnaire, and John Johnstone obtained preliminary pretests. After some revisions, we visited each school, administered the pretest questionnaires to a freshman and a senior class, and carried out group interviews with a small group of seniors and freshmen in each school. These interviews were directed in part to evaluation of questions from the questionnaire, in part to a discussion of the status structure among the students in school. High-school students proved generally to be extremely good informants about the social system in which they found themselves, with those outside the leading crowd in their grade better than those inside it. Casual speculation suggests that this derives from the fact that for outsiders the leading crowd was an important element to be taken into account and, in some cases, to be aspired to. In contrast, outsiders did not have to be taken into account by those on the inside.

At this point, these interviews and the initial pretests gave us strong confirmation that the status system among adolescents in the school was a powerful element in adolescents' behavior and, furthermore, that the status systems differed from school to school. However, it also became apparent that the variation in status systems would not be as great as we had hoped. If this had led us immediately to add schools to the research, thus ensuring a greater diversity of status systems, the research would have gained measurably. Caught up in the mechanics of handling the schools we already had, unfortunately we did not do this. This is one of the many points of the research where the mere weight of mechanical details acted to debar us from carrying out better research.

The summer of 1957 was devoted to hand tabulations of the pilot questionnaires and design of the final questionnaire. It is useful to spend a few sentences discussing the questionnaires because of the special character they assumed. They were intended to study the functioning of ten social systems (more particularly, status systems) rather than a sample of adolescents and, as a consequence, they had to be designed for this. In part, the questionnaire treated the students as *informants* concerning the status system of the school; in part it treated them as *respondents*. Thus, the status system could be characterized both in terms of the way it was perceived by the adolescents within it (as it was in Chapter 3 of the book) and in terms of the way it functioned (as it was in chapters 5, 6, and 7).

A second special attribute of the questionnaire was questions that allowed a reconstruction of many aspects of the informal structure of the school. Questions asking what students were held in especial esteem, what ones were in the leading crowd, what ones were good students, and so forth made it possible to reconstruct the status structure of the school from the questionnaires. Questions asking about informal associations outside class made it possible to reconstruct the structures of association.

Thus, in addition to the usual information from the questionnaire used to characterize the individual himself, there was information which allowed characterization (1) of the school as a whole, (2) of the status and roles of various persons in it, and (3) of the social structure within the school. In fact, the weakest parts of the questionnaire were those concerning attitudes of the respondent—the point at which most questionnaires are strongest.

In the first few weeks of school in the fall of 1957, we administered the questionnaires in each school, using one day at the school for administration. (In this fall quarter I was not teaching, having taught the preceding summer.) Questionnaires were administered in English classes (from 40 minutes to 56 minutes in length) by our own staff. It was emphasized that no one other than our staff would ever see their responses. This and the fact that the questionnaire rather quickly moved into areas that interested them very much (who their friends were, who was the most popular with the opposite sex, who was in the leading crowd) appeared to establish a frame of reference in which they were not answering as *students* but as *adolescents*. If this had not been so, I expect we should have obtained stereotyped responses such as adolescents often give to teachers.

The remaining details of the field work and initial processing of data I will compress into a few points.

(1) Teachers' questionnaires were distributed, filled out, and mailed directly to us by teachers.

(2) Parents' questionnaires were mailed, using addresses obtained from school files, and returned to us by mail.

(3) The large task of copying students' records, including absences, grades, rank in class, and scores on standardized achievement tests, was facilitated by a fortunate circumstance: two undergraduates from Antioch came to the National Opinion Research Center, were sent to us, and after short training in coding were sent to the schools. At each, they spent several days coding this information.

(4) In May of 1958, we returned to the schools for one day and administered spring questionnaires. In each school, we carried out recorded interviews with a small sample of seniors, again using them as informants about the status system, as we had done in the group interviews during the pretests.

(5) The coding, punching, and cleaning of cards was done by a temporary staff made up principally of graduate students.[9] This occupied most of the academic year 1957-1958 and part of the summer of 1958. For carefully precoded, standardized questionnaires such as we had, this is an unconscionably long time (even taking into account the sociometric data, which had to be coded from lists of names); but as long as research like this is a one-shot affair, using amateur personnel for the mechanical details, it will likely remain this.[10]

During the late summer and fall of 1958, analysis was begun in some detail. At that time, Univac I had been installed at the University of Chicago, and we determined to carry out on it much of the sociometric analysis (as well as calculations of correlations among scores such as grades, IQ, and achievement tests and percentaging of ordinary tabulations). In retrospect, the difficulty of programing in machine language and the entrapment that occurs to one when he begins to program meant that a large portion of my own time was wholly wasted.[11]

In the meantime, parts of the analysis had been subdivided; Kurt Jonassohn was doing analysis on parental attitudes and community structure, in addition to continued administration, and John Johnstone was doing analysis on adolescents' use of the mass media. Johnstone's work led to his Ph.D. dissertation, and the part of Chapter 8 that deals with mass media was his analysis and writing. In addition, he carried out aspects of the analysis that appear at various points in the book (the analysis relating to figures 8.4 and 8.5 stemmed from his ideas, and he carried out the delineation of cliques in the sociograms of Chapter 7).

Despite the fact that the analysis proceeded during this period, a curious sort of disorientation occurred. The research had been well planned, the hypotheses were clear, yet I did not know quite what to do. This disorientation lasted until the summer of 1959 when the formal project was complete, Jonassohn and Johnstone had left for full-time jobs, and the project funds were finished. Then, in one fell swoop of about eight weeks, a large part of the analysis was done, and the report to the Office of Education (which later became the book) was written, except for small portions that had been written before.

This curious disorientation at the early stages of data analysis has accompanied all but the simplest research projects I have been engaged in; yet here it was much greater than before or since. In retrospect, it is evident that there is one fundamental cause, and that cause itself had several sources. The cause was that I simply did not know how to carry out the analysis. This, in turn, was partly due to the fact that initially none of the variables at the level of the school was turning out right. Wholly prior to any attempt to measure the pluralism of the different status systems, we had to measure the importance of different status dimensions taken separately; but, depending on what measure was used, the rank order of the schools, especially in the middle ranks, shifted greatly. In an attempt to get confirmation through consistency among different grades in school, from 9 to 12, we found wild fluctuations from grade to grade in several of the small schools. We had hoped to find wider variation in the schools than we did find; the resulting low variation meant that the schools could not be ranked precisely on the importance of different dimensions. Also, there was the continuing puzzle of the one most upper-middle-class school, in which scholastic achievement had from the start appeared to be valued little in the adolescent community, with social activities and sports valued highly. Why should this school, of all schools, appear so consistently low in its adolescents' evaluation of scholastic achievement?

But going beyond the measurement of separate dimensions of status in the schools, we needed to obtain two derivative measures: the pluralism of the status system and the consistency of its values with those of the staff. Though we had not measured the staff's values with enough sensitivity to obtain variations in their adherence to scholastic success as a standard, we could nevertheless approximate the value-consistency measure simply by a measure of the importance of scholastic achievement to the students. Thus, a partially satisfactory measure could be obtained for this variable.

But the status pluralism became extremely difficult. For one thing, a totally unsuspected variable arose to confound the whole issue of pluralism. The pluralism hypotheses implicitly assumed that the dimensions of status were alike in other respects, but in these schools they were not. Most confounding was the difference in ascriptiveness of the different systems: in four schools, status was far more dependent on family background than in the other six. Furthermore, this tendency toward ascribed status was not unrelated to the importance of various other avenues for status. In particular, the importance of scholastic achievement proved to be correlated with the importance of family background.

A second important difference among the dimensions of status was in their source: in some cases, such as athletics, the source appeared largely to be in the solidarity-producing functions the activity performed for the adolescent community itself; in some cases, such as cars, clothes, or good looks, it appeared to lie in the power of the attribute relative to the opposite sex; in others, such as scholastic success, it lay in the direct rewards provided to the successful by the staff; in still others, it lay in the autonomy and freedom from adults the activity implies. A general difference between status-giving attributes or activities was their relation to the two surrounding institutions: the school with its staff, rules, rewards, and so forth and the adult society, with its rules, rewards, and so forth. Some status-giving activities were sponsored by the surrounding institution; others were in reaction to it.

This variation split the problem wide open; unless these other attributes of the status dimensions were held constant, the sources and effects of pluralism could hardly be examined. Yet, whatever pluralism there was in these systems was highly related to the content of the important status attributes.

The embarrassing question is why this point did not come up until that late stage of the research. In the very selection of the schools, this should have been evident at the outset. The reason it was not was that we were very pleased to find small social systems with *different* status systems, and we wanted to make certain that the differences were as great as possible within the limits of the population of schools. We therefore neglected problems that would arise in our study of pluralism.[12]

Thus, the research was stymied, when it came to the analysis, by problems that should have been apparent in the design. Corrections might have been made at the outset. If, for example, a large sample of schools had been studied less intensively, the problem could have been by-passed by holding constant the content of the status systems (and the size of the school) when examining the sources and effects of pluralism. The stymie in this case did not mean it was impossible to study the sources and effects of pluralism, however. Other ways were found of doing so, but this will be discussed later.

The design of the research also helped bring about this disorientation in another way. The hypotheses were stated in such a way that they required statistical comparisons between different status systems but there were only ten schools. The research was designed to do both comparative analysis among the schools and internal analysis of each school. The methodology of the former was quite clear, but that of the latter was less so. What I mean by internal analysis stemmed from the ITU study: translating hypotheses that related to the organization as a whole into subhypotheses, at least one of which could be tested within the organization. Thus, by comparison of the political behavior of large and small locals within the ITU, we were able to test statements about the importance of large autonomous locals as sources of opposition.

Within these schools, there was far less substructure by which such internal comparisons could be made.[13] Thus, I was disoriented by literally not knowing how to carry out the analysis. Only very slowly did a different mode of internal analysis of the system develop: the analysis of roles. In these informally structured social systems there were no a priori designations of roles, so we had, first, to locate the occupants of various roles. We did this from the nominations made by adolescents in the questionnaires. The roles thus determined were:

Members of Leading Crowd

Role Models $\begin{cases} \text{someone to be like} \\[1.2em] \text{someone to be friends with} \end{cases}$

Those with Many Friends
Girls that Boys Would Most Like to Date
Boys that Girls Would Most Like to Date[14]
Best Students
Best Athletes (boys)
Best Dressed (girls)
Most Popular with Opposite Sex (as seen by own sex)

Then slowly, and without our really being aware of it, the analysis became almost wholly an analysis of roles, as an examination of the book will indicate. But because we were still thinking in terms of comparisons of social units (as the ITU study compared shops or locals), it was done slowly and painfully and far less well than it might now be done. Throughout the whole analysis, we maintained the fiction that the principal mode of analysis was that of comparison between schools, while, in fact, coordinate with this we were carrying out an analysis of roles and statuses within the system. Such an analysis of roles is not well formulated in the literature, and I shall attempt later in the paper to describe it.

Still another element was, in part, responsible for the curious disorientation that I experienced in the beginning of this analysis. In the fall of 1959, I presented a paper at the College Entrance Examination Board's annual conference on college admissions and then another at a conference in January.[15] These papers forced me to put aside my own interests in status pluralism and examine the more general question: what were the dominant processes shaping these status systems? This question and the provisional answers I gave to it began to shift my whole orientation in the study. The shift was away from using the schools to examine a problem I was interested in and toward searching out the dominant processes shaping the institution itself. That is, though I began by asking what the sources of status pluralism in a social system are, this widened into the question of what the general sources of the adolescent status system are and what its consequences for education are.

With the preparation of those two papers (and a third prepared in the spring of 1959 [16]), I began to look at the general problem of the structure of activities in a school, the way it affected the structure of rewards, and the way this, in turn, might affect the values attached to an activity. In this manner, the focus of the research shifted from a problem at the center of the research to the problem at the center of the institution. I began to feel that to answer my questions about status pluralism and to leave unanswered the general questions about how these systems developed and maintained themselves would miss the most important elements of the phenomenon I was studying.

That reason, coupled with the frustration in measuring these schools' pluralism, might cause suspicion that I was rationalizing a shift necessary to rescue the study. Yet I think not. I did not even see that the shift was occurring and thus was hardly attempting to rationalize it. In fact, the shift was not occurring in any regular fashion. Instead, it was a period of great disorientation, of casting about with little pieces of analysis which hardly seemed to fit together.

Thus, I think that part of the disorientation I experienced was a result of finding a disjunction between the narrow framework of hypotheses that I had specified about status systems and the dominant processes that were affecting these status systems. The disorientation lifted only when I wrote Chapter 1 of the report (and of the book), which, by setting the frame of reference of the book as the development of an adolescent subculture in industrial society, provided the focus of the book. Thus, the book as it stands is not an examination of the hypotheses of the research about sources and consequences of status pluralism. Instead, it is an examination of the structure of status systems in high schools and the various sources and consequences of the status system.

Is such a transformation a good thing or a bad one in research like this? From one point of view, it is certainly bad, for the problem as posed is shunted aside and partly lost. From other points of view, however, it may be quite good. Instead of forcing the social system under study to give answers to a set of narrowly defined questions (as specific hypotheses must be), it may be better to give the system more freedom, posing only very broad questions—in this case, questions about the sources and consequences of adolescent status systems—rather than hypotheses about the sources and consequences of one attribute of the status system—its pluralism. For if, in studying a social system, one abstracts from it only those elements important to him, he may inexplicably find that his results are weak, inconclusive, and inconsistent wholly because he neglects those processes dominant in the system.

Yet it is important to recognize that the research did remain rather highly focused. It carried out no analysis of individuals, only of these social systems; and it attempted not to describe in full richness the functioning of these systems but only to examine the values that arose within them and the related dimensions of status.

In any case, the shift from my hypotheses to more general problems of these status systems occasioned some part of this disorientation experienced in the analysis. At the same time, the shift brought on some real excitement, for the processes that were shaping status systems in the adolescent community were extremely interesting. Important changes in the total structure of society appeared to be having an impact here, with the weakening of family, the interposition of educational institutions to interrupt generational continuity, the rise of a separate subculture based on age. In addition, a remarkably great disparity was growing: the increasing social sophistication of the young and the freedom and resources to choose their leisure pursuits, combined with a longer and longer period of financial and occupational dependence. I felt I had happened onto some of the most important social changes of the period and, equally important, social changes which could have a powerful impact on the viability of educational institutions.

It should be clear, of course, that the research did not *test* whether these broad changes were occurring in society. Some of the research results appeared to indicate the existence of these changes, but to test such problems requires a different—and difficult—research design. Instead, the research took these points as premises—premises that provided a frame of reference for the research results and helped bring consistency into them.

Thus, these interests and excitements came to supplant those with which the study began. They were forced into the open by the three papers I mentioned above, and at the conclusion of the third there was enough organization of ideas to generate the structure for the book. Chapter 2 had been written before this time, as had parts of Chapter 3; Johnstone had written the section on use of mass media, and many of the tabulations later used in various chapters had been made. Then, after this reorientation was explicit, the other chapters were analyzed and written in about the order of appearance in the book. This analysis and writing, during the months of July and August 1959, were probably the most extended period of intensive, uninterrupted, intellectual concentration I have sustained—a period of complete asceticism. My family was in Baltimore; I was staying in a room at International House in

Chicago, and my only contact with the outside was telephone calls to a machine operator and daily meetings with the girl drawing charts and typing chapters.

This appeared, both at the time and in retrospect, to be a very disorderly way of carrying out research analysis. Is it? Could not the book have been far better if the analysis had proceeded in a more orderly fashion? Certainly this intensive period would have been impossible, had not many of the tabulations already been done, had not most of the mechanical problems already been disposed of.

It may be, however, that such a period is necessary, not in order to carry out analysis, but in order to carry out synthesis. The synthesis necessary in any coherent research is often neglected in our examination of research activity. We can teach methods of analysis; yet any extensive research of the sort that results in a book requires something equally important: an organization or synthesis which provides the essential structure into which the pieces of analysis fit.

Such synthesis is perhaps better done, as it was in the ITU study, over a longer period and as the result of extended discussions between the authors over a period of time.[17] Yet the whole problem of how it is best done and what the interplay between analysis and synthesis should be is an open one.

The remaining activity in making a book out of this crude report was carried out by laborious rewriting, reorganizing, and filling in additional pieces of analysis over the next nine months at Johns Hopkins. The manuscript was given to The Free Press in May 1960 and published in September 1961.

THE METHODOLOGICAL CONSEQUENCES OF THE RESEARCH

The methodological consequences of this research for my subsequent work have been of several sorts. One is what was termed above "role analysis." Analysis of social units based on survey data has ordinarily been conceived in terms of what has been variously called contextual analysis, structural-effects analysis, or compositional-effects analysis. In such analysis, a dependent variable is explained in terms of two (or sometimes more) independent variables: a variable characterizing the individual and a variable characterizing the group, ordinarily the aggregate of the individual independent variable. But most of the analysis in *The Adolescent Society* is not of this sort. It first identified roles and then carried out much of the analysis as an internal analysis of roles. By identifying certain positions of high status (members of the leading crowd, the role models, and those with many friends) and seeing what kinds of persons arrived at those positions, it was possible to learn both about the value systems in the school and about the impact of the external social structure. For example, persons characterized by other roles (best students, best athletes) were examined to see how often they were in positions of status, and from this analysis of the relation between roles one could infer the status and power that derived from various activities in the school. Persons were characterized in terms of their background and examined to see how they appeared differentially in positions of status, thus showing the penetration of ascriptive background factors into the status system of the school.

Again, it was possible to examine some of the effects of the system directly by such analysis of roles through scrutinizing the attitudes held and the activities engaged in by those in positions of status. As one example, the strong psychological impact of the system was apparent in the far more positive self-images held by those with high status and in the differentially positive images held by those in various roles (best student, leader in activities, most popular). The differential attitudes of those in the role of athlete toward the role of athlete and of those in the role of best student toward the role of best student showed the impact of the system on the values of students.

It was still the case that much comparison among schools was necessary, but this comparative analysis was based on the role analysis discussed above. For example, the relative status of athletes and scholars differed from school to school, and that difference was used to characterize the value difference between the schools.

It would be valuable to have an explicit methodology of such role analysis, for it seems that much can be learned about the functioning of a social system in this way. What can be learned differs radically, according to whether roles are filled through popular consent (as in this research), by authoritative appointment, or by other attributes. Where the roles are filled by popular consent, for example, their occupants may be studied to infer the values of the members of the system. In any case, there seems to be a field waiting for definitive methodological treatment. The most exciting point in such methodology is its intimate connection with sociological substance. For it is clear that if analysis of a social system is to proceed very far, it must discard the study of individuals qua individuals and let attributes of the system such as roles, norms, activities, values, and statuses constitute the very units of analysis.

COMPUTER TECHNIQUES AND DYNAMIC SYNTHESIS

The research, and the experience with computers it entailed, convinced me that it was possible to analyze social structures with computers in ways that had never been possible without them. This conviction led to an extensive research project in the use of computers for the analysis of social systems. The research is beginning to bear fruit, but it is still more nearly a promissory note than an accomplished fact. Although several partial analyses have been carried out, progress has not met my hopes. I believe the reason is that such an analysis of a functioning system requires techniques totally different from our usual statistical analysis. In order to carry out the analysis, some parts of the structure must be modeled; there must be a more intimate mixture of analysis and synthesis than occurs in our old techniques. Yet I suspect that when we come to have these methods well developed, the old ones, in which the analysis is done by machine tabulations and the synthesis according to a vague and unreliable model in the researcher's head, will show their crudity.

It is likely that the requirements of studying a social system and the possibilities afforded by a computer will begin to influence radically even the kind of data collected. Let me return for a moment to the mention that was made earlier of the stream of behavior and how it is lost when we turn to quantitative analysis of a social system. What seems to occur is this: the roles, institutions, and other parts of the system are identified, and then an account of the stream of behavior that flows among them is given, mostly through example of concrete cases drawn from observation or interview. This has both structure and dynamics that are missing from most quantitative analyses of social units based on survey data. A good role analysis of the sort described above puts into the analysis the necessary structure. But it still lacks much of the process which constitutes the functioning of the system.

One means of reintroducing the processes of the system or organization is through system simulation based on data rather different from data we usually collect in quantitative analysis. Suppose that we first identified the major roles and role relations in the system, sampled these, and then obtained data on the types of response made by a person in a given role when faced with a given situation. This might be done quite precisely or quite loosely, but the important point is that the result would be an inventory of contingent responses for each role. These results then become the critical set of data necessary for synthesizing a model of the system; in ways we conceive only vaguely now, it becomes possible to pose questions about the system's functioning—questions that cannot be answered now either by our quantitative analyses of systems or by qualitative ones. Such work must at present be stated in very tentative form, but it indicates the direction of ideas first set off by the disorientation that this research produced.

THE SUBSTANTIVE CONSEQUENCES OF THE RESEARCH

Some of the consequences of this research for my own intellectual development are evident in the passages above. I shall indicate the influences of the research on my subsequent work, but prior to that I should mention the fate of my principal initial interest: pluralism of status systems and political pluralism generally.

I continue to believe that the most serious problems facing societies of the future will be those of maintaining social and political pluralism, the ability to generate political innovation without revolution and social innovation without explosive fads. Sources of diversity exist, of course, in the differing interests of different groups; whether a stable political system can be built on such consensus and cleavage, in the absence of long-standing value and cultural differences, remains to be seen. Certainly it is not evident from the research I carried out, for in none of the schools was there a stable pluralism. In some of the schools, the conflict of values (say, between scholarship and athletics) was resolved by cooptation: the leaders of the system were those who achieved in both areas—a very different matter from having two sets of elites, each of which achieved in one area. In one school, it was resolved by a kind of unstable monolithic system: in one grade, athletes would be leaders and popular heroes; in another, scholars. (These examples indicate what was suggested earlier: how it was possible to study the question of status pluralism by analysis of roles within the schools.)

It is evident that the appropriate political system can help ensure the proper balance of consensus and cleavage in the society—contingent, of course, on an appropriate crosscutting of interests in the social and economic structure. Thus, the problem of political pluralism appears to me less tied than before to the question of a pluralistic status system, more tied to a wide range of matters in the social and political structure. It appears to me a problem not amenable to direct study until we can more completely analyze the functioning of a social system.

More generally, this research has shifted sharply my interests in the functioning of social systems. My interests in status systems are now considerably broadened. In particular, schools constitute an example of a social system that develops from an unstructured mass captive within an authoritative institution. As such, it is one of a general class, exemplified by jails, mental hospitals, camps, factories. My interest has centered on the relation between the kind of environmental conditions the institution provides and the kind of status system that grows up within it. In particular, it has centered on the questions of autonomy. Under what conditions can the captive system become a self-governing one? And under what conditions will this self-government be nonoppressive for members and allow them to be self-governing as individuals?

This turns back immediately to the general problem of socialization, for perhaps the major problem that must be faced in socialization of neophytes into responsible members of the system is how and when to give over autonomy and responsibility to the neophyte. It is this problem, I propose, that is the most difficult one educational institutions of industrial societies have to solve today.

This has led in a direction in which I would never have anticipated going: toward experimentation in schools. The research in which I am presently engaged involves the construction of social and economic games to be used in schools. For a variety of reasons, games appear to me to be extremely interesting socializing devices: the reward structure they furnish, the self-governing system they establish, the autonomy they provide without dire consequences, the lack of necessity for outside judges. The aim of the research is simply to explore the role of games in adolescent socialization: their effect on individuals and on the social system of adolescents. The more general aims of the research are related both to games and to broad problems of socialization. I see games as an important device for the construction of social theory. The essence of a good game is that it involves Hobbes's problem of order within its restricted environment: each man pursues avidly his own self-interests, yet the system continues to function without a war of all against all. As a consequence, games seem particularly useful as a mode of testing the viability of a given structural system.

The broad problem of socialization, of course, I see as that stated above: how and when to give over autonomy to the young and how to structure their position in society so that the social subsystems within which adolescents socialize themselves have both autonomy and responsibility.

Thus, my present research is a direct outgrowth of the research that began in the winter of 1954-1955 through an interest in political pluralism. That it bears only a faint resemblance to the ideas that formulated that earlier work I take as an example of the impact of research on ideas.

· NOTES

1. S. M. Lipset, Martin Trow, and James S. Coleman, *Union Democracy* (Glencoe, Ill.: The Free Press, 1956).

2. I have always been very sensitive to "what was expected of me" and thus have been able to find freedom only by placing myself in those situations that rewarded me for those things I wanted to do. This sensitivity probably accounts for a kind of bias found in *The Adolescent Society* toward the demands of the adolescent system. Some adolescents are sufficiently impervious to the expectations of their peers so that they can pay little attention to the society of peers that surrounds them.

3. Published in Robert K. Merton and Paul F. Lazarsfeld, eds., *Continuities in Social Research* (Glencoe, Ill.: The Free Press, 1950).

4. Published in William McPhee and W. A. Glazer, *Public Opinion and Congressional Elections* (Glencoe, Ill.: The Free Press, 1962).

5. Bernard Berelson, Paul F. Lazarsfeld, and William McPhee, *Voting: A Study of Opinion Formation in a Presidential Campaign* (Chicago, Ill.: The University of Chicago Press, 1954).

6. That proposal was never accepted. Neither the Ford Fund for Advancement of Education nor the Carnegie Corporation was interested, and since these were the principal sources of funds for such research at that time, the plans for the research had to be set aside. The spring and summer of that

year I spent with Herbert Menzel in analysis of the study of doctors introducing new drugs.

7. William F. Whyte, *Street Corner Society* (Chicago, Ill.: The University of Chicago Press, 1943).

8. Beginnings in this direction are made by Roger Barker and his associates. See *Big School–Small School* (Palo Alto, Calif.: Stanford University Press, 1963).

9. We debated at several points whether to do the study, or some parts of it, within NORC where I had an appointment and had worked on another project. The additional overhead that would have been necessary beyond that already taken by the university led us to decide against it. This made for few additional problems, but only because Kurt Jonassohn, as assistant project director, was extremely efficient and meticulous in supervising coding, key-punching, and initial runs. Nevertheless, the whole of the academic year 1957-1958 was occupied by the mechanical details of getting the data ready to analyze.

10. Research organizations with which I am familiar, NORC and BASR, are little better, since they, too, operate with amateur personnel gathered for temporary work. In my experience, it is only in the IBM machine work, the upper administrative levels, and interviewing that the research organization provides professional aid for the academic researcher—though these elements are themselves important.

11. The programs did do nearly all we hoped, but the cost in my time was extremely great. On the other hand, it was this experience as much as any other that led me to realize how remarkably suited a digital computer is for statistical analysis of social systems, in contrast to the usual IBM card machines. This has led me much further into the development of methods for studying social systems with computers.

12. We tried to introduce it by including some schools with many extracurricular activities, some with few; but this was of little aid because in some schools, like Elmtown's, the great number of extracurricular activities had no impact on the status system and were, in fact, sometimes disdained.

13. All but one had four grades, which were to some degree separate social systems, and, in retrospect, the analysis would have been greatly aided by treating each grade as a unit. This would have given 39 cases for boys and 35 cases for girls, rather than 10 and 9.

14. This role and that of the girls (above) were not used in the analysis, though they were measured.

15. These are published as "Style and Substance in American High Schools," *College Admissions*, VI (1959), 9-21; and "A Sociologist Suggests New Perspectives," in Franklin Patterson *et al.*, *The Adolescent Citizen* (Glencoe, Ill.: The Free Press, 1960), pp. 288-311.

16. J. S. Coleman, "Academic Achievement and the Structure of Competition," *Harvard Educational Review,* XXIX (1959), 330-351.

17. Equally important in that research was the fact that the synthesis was already far more fully developed by Lipset at the beginning. The research did not drastically modify the original focus but elaborated and confirmed it.

9 *Great Books and Small Groups:*

An Informal History of

a National Survey*

JAMES A. DAVIS

This is, within the narrow limits imposed by perceptual defense and criminal libel, my recollection of how National Opinion Research Center Survey No. 408, the Great Books study, proceeded from its inception in the summer of 1957 to the publication of a book in 1961. Although I tend to view the chronicle as the struggle of a brave study director against time, money, clients, winter weather, Texas Great Books groups, and NORC's business staff, these events may better be viewed as a reasonably typical case study of how modern social research proceeds in a large nonprofit research organization.

This brings us to the subject of money. I think it may be stated as a matter of indisputable fact that there is no money available in the contemporary United States for unrestricted support of large-scale social research. On occasion, a professor whose work is fashionable or whose years of loyal back-scratching on professional committees is deemed worthy of a reward will receive the munificent unrestricted sum of $1,000 or $2,000, most of which he passes on to subsidize graduate students, but private donors prefer to see their names on university dormitories; the association of sociology and socialism is graven in the minds of congressmen; and foundations have (let us face it) retreated from social research. No one is going to give NORC or similar private institutions the wherewithal to pursue their own research interests at

* James A. Davis, *Great Books and Small Groups* (Glencoe, Ill.: The Free Press, 1961).

$50,000 per interest. The citizens of Michigan do subsidize the Survey Research Center, and its senior staff members have sabbaticals and swimming pools and who knows what else; but the private, nonprofit research center is in there hustling in the market place along with Ford, General Dynamics, and Joe's Drugstore.

Thus is born "the client," typically a large foundation or a government agency with a particular research question which it feels is worth the exorbitant costs and personal frustrations involved in commissioning research. And with the birth of the client comes the eternal triangle of client, organization, and study director. It is the operation of this triangle which is the key process in the poignant histories of surveys.

Let me begin, however, with a few kind words for clients. As a matter of fact, the client for this study, a foundation executive in one of the many progeny spawned by the Ford Foundation, is a fine guy, and at least prior to his reading of this document, I consider him my friend. But, as we know from introductory sociology, personalities and roles are two different matters. Rolewise, to be a client is somewhat like being a sugar daddy responsible to a board of directors. It is an extraordinarily expensive business, the satisfactions are occasional and fleeting, there is the distinct impression that one is being ruthlessly exploited, and all of this has to be justified at the annual meeting.

At the same time, those in the humanistic studies who are so enraged at the funds they see flowing into social research might momentarily consider how it would be to receive an enormous commission, most of which disappeared into $25 checks to unknown ladies in New Jersey and a mysterious maw called "overhead," and to have the Medici Fund tell you that you could paint anything you liked as long as it matched the rug in their private audience room.

All of this would work out cozily, as it does in business, were it not for the motivations of study directors. There are, I would guess, no more than a hundred people in the country today leading the lives of noisy desperation characteristic of study directors, but they fall into two types. Historically, relatively few study directors in market research and in nonprofit organizations came from graduate study in sociology or social psychology, a Ph.D. in sociology being no more necessary for competence in this area than a degree in electrical engineering. Among NORC's senior study directors, for instance, are non-Ph.D.'s trained in history, anthropology, and undergraduate liberal arts. Into this occupation, however, like locusts, have come the Ph.D.'s.

They tend to be ambitious, steely-nerved young men who have worked out the implication of the following propositions: (1) academic success is contingent on research publication, regardless of the topic; (2) young men seldom get research grants on their own; (3) people come to research centers and give them the wherewithal to do large-scale studies.

While the two types of study directors appear indistinguishable to the naked eye, they vary considerably in their view of research and of their jobs. The "old-line" staff tend to identify with the research organization and to gain their rewards from pride in craftsmanship and budgetmanship, reputed client satisfaction, and the feeling that they have contributed to the success of the organization. The new men, however, while often willing to deliver a thorough and honest piece of work for the sponsor, find their major satisfactions in milking the research for journal articles or publications to throw into the potlatch of academia, whence cometh their eventual reward: a research professorship.

I am, in truth, accentuating differences that are far from polar, for most people in research work find their major rewards from intellectual challenge (as well as salaries superior to teaching), and "applied" research is generally more challenging intellectually. That is what I said: applied work is usually more challenging—because there are more definite standards of accomplishment. In social-science theoretical work, the feeling that it "sounds right" or has the requisite polysyllabic mumbo jumbo is the typical yardstick; in "pure" empirical research, if *any* significant correlations can be wrung out, the material is generally publishable; but in applied research, there are rather precise questions at issue and the failure to answer them is painfully apparent. In addition, applied research in government agencies, the larger commercial firms, and centers like NORC is characterized by superior probability sampling, larger samples, better interviewing, more careful control of coding and tabulations, and informal monitoring of the work by colleagues who are specialists in the same area. One wonders why the Ph.D.'s have continual intellectual dissatisfaction in their jobs.

The root of the problem, I think, lies in the difference between generality and specificity. Clients commission research because they are interested in something specific: who has health insurance, whether enough people are training for careers in biochemistry, how much scholarship money is available to graduate students, to what extent people near airports are bothered by jet noise, and so on. Sociology is, however, the enemy of the specific. Even though the facts of social life in modern America are less well documented than the facts of marine life at the bottom of the ocean, the academic sociologist (the ultimate judge, employer, or journal editor whom our young Ph.D. wants to impress) has a phobia against research which "merely" describes. This is "nose-counting," "dust-bowl empiricism," "trivia," and so forth and is not part of the grand scheme for building the science of sociology. That the history of natural science is in the reverse order—theories having been developed to explain facts, rather than facts gathered to ornament theories—weighs little against the pressure of intellectual tradition. Therefore, the academically oriented study director is faced with a dilemma. If he completes his research in such a fashion as to satisfy the sponsors, it will lack academic glamour. If, on the other hand, he completes a piece suitable for academic publication, it will probably tell the sponsor nothing about the questions which led to the research.

If one has attained sufficient eminence, one proceeds to conduct the study as one pleases, considering the client lucky to have his problem studied by an important person, even if in the process the client's problem disappears. For younger people and the struggling research organization, this is a dangerous tactic, and the natural strategy is to attempt both tasks: a specific descriptive report "for the client" and a high-brow article or monograph for the study director's self-aggrandizement. Thus, as well as a description of who gets scholarships comes a test of the theory of relative deprivation among graduate students; along with the descriptive materials on whether poor boys go on to college comes a paper on status crystallization and career choice; along with the statistics on what doctors prescribe brand-X drugs comes a paper on sociometric aspects of innovation; and so on.

It must be made clear to the reader, however, that these theoretical forays have little or no connection with the specific research questions. As currently developed, sociological and social-psychological theories are almost useless in predicting a dependent variable, either because they are stated so abstractly and vaguely that it is impossible to translate them into research operations or, if stated in usable terms, they are often wrong or account for only a negligible portion of the variance when compared with the "trivial" things such as age, sex, education, and marital status. I, personally, hold great hopes for the theories now being developed under the general rubric of "dissonance" and for much of George Homans' work, and I have to admit that a little exposure to theory suggests some interesting intellectual problems for research; but my reluctant conclusion is that in 90 per cent (plus or minus 20 per cent) of the research work, "theory" is dragged in only as a status symbol or to improve the eventual merchandising of the results.

Considered, then, as roles, social research is typically conducted by (1) a study director, who may be willing to do what he is paid for, but is more interested in wresting an academic article from the remains, (2) a sponsor, who stokes the fires with money and hopes vaguely that the evasive, fast-talking young man will complete within his lifetime a report bearing vaguely on the topic, and (3) a research organization, beset with financial woes and firmly aware of the fact that the study director (who gets no profits when a study makes money and pays no refund when he runs it into the red) is capable of spending the organization into the poorhouse without shedding a tear and in the process alienating the client beyond the point at which he can be persuaded to pony up the deficit.

Let us now see how these three archetypal characters proceeded to produce NORC Survey 408.

I did not have much to do with the Great Books study until it was about six weeks old. During the early summer of 1957, I was away from Chicago doing field work in Ohio for a community study directed by Peter Rossi. Why? Because I had been hired by NORC in the summer of 1957 to direct a study of physicians, which never took place. In order to keep me busy, I had been sent into darkest Ohio as gunbearer for Mr. Rossi, who was stalking community leaders there.

The infant Great Books study had been ushered into the world by Clyde Hart, NORC's director at that time, and the staff at the Fund for Adult Education, my role being that of pediatrician rather than obstetrician.

Had I been in Chicago for every moment of the initial negotiations, I would probably have little more to add to this chronicle than I can from my observation post in a commercial hotel in "Mediana," Ohio. Indeed, the exact origins of this survey, as for many, are a mystery. My hearsay version goes as follows.

The Fund for Adult Education, a subsidiary of the Ford Foundation, had since 1951 been supporting diverse activities in the area of adult education by grants to ongoing study-discussion programs, continuing educational centers, educational-television experiments, and so on. The fund was oriented to action, not research, and had commissioned little or no professional research prior to the Great Books study.

I am told that it came to pass that from within the parent Ford Foundation came word that the time had arrived for the Fund for Adult Education to render an accounting of its stewardship and that the conventional medium for such an accounting was "research." The Fund for Adult Education, not unexpectedly, proceeded to commission a number of studies, of which Great Books was one.

Now, if the Fund for Adult Education was bemused to find itself bank-rolling a statistical survey, the object of the inquiry—the Great Books Foundation—was flabbergasted. The foundation, an independent, nonprofit corporation with headquarters in Chicago (which has no connection with Great Books of Western World, a commercial publishing venture), coordinates the national program of Great Books, using a small professional staff and a large number of volunteers. As intellectual types, the personnel of Great Books stand somewhere to the right of Jacques Barzun and Arthur Schlesinger, Jr., in their opinion of sociological surveys; but it is amazing how persuasive a large foundation with a history of generosity can be, so eventually the foundation was persuaded to cooperate. I think it would be fair to say that, while the foundation did provide the requisite liaison to complete the study, its stance was of one about to be photographed with a midget on his lap at a congressional hearing.

At this point, the following parties are involved: NORC, a research organization fully aware that evaluation studies usually make the client look bad; the Fund for Adult Education, an action organization already persuaded of the merits of Great Books, but hopeful of gaining concrete evidence of these merits; and the Great Books Foundation, already persuaded of the merits of its program, but quite doubtful that surveys can measure them.

Here ensued a number of conferences in Chicago and New York, during which the basic framework of the study was established. The only firm agreement prior to that time had been that the study was to be concerned with participants rather than the operations of the foundation and that we were interested in "the effects" of participation in Great Books.

For those of you who have not had the opportunity to read *Great*

Books and Small Groups, the Great Books program in 1957 was roughly as follows. In 1957-1958, it consisted of some 1,960 discussion groups dispersed through the United States, with some additional groups in Canada and overseas. Each group meets every other week from September to June, and at each meeting the members discuss a specific selection which they have read before the meeting (e.g., Milton, *Areopagitica;* Tolstoy, *The Death of Ivan Ilyich;* Rousseau, *The Social Contract*). The readings are organized into blocks of one year each and, in theory, should be read in sequence. The groups vary in size (from around five to around thirty-five, with an average of eleven in our sample); in sponsorship (most are affiliated with public libraries, but a number are sponsored by churches, business firms, and individuals); and in leadership (some have a single leader, most have two leaders, a few rotate the leadership each meeting). The leaders are not formally trained teachers, but a number have had brief training courses conducted by the foundation. The members do not pay any tuition or get any certificate for completing the program. In fact, no one can complete the program, as additional years of reading are always available. Members are encouraged to buy the inexpensive readings from the foundation but are not required to do so.

It was this program which was to be evaluated, to the end of discovering whether the effects on the participants were such as to justify the continuation or expansion of Fund for Adult Education support. As a separate operation, the Fund for Adult Education commissioned a management-consultant firm to assess the organization of the foundation, market potential for Great Books, and similar internal affairs.

The design of such research falls naturally into two parts, which can be thought of as sampling, in the sense of deciding which people are to be studied and in what numbers; and questionnaire construction, in the sense of deciding what measures to use on the sampled respondents.

Of the two, sampling presented the fewest problems. It so happened that this is one social-science situation for which there is a clear-cut textbook sample design. According to the course I teach in research methods, one should collect a large number of people, arrange for a random subgroup to participate in Great Books, prevent the remainder from participating in the program, and measure both groups on the dependent variables before and after the experiment. (Technically, if you have done it perfectly, you do not need to measure both groups before.) It also so happened that, as usual, the textbook design was out of the question. Such an experimental study would be "possible," although there would be an enormous number of difficulties—making sure that the controls do not get Great Books or equivalent experience, establishing community programs which mask the mechanics of the sample design, and so forth. The major obstacle turned out to be time. We began active work on the study in the late summer of 1957 and had to deliver a report by fall 1958. It would have been plainly impractical (as well as quite expensive) to get a field experiment organized in two months before the 1957-1958 Great Books year got under way and results of a spring-1958 follow-up assessed by fall 1958. In addition, we all agreed that if the program's effects were as expected, some of them might not show up until after several years of exposure to the program.

It was this idea of long-term effects which enabled us to find a compromise design. If it is correct that, unlike indoctrination movies or television debates, the effects of Great Books require a long, long time for their appearance, then beginning participants should make a reasonably good "control group," particularly if the field work could be hurried so that first-year members were reached before they had attended more than one or two meetings. Because, in addition, both beginning and advanced-year members are equally self-selected, this design even has the advantage over a control group of nonmembers in that the latter would necessarily be suspected of less motivation or interest in joining Great Books.

The great problem with this design—and a problem which remained one of the main issues of the research—is that, as compared with a true experiment, the design left open the possibility that advanced-year members differ systematically from beginners, and it would be their other differences which produced any differences in the dependent variables. For example, almost necessarily, advanced-year members are older than beginners, and if age were related to the dependent variables, spurious differences would emerge. While this problem would give nightmares to an experimental purist, by and large it did not bother me too much. I knew that we were going to have sufficient cases so that by cross-tabulations we could control for any differences in gross variables such as age, sex, education, occupation, and so forth. A major pitfall, however, was the problem of retention. Nobody really knew the dropout rates in Great Books, but from all that is known of volunteer organizations, they had to be high. Furthermore, it made plain common sense that retention in the program would be correlated with the dependent variables, for people who were not getting the "effects," whatever they may be, would be prime candidates for dropping out, as in any educational institution. Even worse, there was no way of controlling for dropouts in a sample limited to current members, since we would have to introduce as statistical controls events that had not happened yet. The best we could do was to introduce some questions about intention to continue and use these for controls in comparing beginning and advanced members.

Looking back now, I wonder why we let it go at that. We could have interviewed some ex-members without too much difficulty, but I do not remember that this was seriously raised as a possibility. Perhaps at that time I already had the germ of the idea of a second study to determine actual dropouts; perhaps I did not. At any rate, I am glad now we let it go; for without the continuing problem of attrition, the study would have begun and ended as another evaluation project.

The net result of all of this was a decision to sample from existing discussion groups, stratified to oversample the advanced-year groups wherein lay the pay dirt, if any. At this point, another vital, but not deliberate, decision was made. The Great Books Foundation has no individual membership rolls but merely a rather loose file of group registrations. Given the lack of coercive structure for the program, a number of groups exist without the official blessing of the foundation; and at least in 1957, rather than groups' petitioning Great Books for the right to exist, functionaries of the foundation continually scanned press clippings to find their groups, which were then sent registration materials. If a file of individuals had existed, I do not know whether we would have used it; but having no choice, we sampled groups—a very important decision, as it turned out.

The final crucial decision was an economic one. Because of the money to be saved, we decided to raise the case base by asking entire groups to fill out a self-administered questionnaire at their meeting, gambling that the members would be literate and thus capable of filling out a questionnaire and hopefully sophisticated enough to cooperate with a research project. We also had the naïve hope that the foundation would put a little heat on the groups to increase cooperation.

In sum, while from an abstract point of view we should have had a sample of individuals both in and out of the Great Books program, because of a series of unwitting decisions based on practical exigencies we ended up with a sample of discussion groups within the program. We began with the Great Books, but the Small Groups got into the study by default.

While limitations of time and budget usually provide enough restrictions so that sample designs for national surveys amount to choosing among a number of restricted possibilities, when it comes to writing a questionnaire, the sky is the limit; or rather, one's guess as to how lengthy a document the respondents will complete without rebellion is the only boundary. Within this area—and it is amazing how much respondents will actually do for you if approached correctly—we had all of Western culture from which to pick items. It is precisely the major advantage and the major problem of surveys that an enormous amount of information can be collected, the marginal increment in cost for an additional item being very small when compared with the fixed costs of sampling and contacting respondents. At the same time, *the* intellectual challenge in survey analysis is in ordering and synthesizing the diverse information—in this schedule ranging from father's occupation to the respondent's opinion as to whether the course of history is capricious, purposive, or mechanistic. (In case you are curious, 25 per cent thought it capricious, 48 per cent thought it purposive, 13 per cent thought it mechanistic, and 13 per cent were "no answer.")

A certain amount of disagreement arose among the parties: the Fund for Adult Education backing two horses, the Great Books Foundation a third, and the study director a dark fourth horse. The first horse was "community participation," a matter of considerable interest to the Fund for Adult Education, which was convinced that participation in Great Books should and maybe even did lead people to become more active in community affairs. The general idea was that after reading, say, the Greek philosophers on the nature of the good society, the Great Books members would be impelled to remodel Toledo, Ohio, and Minneapolis, Minnesota. The idea met with polite skepticism on the part of Great Books and me.

The Great Books Foundation maintained, and with some justice, I think, that their program did not have any purposes at all, at least in the sense of the sort of thing that can be listed and translated into surveys. With the naïveté of a young man who had read all the texts on evaluation research, I kept hounding the foundation to list—1, 2, 3

—the purposes of their program so that they might at least be tried on charges of their own devising. In the face of such pressures for oversimplification, all the foundation staff could come up with was the denial that participants were expected to become more active, passive, liberal, conservative, or anything that directional. Rather, they were expected to become more sophisticated, more critical in their thinking, broader in their approach, and so on, whether or not they chose to favor the left, right, center, or to refrain from community life. In this the participants in the actual survey agreed, the bulk opting for the response, "The Great Books provide an intellectual understanding (of specific social and community problems), but few or no keys to plans for action," as opposed to "Great Books provide both an understanding of the problems and a key to plans of action" or "Great Books are not applicable to specific social and community problems." While the foundation opposed the community-participation stress on the basis of intellectual ideology, I was against it on the practical grounds that I doubted we would get any effects. To begin with, I thought that Great Books members were typically fugitives from community life, rather than involved. I was dead wrong, but more on that later. In addition, from all that I knew of the literature, I was sure that class, sex, political preference, for example, played such an important part in community participation that exposure to a reading discussion program could not make much difference. I was perfectly willing to let the facts speak, but I was not going out of my way to disappoint somebody who had given me $40,000 to do research.

At this point, the inevitable answer occurred: to stress both purely intellectual and also community-participation materials. However, this led to a knotty problem of measurement. In assessing community and political involvement, we felt we were on safe grounds, for there is rich experience in survey measurement of such phenomena, and the content is heavily behavioral and hence fairly easy to translate into questions. In the measurement of such things as "critical thinking," "tolerance of ambiguity," "intellectual sophistication," that way madness lies. I was toying with pulling out existing tests of critical thinking and similar measures, and the Great Books Foundation had in desperation proposed that the respondents be given essay examinations which the foundation staff would evaluate, when the Fund for Adult Education split into two spokesmen, with the entrance of the late Carl Hovland, then a consultant to the fund. Hovland, as a psychologist, was highly concerned about the details of testing: whether the tests could be given under standard conditions, timed, collusion prevented, motivation maintained, and so forth. The answer, of course, is that they could not, or at least could not to the satisfaction of a testing specialist. Hovland felt that we might get away with it but that the results could never be sold to a really hostile critic who was oriented to psychological measurement.

The compromise decision actually became a sellout of the Great Books Foundation's position, for on the basis of Hovland's technical doubts, critical thinking and open-mindedness essentially disappeared, except in the form of some very simple attitude and opinion items in which, for example, the respondents were asked to name "any particular authors or schools which you once disliked, but now find more acceptable." Rather, we began to search for some sugar-coated test-oid materials to get at more superficial things. We actually found two good ones, a set of cartoon items (e.g., a gentleman in a nightshirt nailing a paper on a door, to denote Martin Luther posting his theses; a child in diapers composing at the piano, to denote the young Mozart, etc.), which had originally appeared in *Life* magazine, and a marvelous poetry test,[1] developed in the early 1920's by M. R. Trabue and Allan Abbott. Trabue and Abbott presented the original version of a well-known poem, along with versions systematically deformed in aesthetic content, and the respondent was asked to pick the one liked best. We never did find a complete set of the poetry-test items, even by long-distance calls to Prof. Trabue, by then a sprightly emeritus, but we got enough to go into print. In addition, we packed in voluminous materials on reading quality and musical taste and picked a number of items on philosophical points from the work of Charles Morris.

Note what was happening here. What should have been a series of technical measures of cognitive functioning became a set of crude information measures along with considerable materials on aesthetics and ideologies. Part of the shift can be explained by Hovland's technical qualms, part by the inability of the Great Books Foundation to come up with neat objectives for which nice tests exist, but a good proportion came from the wily maneuvers of the study director.

What was I up to? Viewed from the perspective of getting some academic yardage out of the study, Great Books appeared to have three possibilities, of which I picked the wrong one. The first possibility would have been to conduct an evaluation study of such methodological luster that it would attract attention despite the offbeat character of the sample and the stigma of "Adult Education." I knew in my bones that because of the design limitation and the fuzziness of the measures, this one could not make it. Now, get me right. I felt all along that we could deliver useful and valid information to the client on the effects of Great Books (and I think we did), but I doubted that we could get out a study which would be cited as technically outstanding or particularly convincing to a *resistant* reader. This meant that while for report purposes the materials had to yield an evaluation study, for academic purposes I had to find an analytical theme such that the whole sample could be treated as a single group and differences in exposure to Great Books (our analogue of experimental and control groups) ignored.

Looking back now, I think there were actually two possibilities. It turned out that Great Books members are phenomenally active and involved in local community affairs and that we had the makings of a detailed study of community involvement among young, educated, middle-class Americans. Actually, Vickie Beal, who was on the staff, managed to get a good Ph.D. thesis on this theme, but for some reason I did not pick that tack. The third possibility was a detailed analysis of the members' intellectual lives, their tastes, ideologies, philosophical positions, and so forth. I had done a Ph.D. thesis on taste and status symbols, and I have always been fascinated by the writings of Russell Lynes, David Riesman, Eric Larrabee, etc. In the back of my mind, I envisioned the development—from data, mind you—of typologies of kinds of intellectual stances and styles within the Great Books members.

My own proclivities here were reinforced by a marvelous but misleading field experience. Because of the timing of the program, there was no way to see a Great Books group in action before the questionnaire was completed. (My record is still perfect. I have never actually seen a Great Books discussion, despite several years of almost full-time work on this project.) However, in the summer of 1957 Great Books was running a summer program at Aspen, Colorado, which I arranged to visit. While my strongest memory is of the magnificent train ride through the Rockies—a revelation to a midwestern boy—professionally I came away with two hunches. The first was that Great Books tends to attract a social type which can perhaps be described as "isolated intellectuals." It seemed to me that at Aspen a large number of the people I met were extraordinarily well read, very serious, and given to the construction of homemade philosophical systems, without being hooked into the orbit of academic intellectuals or professional creative artists. I do not mean they were screwballs, but they did seem to be the sort of people whose opinions on whether the course of history was purposive might be interesting. Apparently the people who attended that summer conference were quite unrepresentative of the Great Books membership, for the actual survey showed the Great Books participants to be a pretty clean-cut group of PTA joiners and *Time* readers, not much given to the construction of homemade philosophical systems.

I came away with one other impression which did have a pay-off, although indirect. While I have never seen a legal, in-season Great Books discussion, I did watch a number of the sessions in Aspen and tried to make some guesses about the group dynamics beneath the surface of the discussions. It seemed to me that what was going on was a sort of political process in which people would advance ideas, allies would rally to them, and enemies would muster forces against them and that the course of the discussions was heavily influenced by latent attractions and antagonisms among the participants. Mulling this over later in Chicago, I decided to insert some sociometric questions to see whether a member's isolation or acceptance by others in the discussion group affected his reactions to the program. In the back of my mind was the idea that perhaps the discussions worked best when there were fairly "even sides" in the ideological and interpersonal teams. However, the Great Books Foundation vetoed these questions on the grounds that it might produce complaints if members were asked to name their friends and enemies.

The key items in the final study—a series of questions about functional roles in the groups—were actually devised as a substitute. The items asked each member to rate himself and others in the group in terms of such roles as "joking and kidding," "making tactful comments," "providing 'fuel' for the discussion," and others. The logical structure of the items was suggested by the work of Freed Bales. I have never had a course in small groups, but Bales's influence is very strong at Harvard, where I did my graduate work, and a sort of Balesian, functional approach to roles and group dynamics, absorbed by osmosis during graduate-school days, was the only one I knew.

This is the intellectual history of the questionnaire, a lengthy document bearing the stamp of the Fund for Adult Education's interest in community participation, some vestigial traces of the foundation's interest in "critical thinking," a lot of my own penchant for materials on aesthetics and ideologies, and a good bit of information on functional roles, inserted as a substitute for the excised sociometric items.

Were I to recount in detail the field work, coding, key-punching, and card-cleaning which took up the next ten months, this essay would become a book about a book. Let me merely say that with the superb help of Grace Lieberman, Mary Booth, Ursula Gebhard, Joe Zelan, and a charming group of Antioch College cooperators we came into the possession of schedules from over 90 per cent of the sampled groups and coded the data from the 1,909 individual respondents onto some dozen decks of IBM cards.

I am also going to gloss over the first report, a bulky, 256-page, single-spaced mimeographed document completed in August 1958. It consisted of (1) a description of the Great Books participants, (2) analyses of the members' reported effects of participation in the program, (3) comparisons between beginning and advanced members in terms of dependent variables, and (4) data on role structures and social correlates of role performance. Vickie Beal, who had joined the project staff in the spring of 1958, analyzed the materials on community participation; Ursula Gebhard had done most of the role analysis, and I did the rest.

Because the report was done under great pressure, the entire analysis and writing being concentrated in a period of about three months, the document was overly long and underly organized. In essence, however, we showed (1) strong differences in knowledge between beginners and advanced-year members, (2) slight differences in attitudes, consistent with the idea of increased tolerance or open-mindedness, and (3) few differences in behavior. Taken together, I feel that the material pretty well showed that exposure to Great Books does add to the intellectual depth and perspective of the members but does not produce striking or consistent effects on behavior, which is, after all, just what the results should be if the program does what its organizers think it does. We also found that, rather than being socially marginal, the Great Books members were highly educated, highly involved in their communities, and *less* upwardly mobile than comparable college graduates. While this may be a comfort to the friends of Great Books, it "did in" my plans for analyzing their intellectual lives. If the mem-

bers had turned out to have a considerable proportion of homely philosophers, upwardly mobile people, or members of strange cults, the
results might have been quite "marketable" because academic sociologists gobble up materials on strange cults and deviant people; but
since the participants turned out to be quite typical suburban types,
without being statistically representative of suburban types, it would
be hard to justify detailed analysis of their tastes and ideologies.

The Fund for Adult Education received the report and must have
been clairvoyant, for they gave Great Books a large-scale grant about
a month before they saw the results, such being the crucial role that
social research plays in decision-making in modern America.

The project was far from over in fall 1958. During the preceding
winter and spring, while coding and key-punching were going on, I
had been brooding about the dropout problem and came up with an
idea. It occurred to me that if we could collect data in 1958 to determine which of our respondents had actually dropped out of the program during the year, we could subtract them from the tables and
make comparisons only among members who were "destined" to continue. The following dummy table illustrates the idea.

PER CENT HIGH ON DEPENDENT VARIABLE

1957 STATUS	1958 STATUS		
	Dropped	Continued	Total
Advanced Year	$45_{(200)}$	$70_{(300)}$	$60_{(500)}$
Beginning Year	$10_{(200)}$	$60_{(300)}$	$40_{(500)}$

In this hypothetical example, 60 per cent of the advanced members
are high on the item, in contrast to 40 per cent of the beginners. This
would suggest that exposure to Great Books has a positive effect. However, it is seen that the item is strongly associated with retention, and
when the beginners who continued are compared with advanced-year
members, there is no difference, each group having a percentage of 60.
Thus, the original difference can be explained because of selective attrition. Such a design, of course, could not catch spurious differences
due to "historical trends," such as a decrease over the years in the proportion of new members possessing the trait in question, but it looked
as if it would help a lot in solving the problems raised by dropouts.
In addition, the data-collection problems were minimal, as no new

sample had to be drawn and no information collected beyond continuation status.

In the summer of 1958, we proposed such a project to the Fund for Adult Education, and although they had not yet seen any pay-off from the initial grant, they were kind enough to support a follow-up that year. Ursula Gebhard superintended this operation, as by then I was working on another study, a survey of the financial problems of arts and science graduate students.

Because the follow-up data were simple, we were able by February 1959 to submit an edited version of the original report with dropout controls introduced into the key tabulations. (By the way, they did not greatly change the results, but I felt much more confident with these materials to support our findings.) This report, *A Study of Participants in the Great Books Program*, was printed and distributed by the Fund for Adult Education, and although it has never received any attention outside of those immediately concerned with Great Books and not much among those who are concerned, I think it is a reasonably good nonexperimental evaluation study.

Enough budget money remained to support the substantive analysis of factors associated with retention in Great Books. Thus, *Great Books and Small Groups* has its operational origins in the evaluation study completed in February 1958, but its intellectual roots lie elsewhere— a substantive root reaching to Harvard University and a methodological one to Columbia.

Substantively, the major idea in the new study came from viewing the data, not as "why some people quit Great Books," but as "why some small-scale social systems lose the commitment of their members." The research problem is clearly structural-functional and comes simply because that was how I was indoctrinated at Harvard. My courses with Talcott Parsons and my informal contacts with the students working with Freed Bales had steeped me in the tradition of viewing social behavior as functional or dysfunctional for a given social system. In particular, I had been much impressed with an article, "The Functional Prerequisites of a Society" by Aberle, Cohen, Davis, Levy, and Sutton (*Ethics*, IX [1950], 100-111). I was not actually so much impressed with the content, which is highly speculative, as with the marvelous euphony of the names, and I remember that Parsons once mentioned in class that the paper had been done while the authors were graduate students at Harvard. It had never occurred to me that graduate students could make actual contributions to sociology, and the knowledge that Aberle, Cohen, Davis, Levy, and Sutton had made them served to buck up my spirits in many dark hours in Cambridge.

George Homans was involved, too. While I was at Harvard, I had few contacts with him, and my strong attraction to Homansian theory has come in the last few years. In rereading *The Human Group* while preparing a lecture, I came across his astringent remark, "If we turn to history for help, it is astonishing how few societies have failed to survive . . . [but] small groups are breaking up every day. . . ." Thus was born the idea of studying the functional prerequisites for the survival of small groups. To be absolutely candid, there is also a less cerebral aspect to all this. My own work and interests diverge considerably from that of my graduate-student friends and the intellectual climate of the Harvard department, and I think in the back of my mind was the hope that such a study would show "them" that I was not a naïve empiricist but was capable of wrestling with the high-brow problems of sociological theory with a big T.

The methodological contribution of the book comes from a statistical technique for "contextual analysis" which Joe Spaeth, Carolyn Huson, and I worked out during 1958 and 1959. Joe was a senior assistant in the Graduate Student Finances study, and Carolyn was the research assistant on the Great Books continuation, Vickie Beal and Ursula Gebhard having moved on to Grinnell College and San Francisco, respectively. The statistical technique is fairly complicated, and I am going to have to assume that the reader is familiar with it, as a detailed explanation runs to ten or twelve pages. In essence, the contribution lies not in the idea, which can be traced to Paul Lazarsfeld and Patricia Kendall at Columbia University, but in spelling out in a mathematical form the logical possibilities which can occur. What is new here is the idea that there are specific "kinds" of "structural effects."

The "influential" in the flow of influence here is Peter Blau, a Columbia Ph.D., now of the department of sociology at Chicago. As of 1959, I was quite unaware of "contextual analysis," although some had rubbed off on me from Jim Coleman when he was in Chicago, and I had read but not fully appreciated Lazarsfeld and Kendall's discussion in their classic article, "Problems of Survey Analysis." However, the longer I was around Chicago, the more I heard my graduate assistants talking about "structural effects," as discussed by Peter Blau in his courses. One day, Joe Spaeth came back from a Blau lecture and

230 • *James A. Davis*

started talking about structural effects. I remember asking Joe what they really were. He tried to explain them to me in Blau's formulation, but I did not understand it. When I do not understand something statistical, I try to work it out with dummy data, and I remember sitting around with Joe, trying various examples in order to pinpoint the idea. The result was that it occurred to both of us that there were probably several kinds of structural effects, and we instantly got the idea of trying to systematize them.

The trick lay in how to systematize them. I have never had any mathematics training beyond college algebra, and I try, without much success, to teach myself by reading on my own. At that time, I was reading E. F. Beach's *Economic Models*, so linear and nonlinear functions were on my mind. It was only a short step from there to seeing that when the contextual attribute is expressed as a proportion, its variation could be treated as quantitative and the contextual effect could be described as a mathematical function.

There were two different versions. In the first, the contextual effects were treated as deviations from a regression line. This version was sent off to *Sociometry*, which was quite unenthusiastic, one reader saying that structural effects were very important but we did not really understand them, the other reader saying that although we probably understood structural effects, they were not worth analyzing.

Although Joe and I were quite discouraged by this, we picked up the idea every so often and in spring of 1959 came up with the version which finally appeared in print. Carolyn Huson did most of the work on the section regarding tests of significance, so the final paper was authored by Davis, Spaeth, and Huson. An interesting case of independent invention soon turned up. At the fall-1959 meetings of the American Sociological Association, I read the paper and learned that two study directors at the Bureau of Applied Social Research at Columbia, David Caplowitz and Wagner Thielens, Jr., had worked out a similar scheme, which, however, they had dropped to work on something else. Apparently it was an idea which would come to light inevitably, once the problem was seen. I was convinced that this version would "sell," so I sent it off to *The American Sociological Review*, which accepted it and printed it without unseemly haste in April 1961.

The rest was just hard work. During the spring and summer of 1959, the statistical technique was applied to data on dropouts from the fall-1958 follow-up. Carolyn Huson analyzed the materials on group leadership and its effects; Herbert Hamilton, an advanced graduate student, came on the project and analyzed the individual level differences; and Ursula Gebhard's scholarly master's thesis on correlates of role performance was reworked in my "breezy" prose as background for the materials on the relationship between role structure and retention. It was immediately clear that the strongest effect in the data was that membership retention increased considerably in groups where a high percentage of the members had some role (i.e., were named by people in their group as active in some phase of the discussion), but it did not make much difference what the role was. The functional theories which suggest that role content is important did not pan out at all, but they had led us to find the important variable; and, for that matter, my idea of balance of power never paid off either.

The work was complex and tricky. The actual data never quite fitted any of the theoretical models, and the tabulations for partial relationships when the same characteristic has to be held constant twice (at the group level and at the individual level) were nasty; but, once it became clear that role systems were the key to the whole thing, the parts fell into place, and a final report was completed in November 1959.

The report was submitted to The University of Chicago Press, which turned it down because they said it did not have any sales potential, and to The Free Press, which also said that it would not sell but agreed to print it, apparently just for the fun of it. During the 1959-1960 academic year, I was tied up with the final report of the Graduate Student Finances study (a story which would make quite a chronicle itself, but which chronicle could not be sent through the United States mails), so I reorganized and rewrote the report in the summer of 1960, delivering the manuscript to Jeremiah Kaplan of The Free Press and a waiting world in the fall of 1960. The Free Press was delivered of the completed volume in the fall of 1961, a little over four years after my visit to Aspen.

Were I to draw morals from this history, they would be these.

(1) I think the chronicle of *Great Books and Small Groups* illustrates the tremendous importance of technical methodological developments in the substantive development of social science. It is commonly believed that research technology is a mere servant of substantive or theoretical interests. Actually, research technology makes a direct contribution to the content of the field in the same way that the invention

of the microscope or radiotelescope shaped the content of physical science, or perhaps more exactly in the same way that Whorfians claim language shapes our thinking. We can ask of the data only questions that can be translated into specific research operations, and until such translations exist, the research questions remain purely ruminative. Thus, the existence of a national research center makes it both possible and inevitable that comparative studies of groups will take place. Thus, the statistics of correlation and partial correlation give meaning to the vague concepts of "cause," "intervening factors," "spurious relationships," and such, and thus, the development of techniques for contextual analysis focuses our attention on "social climates." The history of content in social science is the history of fad, fashion, and momentary preoccupations, but the history of research methods is one of cumulative developments which have enabled us to ask increasingly precise and sophisticated questions about human behavior. In this sense, I believe progress in social science is mostly in the ability to ask questions, not in the ability to foresee the answers.

Second, I think that this chronicle illustrates the ways in which survey analysis is much akin to artistic creation. There are so many questions which might be asked, so many correlations which can be run, so many ways in which the findings can be organized, and so few rules or precedents for making these choices that a thousand different studies could come out of the same data. Beyond his technical responsibility for guaranteeing accuracy and honest statistical calculations, the real job of the study director is to select and integrate. Of all the findings, only some should be selected for presentation, but which ones? Is this particular finding so unimportant that it should be left out as confusing to the reader, or so important that it must be reported even though it will make the results appear terribly complicated? Should we emphasize the smashing difference which is, however, "obvious," or should we give play to the puzzling surprise, even though it produces only a small difference? The 101 (an IBM machine used for cross tabulations) takes independent variables in batches of four; and having listed the seven obvious columns for cross tabulations, which "unobvious" one do you choose as a gamble for the eighth? How much attention shall we give to the client's areas of success and how much to his areas of failure? How much of the data shall we present in the report: so much that no one will read it, or so little that the reader cannot check our conclusions against the evidence? In multivariate analyses, one can produce a large range in percentages either by dividing a very few variables into fine categories (e.g., cutting income by thousands of dollars versus dividing it at the median) or by taking a larger number of items and dichotomizing them. Which is preferable? Statistics books will not help you, for the answers must come from the study director's experience and his intellectual taste, his ability to simplify but not gloss over, to be cautious without pettifoggery, to synthesize without distorting the facts, to interpret but not project his prejudices on the data. These, I submit, are ultimately aesthetic decisions, and the process of making these decisions is much like aesthetic creation.

But all this should not be construed as support for the fallacious idea that "you can prove anything with statistics." Short of deliberate falsification, statistical data are remarkably resistant, as anyone who has desperately tried to save a pet hypothesis knows. It is almost impossible for two competent study directors to arrive at *contradictory* conclusions from the same data, but it is almost inevitable that they will differ considerably in their emphases, organization, and selection.

Thus, if survey analysis is an art, it is not an art like sculpture or painting, in which one can make almost anything out of the raw materials. Rather, it is an art very much like architecture, in which it is possible to show disciplined creativity by producing elegant structures while working with raw materials characterized by limited engineering properties and for clients with definite goals and finite budgets.

Balance between discipline and creativity is very difficult in social science. By and large, the fashionable people in sociology are "action painters" who dribble their thoughts on the canvas of the journals, unrestrained by systematic evidence, while at the opposite pole there are hordes of "engineers" who grind out academic development housing according to the mechanical formulas of elementary statistics texts. It is not easy to steer between these courses, and I am not claiming that I did so in this study, but my opinion is that the fun lies in trying to do so.

There is a lot of misery in surveys, most of the time and money going into monotonous clerical and statistical routines, with interruptions only for squabbles with the client, budget crises, petty machinations for a place in the academic sun, and social case work with neurotic graduate students. And nobody ever reads the final report. Those few moments, however, when a new set of tables comes up from the machine room and questions begin to be answered; when relationships

actually hold under controls; when the pile of tables on the desk suddenly meshes to yield a coherent chapter; when in a flash you see a
neat test for an interpretation; when you realize you have found out
something about something important that nobody ever knew before—
these are the moments that justify research.

• NOTES

[1] M. R. Trabue and Allen Abbott, "A Measure of the Ability to Judge Poetry,"
Teacher's College Record, XXI (1921).

10 The Sociability Project:

A Chronicle of Frustration

and Achievement*

DAVID RIESMAN AND JEANNE WATSON

> Yi-ch'uan came again with Chang Hsün during the spring.
> Shao Yung invited them to go for a walk with him on the T'ien-
> mên road to look at the flowers. Yi-ch'uan declined, saying that
> he had never been in the habit of looking at flowers. Shao
> Yung replied: "What is the harm? All things have ultimate
> principles. We look at flowers differently from ordinary men,
> in order to see into the mysteries of creation." Yi-ch'uan said:
> "In that case I shall be glad to accompany you." †

INTRODUCTION

The enterprise which came to be known as the Sociability Project
grew out of collaboration among a number of different persons. The
project was financed by the National Institute of Mental Health
(NIMH), Public Health Service.[1] It was initiated by Nelson N. Foote
and David Riesman, and, though still unfinished, it is being carried to
completion by David Riesman, Jeanne Watson, and Robert J. Potter.
In trying to reconstruct the history of the project, Riesman and Watson
have been ably assisted by Robert J. Potter, who participated actively
in preparing materials for this report, and by the memorandum which
Ariadne P. Beck ("Andi") prepared as a contribution to the chronicle.

---

* A chronicle of research in progress.

† A. C. Graham, *Two Chinese Philosophers Ch'eng Ming-tao and Ch'eng Yi-
ch'uan* (London: Percy Lund, Humphries and Company Ltd., 1958), pp. 109-110.

We regret the fact that Nelson Foote chose not to make any contribution to the chronicle. He would have additional things to say about the Sociability Project and, almost certainly, conflicting interpretations of what happened. So, too, would the graduate students who served for varying lengths of time as research assistants on the project; so in addition would our former colleagues on the faculty of the University of Chicago—Reuel Denney, Anselm Strauss, and Everett C. Hughes— all of whom facilitated and encouraged our work and have continued to be interested in it.

Our discussion of the Sociability Project, like the project itself, relies on the methods of participant observation and detailed reporting of personal experience to produce an account of human behavior. This is not a conventional approach to the discussion of research operations but ours was not a conventional project. Our task was to study the nature and significance of sociable interaction in modern middle-class society. In order to do this, we felt we could not begin with conventional methods of data management and that we had to expose ourselves in a rather more open and intuitive manner to the complexities of our subject. Our emphasis was on exploration rather than the collection of systematic information for the testing of specific hypotheses. We sought to find new ways of thinking about sociability, new ways of interpreting it, new ways of understanding what may be gained and lost in the course of sociable interaction. In immersing ourselves in the study of sociability, we found that unconscious or irrational elements of the personality were brought to bear on the research, as well as conscious and rational ones. Similarly, the wholistic quality of our individual involvements in the Sociability Project gave more than ordinary weight to those factors of personality which generate interpersonal alliances and antagonisms. The wide differences which existed among us initially with respect to training and experience in research, our different and often opposed commitments regarding the processes and preferred outcomes of research, were sometimes aggravated, often assuaged, and always blurred by the urgent need, which we each felt, to discover in our associates congenial partners in research. The emergent ethos of the project, with its emphasis on pluralism in style and method, diffusion of initiative and responsibility, and compulsive attention to the processing of data, seems to flow from many sources: the vagueness and difficulty of the research task; the quality of sociability as a projective, expressive form of interaction rather than an instrumental one; the existence of wide differences in personal and professional style among the individuals who were committed to work together as a research team; and an eventual consensus of temperament favoring permissive and receptive methodologies for both internal and external operations of the research staff. Our chronicle is an attempt to examine the operation of these and other factors as they were manifested in a particular situation. We do not mean to suggest that what happened on our project is in any sense typical or representative of what may happen elsewhere.

About the Sociability Project

The Sociability Project was an attempt to study systematically the interaction which occurs in sociable settings. Two separate studies comprise the project. The first focused on the observation of interaction among college students who were serving on the staff of a summer camp for adults. The second focused on party sociability and included detailed analysis of the reported conversation at twenty-six parties. The general object in both studies was to develop a conceptual framework for describing and analyzing sociable interaction.

Our first step in conceptualizing sociable interaction, or sociability, was to develop a series of distinctions which would differentiate sociable interaction from work-oriented interaction and from familial interaction.[2] In comparison with other types of interaction, sociability has less tendency to be organized in terms of the requirements of an external task and is more free to reflect the needs, interests, and habits of the participants. It has a special contribution to make to both the development of unique individual identities and the development of shared norms and identities which facilitate social integration.

Acting on the assumption that sociability is a form of interaction in which the display, maintenance, and development of identity is of paramount importance, we constructed a unit for the analysis of interaction which would, we hoped, capture the substance of what was affirmed about identity in each conversational exchange or episode.[3] For each episode, we noted the stance which the central participants adopted vis-à-vis one another and the tie which they used as a basis for conversation; we recorded what aspect of the external environment they chose to discuss and what relationship they expressed toward that environment. These four elements of conversation define the structure of the episode. At the same time, they indicate the fundamental affirmations which are being made about the relationship of individuals to one another and to the society in which they live. Comparisons among different subcultures and populations with respect to these four elements of episode structure are expected to yield illuminating information about differences in patterns of social integration.

Our description of episodes of sociability was not limited to the four structural properties noted above but included also many other qualities and characteristics. Some of these are associated with one or another of the structural properties; others are of a kind which add color to the interaction but do not give structure. Taken together, they permit a variety of analyses of the character of sociable interaction.

A continuing interest for us has been the problem of *anomie,* by which we mean to refer to such individual characteristics as purposelessness, alienation, and lack of commitment. Analysis of emotional themes in sociability indicates that anomic and nihilistic themes are related both to the manner in which participation is organized (large groups encourage nihilism; dyads dispel it) and to such social characteristics of the groups studied as status and social class.[4] In general, nihilism is maximized in situations where individuals have least power to influence the events which impinge upon them. Thus, sociability serves both to reflect and reinforce the objective properties of the situation in which participants find themselves.

Customary expectations in sociability about what subject matter will be discussed and what sentiments may be expressed, as well as about the kind and amount of personal exposure to be expected, are governed in part by the type of partner with whom one is conversing. Yet, surprisingly, our data showed no evidence of differential sex roles, except that men talked more than women; with this exception, for the parties which we observed and in the terms of our analysis, we found no significant differences between men and women.

Much more important was the difference arising from variations in types of acquaintance.[5] For example, two contrasting categories of acquaintance are *familiars,* people who see each other frequently as a result of common affiliation with a group based on institutional ties, and *casuals,* people who have had little or no contact with one another prior to the situation in which they are observed. Interaction between familiars is highly routinized and reiterative; interaction between casuals is much more likely to be creative, exciting, and even intimate.

Historical traditions and subcultural patterns influence changes in sociable style. The decline of puritanism, the spread of leisure, the growth of tolerance—these and many other developments in upper-middle-class culture have accompanied a shift toward more relaxed and more egalitarian norms of behavior in sociability. Monopolies of responsibility for the conduct of a party have tended to shift from the host to the guests, from dowagers to demos, from Society to Café Society; and acceptance of mediocre performance and even passive nonperformance has increased accordingly.[6]

We cannot be sure that these brief and perhaps cryptic references to some of the highlights in the papers we prepared for publication or for presentation at professional meetings will give the reader any adequate sense of the range and variety of concerns of the Sociability Project. Each of us came to the study of sociability with no specialized preparation in that particular field or in anything quite comparable to it; we came instead with diverse questions and interests, as well as with the experiences all of us have as amateurs of sociability. There was no one line of inquiry which we wished to pursue, no established tradition of research upon which to build. Georg Simmel's scintillating essay on sociability,[7] for example, was a stimulus to reflection and speculation but scarcely a program for research. Indeed, our project was an effort to be systematic and even quantitative in areas that, on the whole, had been dealt with previously in diffuse and theoretical fashion. Under these conditions, the project was shaped by the intellectual preoccupations and curiosities of the principal investigators. It is with a description of these that our chronicle begins.

I. A CHRONICLE OF THE FIRST YEAR

Intellectual Origins

The immediate origin of the Sociability Project can be traced to a seminar on leisure and play given by Foote and Riesman in the academic year 1954-1955. Prior to that time, both men had interests pertinent to the matter at hand. Foote had a long-standing interest in play, and part of his interest in sociability was an interest in the character of fun. What, he asked, makes sociability fun? Beyond this, Foote tended to see individuals as forever changing; he was skeptical of the permanency of character or nature, and he felt that sociable interaction held part of the key to the process of change. He was interested in natural and staged encounters which would challenge the individual to alter in some significant respect his view of himself or others. With Leonard Cottrell, Jr., he was writing a book on identity and interpersonal competence,[8] and he was supervising research on the development of such competence. He was particularly interested in the processes of identity formation and development among young adults.

Riesman was less concerned with change and more with form and process; if not less interested in individual careers, he was less clinical and more preoccupied with the relation between careers and cultural patterns. In his own words (from a letter):

> I have had an image of our society as one in which people would soon be forced to play and consume beyond their capacities. With the relaxation of hunger and gain in large segments of the middle class, I was curious about the emergence of new potentialities for abundance and relatedness. Before coming to the Sociability Project, I had worked for a time on a study of middle-aged and aging persons and like many people had been troubled about the leisure of men who no longer worked but were forcibly retired.
>
> I also brought to the Sociability Project a long-standing concern with friendship in cross-cultural perspective. Having known many refugee or visiting Europeans, intellectuals, and artists, I was of course aware that they judged American friendship to be superficial. In a seminar of Harry Stack Sullivan's and in his writings, I had been stimulated by his concern with the chum, the age-mate or peer, as an essential relationship in personal growth. So, too, I had been influenced by sociometry, especially by its method of using friendship clusters as a way of studying personal and political influence, or isolation.
>
> In still another perspective, I brought to the project my sense of the fragility of civilization and especially of the more intangible aspects of the human enterprise. For thirty years I have been interested in the way negotiations occur or break down among great powers, both inside and across national boundaries, and in the role of lawyers, go-betweens, or brokers or diplomats in mediating—or exacerbating—conflict, compromise, and cooperation. As a participant-observer in negotiations, parties, seminars, conferences, conventions, and other such occasions, I have been keenly aware of the resiliency, resonance, and shakiness in these affairs, sensitive to the constructive and destructive, sadistic, autistic, and benign, the inventive and the stereotyped responses. When we came to study sociability, my interest in the blockages and impasses which destroy and the imaginative inventions which can salvage a social situation was mobilized. In addition, Max Weber's dichotomy between the charismatic and the bureaucratic, along with other similar dichotomies, seemed relevant to an understanding of "presence" and "absence" in social situations. My concern here was less with ritual, elegance, or

drama for its own sake than with the way in which people consciously or unconsciously prospect for one another, recognizing friends or enemies, partners or people who "do not count," within and across lines of sex, social class, and personal and professional style.

The seminar on leisure and play examined such matters as the social definitions of work and play, historical and ecological shifts in the availability of leisure time, and the images of play presented by the mass media. Riesman brought his friend and close colleague Reuel Denney into the seminar and into the first stages of the Sociability Project. Denney shared an interest in these matters and in the history of sociable styles in American life, particularly in terms of frontier traditions, regional differences, and changing ethnic patterns. In discussing differences between work and play, the seminar sought to understand contrasts between sociability, interactions at work (such as those studied by George Homans),[9] and the closer relations of families and other intimates. It explored some of the ideas about play and sociability drawn from the writings of Piaget, Simmel, Erikson, and Huizinga.

Operational Origins

At the conclusion of the seminar, Foote suggested to Riesman that they submit an application to the National Institute of Mental Health for funds to explore further some of their ideas about leisure and play. Riesman was doubtful about whether these ideas constituted a basis for research and yet was also skeptical of his own qualification to judge what was or was not researchable. However, he agreed to cooperate with Foote, leaving to the latter the mechanics of the proposal: the statement of the problem, the amount and phasing of the money, the number and specification of others to be recruited, and, indeed, the general shape of the enterprise.

The proposal envisioned a staff of one full-time project director and four half-time graduate students. Young adults were to be recruited from the area around the University of Chicago. Half of them would participate in role-playing groups, similar to those currently meeting as part of the project on interpersonal competence, with the object of engaging in "staged spontaneous" sociable encounters. All of them would be exposed to beginning and end observation of sociability and measurement of relevant aspects of identity.

Riesman viewed the proposed Sociability Project as an opportunity for him to learn about methods of doing research still new to him. He writes:

My knowledge of social-science research methods before working on *The Lonely Crowd* [10] was confined to very limited experience with survey techniques, interviewing, and "projective" analysis of protocols. I wanted very much to take part in a large research project in which I could learn something and make up, as it were, retroactively for what I had missed in not attending graduate school. It was this that led me to join the Kansas City Project several years earlier (and to be disappointed there, being for a time put into an unwanted position of partial leadership rather than being given the training I sought).

I felt great admiration for Foote and was especially impressed with his courage to try hard work on "soft" data. I was doubtful about his suggestion that the Sociability Project focus on a study of identity but felt too unsure of my own research abilities to propose an alternative. I had been greatly drawn to Erikson's work, and in the Kansas City Study I had witnessed an effort to make operational his stages-on-life's-way. I felt that this had been too gross and heavy-handed a treatment of a work of poetic nuance and evocativeness. In the Kansas City Study, I had wanted to look for people who exemplified aging in its most creative ways rather than just random people.[11]

I supposed that the same design in the Sociability Project would have meant looking for people who, on some a priori ground, were going through a decisive shift of identity. But how could one discover that? No mechanism for staining invisible tissues appeared to me likely for pursuing such subtle personal processes, and the hope of literally applying the Eriksonian stages seemed to me farfetched. Furthermore, I doubted whether graduate students possessed, in the main, the sociable experience or clinical gift that would enable them to catch on the wing the fleeting experiences of self-realization in a group. Still, I had confidence in Foote and hoped that he would be able to find a way to do research on some of these matters.

My previous experience with applications for research funds had not been encouraging. I had been turned down in several efforts to raise money for my own research, in some cases falling between the jurisdictional lines of the foundations. I had actually very little expectation that money would be granted for the Sociability Project. With Reuel Denney, I had made an application to the Behavioral Sciences Division of the Ford Foundation to set up a Center for the Study of Leisure at Chicago and had also begun with Mark Benney a small-scale study of the interview under a grant from the Foundations' Fund for Research in Psychiatry. What happened then was that all these ships came in at once, famine being succeeded by indigestion.

Perhaps it should be added that I grew up as a social scientist in an epoch when famine was the customary lot of the researcher. The vote of confidence that NIMH placed in Foote and me imbued me with a feeling of responsibility and obligation that younger researchers may have in lesser degree, since their experience is not of chasing grants but of grants chasing them; they assume that whatever they want to work on should and will be supported.

Moreover, the fact that NIMH is a government agency influenced my attitude. I regarded NIMH as freer and more courageous than most private foundations but was aware that "mental health" is regarded as a subversive concept in some quarters and that research done at government expense involved congressional and other constituencies that could not be foreseen. I was anxious to deliver a product which would justify all expectations.[12] At the same time, I lacked confidence that anything we could discover would be relevant for mental health. For while the relation between sociability in our society and individual development could be assumed in a general way, it was another matter to discover anything not already apparent, anything concrete that was something other than a how-to-do-it manual for sociable hosts and guests. It was here that Foote's confidence overcame my own misgivings.

When the grant came in from NIMH, requiring that its first installment be spent by the end of the calendar year 1955, Foote and Riesman began to look for a project director and for graduate-student assistants.

At this point, one hurdle that besets so many projects with headquarters in a university made its appearance: to secure a first-rate person, it would be desirable to have a departmental appointment or at least some attractive entree into a department. We approached several younger social psychologists who, while interested in the project itself, were drawn away from it by more promising academic futures elsewhere. Both Riesman and Foote were pleased when they were able to persuade Watson to come to the project; they felt that she was admirably qualified for the work.

However, concerning the search for graduate-student assistants, Riesman had some doubts; he felt that the project was not yet ready for the leap into actual operation. At this point, he did not feel prepared to direct graduate students; he was seeking exceptional apprentices with whom he could work as colleagues on the basis of parity and joint exploration. In considering what was needed, he looked to the intellectual and personal quality of the candidates as one way of shaping the outcome of the project itself. He did not want to see all the slots filled too hastily, fearing that this would foreclose possibilities of reshaping the project as it progressed.

But the academic calendar ran against him here. Good graduate students were scarce, and other projects at Chicago were competing for them. Therefore, he reluctantly joined in the process of recruitment, going, for example, to New York to talk with one potential candidate for the Sociability Project and with others who might be brought to Chicago. He participated in hiring two of the assistants, but the other two were hired despite his reservations about whether they could add anything to the project.

There were several reasons for the haste in hiring assistants. Not least important was the timing of the grant itself. As already stated, money given in the fall of 1954 had to be spent by the end of the calendar year of 1955 or else returned to NIMH.[13] There was some expectation that the Sociability Project would be a continuation, in a sense, of the project on interpersonal competence which was currently in progress at the Family Study Center, so that it was reasonable for assistants for the new project to be trained by being given assignments on the old project. To get the best graduate students, it was necessary to corral people early and give them good pay; in addition, each was promised an opportunity to gather data which could be utilized for a thesis—M.A. or Ph.D., depending on the academic status of the student.

The four assistants who were eventually hired included two men and two women. Of these, one man and one woman were working for an M.A. and the other two, both with considerable research experience, were working for a Ph.D. They came from different fields—sociology, human development, and educational psychology—and they differed widely in background and training. Foote hoped that, by composing a varied staff, it would be possible to maximize the range of sociability to which they would have access. One of the men (Al) had had experience working in a group program of retraining for young male delinquents. The other man (Bob) had an M.A. in sociology from the University of Colorado and several years' experience as a pilot in the Air Force. Of the women, Andi was essentially an amateur in social science and research, and Joan had worked extensively with Thelen on studies of group dynamics.

An early view of the Sociability Project as it was presented to the assistants is given by this quotation from Andi.

January 1955:   Riesman and Foote get grant. I arrived in Chicago. Am interviewed by both, but project is presented to me by Foote. His view of project emphasizes development of identity and competence, and the role of sociability in developing these. . . .

Summer 1955:   Riesman was a little more in our presence though he didn't meet much with us. He was very much there for me though in his standard—several times during that summer he told us that he thought reading the literature in the general area was not desirable. He thought that our lack of knowledge in the research aspect of the work was an advantage—maybe we would come up with a fresh point of view. . . . At that time, though, it appealed to all the wrong tendencies in my personality—wrong for that period in my training and professional development—and, so far as I can see now, it was the primary factor that determined my eventual problems with the project and with Riesman in particular. In other words, on some level or other I accepted as our task that we were to develop, in a new and creative way, this unexplored field.

The initial perceptions and response of the project director, Jeanne Watson, were somewhat different. She was hired in February 1955 with the understanding that she could not assume full-time responsibilities until September 1955. Her comments follow.[14]

ON MY DECISION TO TAKE THE JOB.   Any commitment I make to this project must be to the possibility of doing some kind of research relevant to the concerns of Foote and Riesman, not to the particular project outlined in the proposal. The proposal strikes me as deceptively specific; it appears to give specific plans for recruitment of subjects—at least six groups per year, each consisting of eight persons in the eighteen to thirty age range, for measurement and observation and for "staged spontaneous" activities for half of the subjects. But close examination of the proposal convinces me that it is hollow; it does not indicate what the object of the research is or what is to be achieved by the measurement, observation, and experimental interaction.

My impression is that others share these doubts. Foote and Riesman both seem willing to discard the specific proposal if a better one can be devised, and they say that NIMH, also, encourages grantees to do what seems reasonable and not to be bound by what they have put in writing.[15] The present proposal represents one effort to show how the present concerns of Foote and Riesman can be advanced by empirical research, but the reason they want to hire me is because they want someone with training in research, and they are willing to have me work out a different research design.

The real appeal of the project lies in the opportunity to work closely with Foote and Riesman. They have been described to me as two of the most creative social scientists in America today, and certainly their interest in play, leisure, and sociability represents an exciting departure from the stereotyped concerns of their colleagues.

I have not thought much about leisure, play, or sociability myself. However, I think my training and background are relevant for the kind of research which they want to do—my experience in observation and analysis of face-to-face interaction, my training in the dynamics of groups, my thesis on some relationships between personality and interpersonal behavior, and my personal experience with psychotherapy. Probably the statement which most interested me was Riesman's, to the effect that he thinks of sociability as a form of projective behavior comparable to dreams, in which the salient concerns of participants are reflected indirectly and symbolically.

I do not really understand Riesman's statement that he has no interest in cause and effect but is interested only in description.[16] It is a conception of research which is new to me, but it should be interesting to see what it entails. I think Foote's conception of what is meant by research is much closer to mine than Riesman's.

Spring 1955: We are using the spring quarter as a period of preparation. Foote is moving ahead with the hiring of assistants, and he had two persons ready for me to interview on my first visits to Chicago (Andi and Bob Potter). Both have been hired and put to work at the Family Study Center. I don't feel able to participate actively in the hiring of assistants, since I do not know which persons are available and I am not clear about what their jobs will be; but I am willing to accept Foote's recommendations.

Foote has organized a seminar which will review research relevant for the Sociability Project. It is focusing particularly on studies which involve direct observation of behavior, and I am glad to have a chance to review this literature.[17] Also that seminar helps to make clear certain ground rules for the project,[18] i.e., the data will be gathered by *direct observation* and the units of observation will be *episodes of sociable interaction.*[19]

I am commuting to Chicago one day a week in order to attend this seminar and engage in preliminary discussions about the project. Foote, Riesman, and I generally have lunch together before the seminar. These lunch sessions are difficult for me because, instead of trying to work on plans for the project, we seem to spend our time in formal sociability. Foote and Riesman talk about persons they have seen and things they have read. This is a kind of conversation which leaves me feeling at a loss.[20] Occasionally, of course, we talk in general terms about the project, but they do not seem to feel any pressure to get work done.

Yet this very difference in our approach to work makes it clear what my role on the project will be. Foote and Riesman do their thinking by allowing their minds to roam freely. It would be both hopeless and inappropriate for me to ask them to focus their conversation on systematic planning for a research enterprise. That is my job. I spend my time listening to them, trying to sift from their wide-ranging comments some ideas about what questions could be advanced by empirical research. Yet it is obvious to me, and I think to them, that not all questions call for empirical research.

I am working hard to learn to understand Foote and Riesman. It is a little surprising to me to find how extremely difficult it is simply to understand what they are saying. I guess every social scientist develops his own way of thinking, and each new acquaintance means that one has to learn a new language. Right now it is quite a strain.

There is something else which contributes to my sense of being cast adrift, without moorings. I come to this project after eight years at the University of Michigan. During this time I acquired my Ph.D. and virtually all of my professional training and experience. Before coming to Michigan, I spent one year at Columbia University attending classes and working on an M.A. The classes were large, and I felt almost anonymous. Yet now it appears that Foote and Riesman made no inquiry about me of anyone at the University of Michigan, turning exclusively to Columbia for references. It is as if my entire professional training is being politely disregarded and set aside in favor of some other mystic quality bestowed by Columbia. But I do not know of anything I bring to the project outside my professional training at Michigan; and if that is not relevant, what is? [21]

I am beginning to get some idea of where my professional interests intersect with those of Foote and Riesman. I like the idea of focusing our attention on the connection between sociability and identity, with sociability serving both to maintain and to facilitate change in identity. Foote seems particularly interested in the "change" part of this relationship. I am not sure that we can get good information about change, but, if we want to try, the idea of working with young adults—persons in the process of becoming adult—makes sense. Spurred on by my own need for orientation in and about Chicago, it occurred to me that a group of newcomers to Chicago might provide access to persons whose identity was in flux. (My M.A. thesis at Columbia dealt with change in situation as a spur to attitude change.) We have tabled this for later consideration as a possible gimmick for recruiting subjects.

The First Months of Project Operation

Andi was hired in February and Bob Potter in March; a third assistant was hired early in the spring quarter and fired before the end of the quarter. All three went to work immediately at the Family Study

Center. It was anticipated that their work would provide both direct and indirect preparation for later assignments on the Sociability Project. In addition, they participated in Foote's seminar on sociability and in occasional staff meetings. Watson, hired in February, spent one day a week in Chicago during the spring quarter, working on the Sociability Project.

During the summer, responsibility for the operations of the Sociability Project was transferred to Reuel Denney.[22] Watson remained in Ann Arbor, trying to finish up work on her previous job; Foote was busy directing a workshop given by staff members of the Family Study Center; and Riesman was out of town. A candidate (Joan) was nominated by Watson for one of the research assistantships and was hired in June; in the fall, a fourth assistant (Al) was brought to the project by Foote.

In the fall of 1955, Foote, Watson, and four assistants were ready to give full attention to the Sociability Project, and work began in earnest. Riesman, however, continued to be out of town during much of the fall quarter of 1955 and did not become actively involved in the project until a later date.[23]

Spring and Summer 1955: Getting Ready

During the spring and summer of 1955, attention was given to a variety of preparatory activities. Staff members were trying to get ready to observe sociability, to interview subjects about sociability, and to stage experimental sociable encounters. Preparations for all these aspects of the project moved ahead simultaneously. In addition, there was a continuing effort to gain a clearer conceptual understanding of the problems to be investigated by the project. These early theoretical explorations are summarized in Part III of this chronicle.

OBSERVATION OF SOCIABILITY. Foote's seminar on techniques of observation provided the nucleus for preparatory work on observing sociability. Sociability staff members supplemented this review of what other researchers had done with various observations of their own. During the spring, the project director and research assistants began to write up observations of sociability. Sometimes these were single incidents which had occurred in the course of a larger event, sometimes they were capsuled accounts of parties or other occasions, and occasionally they reported in detail on an entire event. These reports were generally viewed as exercises, the object being self-training rather than collection of data for analysis. Most of the reported observations were received into the files without comment. One party reported by the project director was made the subject of discussion in several consecutive staff meetings under the leadership of Reuel Denney. At the end of this time, approximately three pages of the report had been discussed, covering perhaps half an hour of elapsed time, and Denney had led the way to several exciting discoveries about sociability. Then the analysis was discontinued because of the pressure of other events.

It was soon apparent that the observer of sociability had to become, first, an observer of self-in-interaction. In a natural setting—whether his own or that of his subject—the observer cannot be a nonparticipant and he cannot follow a standardized pattern of interaction. He is first of all himself, and the qualities which he brings to the interaction have much to do with what is evoked from others and with what he himself is able to see. To an extent unusual even for social science, the observer helps to give form to his own data.[24] To be sure, at large sociable occasions, the participant-observer can observe the way people distribute themselves, even when he is marginal to the scene. But in smaller gatherings devoted to conversation, such as those on which we came to concentrate, the participant-observer is not at all invisible, and his personal qualities, though hardly decisive, color what is said and done. As a first step toward objective self-awareness, each staff member was asked to write a history of his own sociable career.

During the summer, the assistants worked with Reuel Denney on the observation of sociability at the Family Study Center workshop. They were active participants in the workshop as well as observers of sociability and found that they had little time available for observation. Accordingly, they supplemented their workshop activities by seeking other opportunities for observation. Denney took the entire staff to see the movie *Marty,* after which they held a detailed discussion of the interaction portrayed in the movie. Once again, excitement was generated about what could be learned from a study of sociable interaction.

Later in the summer, Denney led the staff in a series of discussions which attempted to set down on paper all the things for which an observer might look in an episode of sociability. The list ranged from such ecological matters as the number, age, and marital status of participants to the question of whether they felt private or exposed; from observations about such dimensions of cultural context as place, timing, class-style, and mobility direction of participants to normative processes which could be observed at the party—performing and "audiencing," handling of status differences, affective tone.

ROLE-PLAYING TECHNIQUES. The research currently in progress at the Family Study Center involved the use of role-playing techniques to improve interpersonal competence in the specific areas of autonomy, creativity, and empathy. Foote hoped that the Sociability Project would be able to build on these experiences to develop role-playing techniques relevant for staged incidents of sociability. Accordingly, assistants hired in the spring were assigned to help in the coding of role-playing scenes, staged by the three different subprojects on interpersonal competence. The Family Study Center workshop in the summer of 1955 represented a continuation of the work on interpersonal competence, and the research assistants on the Sociability Project were asked to take an active part in it. Each assistant worked with one pair of leaders, joining the workshop group as a participant and also taking part in discussions among the leaders about the group. Thus, they gained some sense of what the workshop was like both for leaders and for participants. This experience would have been of value if the Sociability Project, as anticipated, had moved ahead in the direction of experimentally staged incidents of sociability; but, as it turned out, the Sociability Project took a different course.

INTERVIEWS ABOUT SOCIABILITY. Work on an interview schedule began in the spring, and a few interviews with students were obtained at that time. The interviewer asked the respondent to talk about his views of sociability, his experience in sociability, his friends, and himself. A revised interview schedule was prepared for the workshop but never used. The four stated objectives at this time were (1) to find out what "unfinished personal business" the respondent brought to the workshop, (2) to get a comparison of home and workshop in terms of amount of attention given to each and similarity or difference between them, (3) to find out about his reactions to others present as partners for sociability, and (4) to get some background information on his views of sociability, friendship, and self.

Occasional interviews were obtained as the summer progressed, and there was some further discussion about questions which might be included in an interview. Denney's list of matters which might be observed in episodes of sociability was paralleled by a list of questions which a participant might be asked about his experience of a particular sociable event.

Fall Quarter 1955: The Plunge into Work

With the beginning of the fall quarter, the full staff of the Sociability Project was assembled and more or less ready for work. They differed greatly in the kind and amount of contact they had had with the Sociability Project. Of the four assistants, Bob and Andi had worked for two quarters each on the Interpersonal Competence Project, one working with the team which studied creativity and the other with the team which studied empathy. Joan had joined the Sociability Project in time to participate in the summer workshop, and Al, in the fall, was new to the project. Watson had worked intermittently on the project since March but felt that her move to Chicago in the fall marked the beginning of her term of office. Foote had brought the project to its present state of readiness: he had written the proposal, taken major responsibility for hiring personnel, organized the seminar on observation, and supervised the job arrangements for the research assistants. Denney, who had contributed much to the project during the spring and summer, was caught up in other activities in the fall; and Riesman continued to be immersed in other responsibilities and only marginally involved in the Sociability Project.

Foote was eager to get on with the job. All the preparatory activities which seemed feasible had been completed. The staff was hired and more or less ready to go; the interview schedule was at hand; the research plan had been given in the original proposal. It was time to move into the field and see what we could do. Watson, however, came late to the project and was not ready to move so quickly into the field. She wanted more time to work out a plan for the project and to define the questions which it would investigate. Nevertheless, spurred on by the need to make use of the services of four half-time research assistants, she agreed that the fall quarter should be a pretest period [25] and that work should begin at once.

INTERVIEWING. Each assistant was asked to contact three subjects, using his own resources for making the contacts. In addition, the four assistants and the project director would work with one group of five subjects—a group recruited from the School of Social Service Administration (SSA). The idea was that each assistant would develop his own techniques and body of data with his "special subjects," while development of standardized skills and common language would proceed through comparative analyses of experiences with the common group.

The latter part of September was used to perfect the interview schedule, which was completed on October 4, and to set up interview arrangements. October was a month of intensive interviewing. The interviews lasted several hours and, when transcribed, ran to about twenty pages each. Watson prepared standardized summaries of all the interviews and began planning a code which would work from the interview summaries. However, this line of action was dropped, and none of the seventeen interviews was ever coded or analyzed in any other way. Riesman, on October 22, indicated in a memo that "it is my impression that the report of episodes is a more fruitful road into our problems than the sociability interview." Bob Potter, on December 6, echoed these sentiments and presented a detailed exposition of the thesis that we could learn more from observation than from interviews.

Of the five sets of people who had been interviewed, two continued to serve for some time as subjects. One set of interviews had been with students in the law school. The staff member (Al) who had conducted these interviews continued throughout the year to make observations of sociability in an informal luncheon group to which the law students belonged. A second set of interviews became the nucleus for a project on "Close Friendship and Identity Change."

The other two assistants collected observations of their subjects in several different sociable encounters and then severed the relationship. As for the SSA students, reports were made of several different passing encounters with individuals from this group, and then, on November 1, a party was arranged. Participants included the five staff members and their five subjects. The interaction was relatively strained; nevertheless, it provided material both for practice in writing up observations and later for the study of parties. This was the last contact with the SSA students.

There is, of course, no real expectation that data collected during a period of pretesting and experimentation will be utilized for research purposes. However, there is a definite expectation that they will be used as a means of learning how to proceed with the next stage of the project. In our case, there was virtually no comment of any kind about the interviews, except for the two memos asserting that the interviews were not doing the job we wanted to do. The work was done and the products were filed, but they brought only a sense of failure and frustration with no compensating sense of progress.

OBSERVATION. Meantime, we had been proceeding with work on our alternative methodology, the direct observation of sociability. Episodes of sociable interaction among staff members were staged and observed on October 18 and 28. Both of these episodes fell flat as examples of sociability because the more aggressive participants seized the opportunity to interview their more withdrawn partners about private, nonwork activities. This interviewing was resisted, and in each case the result was a kind of stalemate: one person demanded answers, the other parried or evaded. However, they did serve their primary purpose of providing interaction which we could observe and write up. The SSA party on November 1 provided a more extensive exercise in participant observation. Comparison of the write-ups provided by different observers led each of us to greater awareness of his own biases and preoccupations and to greater understanding of his own patterns of recall and forgetting: each could begin to see which factors facilitated and which blocked recall of his own experiences in interaction.

Considerable attention was given to the problems of *sequence and turning points in interaction*. We felt that each episode of interaction was defined by an opening comment and response to it, and we tried to recall initiations, turning points, and terminations. This proved to be extremely difficult. We found that we could recall with relative ease what happened within an episode but that it was much more difficult to remember how an episode began or ended or where it occurred in relation to other episodes. These difficulties were associated with another problem: the all-or-none quality of episode-recall. Once reminded that a discussion about X had occurred, one could recall many details of that discussion, but it was possible to give a full report of episodes V, W, Y, and Z and never notice any reminders of the existence of episode X.

Later in the project, when we were concentrating on the production of reports from memory, we learned that the real context for an episode (or sequence of episodes) is not the event at which it occurs but the world of events and ideas with which it is associated. Thus, we

found in the days following a party that we would suddenly be reminded of "forgotten" episodes by fresh encounters with the persons
or ideas or places that had figured in the "forgotten" conversation. In
the early days of November, we did not reach this generalization, but
we did conclude that memory was more adequate as a tool for the
study of episodes than for the study of sequence and chronology.

After two weeks of practice in observing ourselves and discussion
of ways of standardizing our write-ups of episodes, we moved into the
field. Each assistant made an effort in November and December to
collect observations of the sociability of his own group of subjects, including the SSA students. Each created his own means of gaining
access to sociable interaction. Bob spent two evenings in the apartment of a group of airline hostesses, noting interaction among the girls
and also between the girls and various young men who dropped in.
Joan set up one evening of bridge and one evening of Scrabble, recording interaction both during and after the games. Al joined the
group of law students who met for lunch, eating with them several
times a week. Andi got herself invited home for Thanksgiving by one
dating couple and, in addition, met separately with them and with one
other subject—defining her activity as that of participant-observer in
the development of friendship rather than as observation-from-the-
outside. The project director reviewed the reports as they came in,
made suggestions to each individual about ways of improving and
standardizing his reporting, and began to think about ways of analyzing the observations.

This list of accomplishments may seem meager, and, indeed, we
were dissatisfied with it. However, it appeared that an immense
amount of time was consumed, first, in making arrangements to be
present at events which would permit us to observe the sociable behavior of our subjects and, second, in writing up reports of the observations. In addition, staff members were still working to transcribe the
long interviews previously obtained; numerous staff meetings were
held in which the theoretical concerns of the project were given attention (see Part III); and demands of the academic half time of the
assistants were implacable. Whatever the reasons or excuses, it remains true that Foote and Riesman were evaluating the success of
the enterprise in terms of the written materials; these were inadequate
in quantity and poor in quality.

II. AN ANALYSIS OF THE PERIOD OF CRISIS

Our report so far has focused on the production of work. This work was, in fact, produced, but it is not the central thing which we remember from the year 1955-1956. What we remember is that it was a very bad year. Each of us was experiencing great stress on the project and suffering through this experience in relative isolation. There were signs of tension in the summer and fall of 1955, but things did not really come to a head until November or December. From then until late spring, the project was in a state of crisis. It is probably impossible for any of us to give an accurate and coherent account of what was involved in this crisis, but certainly we have each had a try at it—first, perhaps, in fantasy; then in attempts at sweeping corrective actions; and finally, now, in this effort toward a reasoned appraisal.

Our discussion will be presented in three parts: a report on the physical environment of the project; an analysis of problems arising from the nature of our task; and some comments on staff organization.

The Physical Setting

The Family Study Center, along with various other centers and projects, found space for itself in one of several private houses (neither charming nor notably clean) which the expanding university had converted to academic uses. The Sociability Project was quartered in the Family Study Center and, during the year 1955-1956, was the only research project located there, although there were several projects which were affiliated with the Family Study Center and located elsewhere in the city of Chicago. Foote, as Director of the Family Study Center, had his office there and was present most of the time, engrossed in his own work. Watson was present from nine to six every day, and the assistants drifted in and out according to their class schedules and inclination. The building had an empty, waiting quality; there was no nucleus of people to gather for a coffee break or for lunch, just the waiting desks and typewriters.

The Sociology Department, where Riesman had his office, was located in the Social Science Research Building in another part of the campus. Factionalism characterized both the department and the organization of social research at the university, so that the choice of

Foote as a sponsor and the Family Study Center as a base of operations served to cut off the Sociability Project from other research operations on campus. Political and geographical considerations combined to isolate the staff members of the project from their colleagues. It seems likely that such isolation could have been overcome by more self-confident and less preoccupied individuals than the assistants and the project director; perhaps the fact that the latter was a woman in an academic climate which, like that at most major universities, is largely a stag affair did not help matters.

Isolation is sometimes conducive to creativity, sometimes to strong in-group identifications. In our case, the effects were mostly negative. As emotional tensions increased on the project, the world outside receded from view. The project came to be a "tight little island," locked in internecine warfare and unable to relate difficulties encountered on this project to the experiences of colleagues on other projects. Projective interpretations of persons and events substituted for rational ones, and emotional realities took precedence over the objective task.

Watson found in this situation a strong contrast to the one from which she came, at the Institute of Social Research of the University of Michigan. There, many different projects at varying stages of development are housed in the same building. A project director can always find other project directors to talk to—some who have moved through whatever phase the project is currently in, others who are approaching it, all of them with curiosity and interest and competence in research. Similarly, research assistants can find counterparts from other projects; they can gripe about their jobs and their bosses and the idiocy of their assignments and can return to work with fresh perspective on what they are trying to do. This kind of bureaucratic setting produces relatively matter-of-fact attitudes about the mechanics of research: it is taken for granted that a half-time job means coming to work twenty hours per week; that persons in superior positions have greater authority (though of course their claim to greater competence will be challenged by any self-respecting graduate student); that doing research involves performance of a variety of different interrelated tasks, any one of which may present temporary difficulty, but any one of which can be managed—or, if impossible, revised. None of these things was taken for granted on the Sociability Project. Watson felt as if she had to provide in her own person the full weight of a research organization, geared to production and competence; but she was not able to do this.

It seems probable that Foote, like Watson, perceived the situation as one in which it was important to make manifest the realities of a work-oriented organization. He worked hard himself, made it clear that he expected others to work hard, and tried to ignore the emotional tensions which were preventing others from doing work. As Director of the Family Study Center, he was responsible for the immediate environment of the project: the physical equipment and the affiliated persons and projects composing the Family Study Center. He worked to develop an *esprit de corps* among this larger group, establishing regular Family Study Center staff meetings to bring together a staff which was dispersed throughout the city. His strategy was to leave the internal affairs of the Sociability Project in the hands of the project director and to rely on example and expectations to establish a climate of work-oriented colleagueship. Staff members of the Sociability Project, however, found it unsettling to be defined as the central project of a center devoted to research on the family; they had committed themselves not to research on the family but to a study of sociability and identity, and they felt that their natural colleagues were to be found in the university departments with which they were affiliated (or perhaps in the Center for the Study of Leisure) rather than in the Family Study Center.

It should be evident from what has been said here that the environment of the Sociability Project was relevant primarily because of intrinsic factors within the project itself, such as the nature of its task and the personal styles of its staff; the anxieties of the latter made salient the marginal locale in which the work was done. A larger, going concern, such as, for example, the National Opinion Research Center, could create its own institutional environment and operate at a distance from the university quadrangles with less feeling of isolation. A single project does not have the resilience of a research center. The isolation of the Sociability Project permitted its dominant qualities to flourish and become magnified, protecting the project from both the minimum and maximum standards of more bureaucratic enterprises.[26]

The Task of Observing Sociability

Any attempt to understand the crisis on the Sociability Project must take into account the nature of its task, namely, to collect and analyze observations of sociable interaction. This proved to be a peculiarly difficult assignment. In the discussion which follows, we shall draw on the entire experience of the project, not only that of the first few months, in trying to make explicit the difficulties which are inherent in the task of observing sociability.

Access to Subjects

We have indicated that Foote set out to hire a diversified staff, hoping that this would facilitate access to different realms of sociability. In later years of the project, this technique proved successful: research assistants garnered party reports from special populations to which they had access in New York City, in the South, in North Chicago; Riesman was able to contribute somewhat to the range of data by reporting on sociable events which he encountered on speaking trips to each of the coasts, where he was entertained by "tribes" in which he was something of a stranger, yet where he had entree. Friends and colleagues outside the project added further to our data by reporting occasional parties to which they had access.

In the first winter of 1955-1956, however, we had not yet decided to report on our own sociability. The assignment given the assistants was to recruit individual subjects, first to interview them and later to arrange to observe them in sociable situations. This approach proved painful for all concerned. It established the idea that what we wanted to do was to study individuals, in depth. The observation seemed to be a continuation of the probing inquiry of the interview, and subjects could not introduce the observer into their sociable activities without extreme self-consciousness.

One of the assistants met this problem by establishing herself as a continuing friend rather than an occasional observer; another shifted his attention to a group of persons who did not know of his research interests. A third arranged with her subjects to observe them while they were playing games (bridge, Scrabble) rather than intruding into unstructured sociable situations. Thus, all of them found that the basic assumption underlying the research proposal was untenable: we could not recruit individual subjects to respond *both* to interviews and tests *and* to observation.[27]

A logical next step was to try to recruit subjects for purposes of observation only. Al's luncheon group was an example of this technique, except that instead of recruitment he simply "joined" the group. A different version of this technique was to stage a party for persons whom one knew only peripherally. We tried this several times, once with the SSA students at the party already reported and once when the wife of one of the assistants issued invitations to a group of people whom she knew slightly from her place of work. In every case, the attempt to convert a peripheral relationship into one permitting observation of sociability was only partially satisfactory. It became clear that the observer had to do more than obtain access to his subjects; he had to be a natural part of their social environment, part of the same social fabric. Otherwise, he missed much of the submerged part of the sociable interchange; he was ignorant of their particular language of allusion, he engineered the wrong kind of event, or he evoked a pattern of behavior appropriate for strangers but not for "friends."

In short, December of 1955 found us under orders to bring back the "inside dope," remaining outsiders looking in. We could not do it! Later, in the summer of 1956, we managed a more proximate solution: we arranged for two persons to go to an adult camp and join a staff consisting entirely of college students. This solved the problem of access but dealt less successfully with the requirement that sociable partners, to be a good match for each other, must partake of social worlds that are not too disparate.

Eventually, we decided that the only solution was to draw on our own sociable contacts, extending the range of our sociable activity, soliciting and accepting invitations from persons whom we knew only slightly, using brokers to introduce us to parties where we might not otherwise be asked, and in general using whatever resources were available to us as individuals for maximizing the range of our sociable experiences.

During the first year, responsibility for making contacts with subjects rested with the individual research assistants. This seemed to be the most natural means of procedure, since sociability itself is an activity which is guided by individual preference and initiative rather than institutional roles or arrangements. It did not seem appropriate to utilize the procedure often followed in studies of formal organizations, in which senior staff members on the research project or others of high visible status help with initial contacts, working through "gate keepers" in the institutions of interest to pave the way for research activities. However, we found that our policy placed too heavy a load on the junior staff. The young and inexperienced graduate students came to feel that the difficulties they experienced in gaining access to subjects showed incompetence. Accordingly, our next step was to arrange for entree into an institutional setting: the summer camp. At that point, Foote helped to set up the initial contacts and make the necessary arrangements. In the beginning, however, he delegated such problems of "methodology" to Watson. This was a time when she was still trying to design a project, and she did not assume any responsibility for helping the assistants to open up contacts with subjects.

Riesman's way of helping with the difficulties of observation was to propose an alternative approach to the research. He raised the question of whether sociable interaction might be best studied not only by observation in natural settings but also by seeking out people who could be informants about it because their professional work involved sociability as a dominant theme. It was he who had encouraged Potter to make use of his experience as a pilot to establish contact with airline hostesses or stewardesses; and with Reuel Denney, he raised the question as to what other occupations might be accessible to this sort of inquiry. Potter, however, approached the hostesses as subjects rather than informants. In any event, Foote and Watson were unwilling to shift the nature of the inquiry to one which would depend on informants; the questions which interested them could be better studied through direct observation and in the case of Watson, at least, the suggestion of using informants was resisted because she felt unprepared to direct that kind of research. Since Riesman and Denney were also interested in natural settings, the possibility of using informants was dropped.

The Study of Sociability: Legitimate or Illegitimate?

Again and again, we were reminded that sociability is generally viewed as a private affair, with observation and study not welcome. Colleagues among faculty and students at the University of Chicago had been antagonized and alerted to the "bad faith" of the Sociability Project by an exercise in observing sociability which had been carried out by members of an advanced class a year before the project was organized. At the workshop, staff members were challenged as to the legitimacy of their enterprise, and in subsequent contacts with subjects, we encountered both passive resistance and outright opposition. As Potter put it:

> Concerning research methods and ethics, my attitude was that one just does not know subjects. The interviewing I had done (for the first

time) immediately prior to the project confirmed this view. I simply did not know the persons whom I interviewed in the field-methods course. . . . The greatest surprise of all, for me, came when we began actually seeing subjects and beginning to know them. I could have understood unwillingness and inability to talk about or permit observation of sexual behavior on the part of subjects. This would have fitted well into my own definition of privacy, which is based, I suppose, on a Puritan ethic. But to discover that people—particularly my "science-is-godly-pursuit-of-truth" colleagues in sociology *and on the project*—could not and would not talk ad infinitum about their sociable experience seemed impossible. . . . Why should doing what one enjoys, passing the time of day, acting human in the face of one's best friends, be a matter about which defenses against exposure are so salient?

Why, indeed? Answers to this question are surprisingly projective, reflecting perhaps the defenses which an individual uses to maintain his own self-esteem and sense of identity or perhaps the threats which he is defending against. The very tension which surrounds the question of whether the study of sociability is legitimate testifies to its importance, at least for students of the socially based self. But such speculation was of little help to us. Our immediate need was to be sure that what we were doing was legitimate. This meant building in proper safeguards to protect our subjects and ourselves, and it meant referring our procedures to the models of ethical behavior sanctioned by our professional colleagues.

But one must ask: which profession? We discovered later that psychologists are less worried about intruding into private affairs than are sociologists, provided that there is scientific reason for doing so. Psychologists are, in general, quite prepared to deceive naïve subjects, but their ethical code requires not only that there be a strong reason for doing so but that the deception be liquidated by informing the subjects eventually and explaining the whole matter to them. Sociologists, on the other hand, tend to stress the importance of initial honesty in approaching all segments of the society or institution which one wishes to study and of a commitment to disguise their subjects in any published report so that no outsider can recognize them and no insider can gather ammunition for malicious attack.

One of our difficulties was that it made no sense to try explaining the project to people whom we saw only briefly, as in the observation of a single party. If one is working intensively with a few subjects or for a period of time with a panel of subjects, one can take the time necessary to persuade them of the value of the project and win their cooperation. But at a party or other brief event, any attempt to explain the project or request permission to observe can only antagonize. It makes the subjects uneasy, without giving them enough information to judge whether or not they will be hurt by observation. In fact, the anxieties stirred up by discussing our desire to observe, no matter how impersonally, did hurt and were, therefore, undesirable. Nor was it satisfactory to seek to secure the agreement of the host or hostess to our being observers at a party to which we were invited in the normal course of social life: to do so simply transferred the ethical dilemma to them, without ending our own responsibility.

Eventually we concluded that the doctrine of "answerability" to our subjects could best be handled by a dialogue in which we as researchers did not ask permission of our subjects to observe them but did in effect ask such permission internally. We would put ourselves in the role of subjects and take note of their concerns and anxieties and what they would or would not like to have us do. We would do only what we could reasonably expect them to agree to if they had full information and if we had unlimited time and occasion to explain what use we would make of the observations.

ETHICAL
BASIS

We found that each researcher tended to attribute to his subjects the kinds of anxieties and defenses which would be appropriate for himself. Because these were different for each one of us, we each felt that at least some of the other researchers were being "unethical" in their dealings with subjects. Riesman, writing later, declared:

> In the desire of at least one of the assistants to define the study of sociability in terms of observant participation in the processes of friendship development, I felt that there was a danger of introducing a touch of voyeurism of the sort that draws people both into social research and into psychotherapy; perhaps I was especially sensitive since I was aware of such dangers in my own work while at the same time recognizing that motives are almost always mixed and seldom pure or uncomplicated. At the same time, I felt that ethical scruples about "observing from the outside" could amount to a request for a hunting license for one's own "sentimental education" without any responsibility to the project.

Confronting such problems, Joan wrote in a memorandum of December 8, 1955, as follows:

Those of us observing sociability have been somewhat concerned from the start about our role as observers. We have felt guilty about watching other people. We have felt uncomfortable when our subjects have expressed concern over our observing them.

I would like to emphasize the nature of the group structure which has been pointed out [in preceding pages of the memo] as one of the factors which cause tension for both observers and observees. The suggestion was made that unless there is some way of externalizing and depersonalizing interaction—by a game, by conventional conversation, and so forth—there is real concern on the part of both parties. . . . To go one step further, I would suggest that where there is no structure or very little—where there are very few depersonalizing agents, such as in the intimate friends groups which Andi is observing—no observation can be tolerated. Instead, as with Andi, one becomes part of the group and reports on it through introspection.

What this says is that there are some kinds of sociability which are not readily observable . . . those kinds where there is a very intimate relationship between subjects. By intimate perhaps I mean that each person is revealing something to the other about himself in a situation in which there are no depersonalizing agents or structures. . . . Of course for this reason we don't have very many intimate relations. We don't take such chances very often. And certainly we would be uncomfortable about anyone observing such a relationship.

Potter wrote:

The fear of exposure, in ourselves and our subjects, can be interpreted as an other-directed fear, one which would not be present if the inner standard were clear and firm. Subjects were not content with what they chose as their enjoyment; their defensiveness indicates lack of confidence in their own judgment, to the extent that they would renounce it if exposed to the standards of the observer, the project, or the public. This fear of being subjected, in sociability, to an explicit standard constructed by the project is proof of either the absence of internalized standards for sociability or lack of confidence in them. Indeed, this is a main finding of the project.

Andi resolved the problem by giving up the role of objective researcher and becoming a friend and confidante of her subjects. She writes:

At that time, it was an issue for me that everyone but me felt that people would be uncooperative or actually found it difficult to get cooperation. From my view at that time, there was very little genuine interest in the people who were approached as subjects, and they naturally reacted to being "observed like flies."

Several years later, when reports of our project were in preparation, Riesman and Watson found themselves differing marginally but importantly concerning the degree to which it was legitimate to draw upon party protocols in published reports. Watson wrote:

> I was not able to start submitting good party reports until we had a code to work with. The code constituted an assurance that our analysis would be focused on certain aspects of behavior; that we had a workable technique for separating "acts" from "persons"; that our analysis would look for systematic regularities in a large population of episodes rather than "spying" on each party, per se. . . . I think I would state the problem as follows. It generally appears that the role of friend-and-participant is incompatible with the role of observer-and-objective-analyst. At parties, we appeared to be friends-and-participants. The ethical agreement with ourselves was that, after the parties, we would not do anything incompatible with that role; scientific evaluation and appraisal, which (if unsolicited) is incompatible with the norms of friendship, would focus on behavior and not on persons or parties.

Riesman replied:

> I, too, took for granted that we were never engaged in studying the sociability of individuals, as such. But where necessary to give clarity and reality to our understanding of the choreography or phasing of parties, I wanted to make use of illustrations summarizing or collapsed from actual events. These would be disguised as carefully as possible and permission, if necessary, would be sought from anyone who might legitimately or anxiously fear recognition. I would never deny that publication, even when strictly limited in this way, might do harm in the terms which troubled all of us, and especially you, namely, the feeling of people that they had been exploited even if no one else could possibly recognize them (and even if, due to disguise, they thought we were referring to them when, in fact, we were not). Nor would I stress too much what many observers have noticed, namely, that many people when confronted with their own behavior in disguised form do not recognize themselves.
>
> The issue I raised was whether the possibility of doing harm should not be weighed against the value of reporting the (disguised) data to give a vivid, realistic, and human sense of what went on and what we are talking about. At the same time, we must guard against the temptation to show how brilliant and bright we can be at the expense of our subjects. However, we do not solve all of our problems by simply reporting tables and interpretations of quantitative findings. There is then the danger of giving our findings, though based on fugitive and

fragmentary data, perhaps excessive weight. By drawing more freely on our actual observations, we may be able to say both more and less than we can say on the basis of extrapolation from our tables. In full descriptive reports it would seem possible to say something about the ways in which people often frustrate themselves by efforts to have a good time in sociability.

Watson rejoined:

> I think the kind of analysis which you wish to do precludes disguise, at least insofar as participants themselves are concerned. If we are to learn from detailed examination of actual events, it is necessary to describe behavior so explicitly that the event must be recognizable to participants, even if their names and the names of locales and professions are changed. We may decide that we wish to report our data in this way, but, in that case, I don't think we should make any claim to more than nominal disguise.

In the preceding quotations, we have tried to illustrate the way in which each person tried to build into the project his own kind of protection for subjects and to be troubled by the kind of protection—or lack of it—which was offered by others on the project. These matters were always viewed as questions of ethics. Each person believed that the kind of safeguard which he advocated was both necessary and sufficient as a condition for ethical relations with subjects, whereas the "evasions" advocated by others would violate ethical standards or would unnecessarily restrict the scope of the project.

It seems to us now that more was involved in these differences than disagreements about abstract ethical principle. What was involved, for each individual, was his own personal code of behavior: the set of axioms, ego defenses, and imperatives for behavior which shaped his relations with others and his own self-image. On these he could not afford to yield. Neither could he afford to tolerate violations of his code by colleagues on the project. Anxiety was high; defensiveness was created by the many criticisms of the Sociability Project from colleagues both inside and outside the project, as well as from subjects. Each person felt that he had been placed in an untenable position by the actions of others on the project, and the resulting tensions and antagonisms within the project staff were disruptive. Eventually, alliances developed and factions became established. These reflected, not professional training and associated patterns of ethical ideology, but, rather, matching in terms of personality: persons aligned themselves with others who could make them feel less anxious and avoided persons who made them feel more anxious. In effect, we are suggesting that ethical dispute was used to camouflage and channel individual anxieties.

And yet this is only part of the story. The study of sociability poses problems with respect to relations with subjects which, although not unique to the study of sociability, are more forcefully presented there than elsewhere. One set of difficulties arises from the fact that the rules of behavior which regulate sociable behavior are largely covert and implicit; there is no standard code of behavior to which disagreements about appropriate procedures for interaction may be referred. In this, sociable situations differ from the institutions and organizations which comprise other segments of our culture, such as the institutions of work, family, therapy, and so forth, and also from the experimental situations created in laboratory studies of behavior. The question of how researchers will relate to subjects must always be answered within the terms of the institutional framework in which they meet; and when they meet in a sociable situation, where interaction is largely governed by the codes of behavior which the individual brings with him to the interaction (codes derived from both personal experience and group memberships), then it follows that researchers, like their subjects, will bring to the situation diverse and individualized expectations about what constitutes proper behavior.

A further difficulty is that sociable interaction takes the form of conversation among friends. Regardless of whether the interaction has been deliberately staged by an experimenter or occurs in a natural setting, the researcher finds himself responding to his "subjects" as friends and allies. When, in addition, as on our project, he is separated from his colleagues by barriers of mistrust and anxiety, he finds himself in the peculiar situation of being identified *with* his subjects and *against* his colleagues. The "ethical" problem is not so much that subjects refuse to trust the particular experimenter but that the researcher himself does not want to "betray" his friends into his colleagues' hands.

For all these reasons, the business of building legitimate relations with subjects on the Sociability Project was inextricably intertwined with the business of building good relations with fellow staff members. Difficulty encountered in one area aggravated problems in the other, and every attempt to "talk things out" only made things worse by emphasizing the extent of individual differences among staff members and the rigidity with which positions were held.

Discussions on the Sociability Project made little progress toward a statement of an ethical code to govern relations with subjects, largely, as we have indicated, because discussions about ethics were made the vehicle for covert conversations about matters of even more immediate emotional impact. Each individual was left to work out his own personal version of an ethical code. In this chronicle, however, it may be pertinent to try to examine objectively the question of what possible damage to individuals or violation of ethical principle might be involved in the observation of sociability. We shall not consider the possibility of damage resulting from malicious or salacious intent but shall assume that the research is motivated by genuine scientific curiosity, combined with good will toward subjects.

Perhaps it is too easy to stipulate that the motives are all honorable; after all, how can one be *sure* that voyeurism or hostility or masochism is not among one's motives? He cannot; but so long as such "illegitimate" motives are kept under control and do not govern the conduct of the research, they need not be damaging.

The primary ethical consideration is that harm shall not come to subjects as a result of one's activities. Perhaps there is duplicity in the role of a secret observer of sociability; he presents himself as a friend or potential friend, and yet he has another objective, namely, the gathering of data. But is the existence of hidden objectives harmful or unethical? Many people come to social gatherings with more than one objective; they do not announce everything they have in mind, and, in fact, we suspect that there are many hidden objectives in sociable situations which are far less defensible than ours. The important thing is not the presence or absence of these other objectives but the question of whether the manifest role—that of friend and partner in sociability—is misrepresented. It is exploitation if the guise of friendship is assumed for an unfriendly purpose; but if the friendly interaction is entered into in good faith, with other objectives being *in addition to* and not *instead of*, we see no serious violation of ethical standards.

One may, perhaps, wonder if he is giving "full value" as a guest; perhaps he is so busy memorizing or watching that he cannot carry his share of the action. Some of us who are inclined anyway to feel socially inept may be particularly vulnerable to this form of guilt. But, on balance, this does not seem to have been a significant problem for our observers; often they were carrying a disproportionately large share of the interaction or performance. Even if it were to be a problem, it can hardly be considered a cause for ethical protest—any more, at least, than is the case with a person who comes to a party sick or tired or otherwise preoccupied.

But perhaps the guilt is not so much about one's psychological "absence" as about an active seeking after data. Thus, the observer may feel pressure to intrude himself into situations which he would not normally enter; to join groups of strangers and to watch persons who believe that, if they are being observed, it is out of interest in them and not for purposes of research. And yet such pressure exists primarily at large and impersonal parties—for example, at a huge wedding or similar reception—where people are displaying themselves to others and facilitating, if not actually inviting, observation. At small parties of friends, if one conducts oneself as a friend and does not permit the concerns of the research to intrude during the party, there is no occasion for this kind of effortful observing.

Perhaps the greatest problem is associated with the fact that reports of parties are written down as nearly verbatim as possible and made available to others. In the process of writing down, observers altered names and censored out other tags that would identify individuals. Even so, we treated our party reports as confidential, locking files even when going out to lunch and restricting the staff to persons who we thought could analyze data simply for what it had to say about sociability and not about particular participants. Yet the decontamination of the data could not be entirely complete, for we knew the identity of the observer and thus gained vicarious access to his sociability and that of his circles. Individuals often behave one way with friends and intimates and another way with strangers. It may be considered a violation of privacy if their "private" behavior is made known to outsiders, even if the latter do not in fact know who they are. This is all the more a problem if the "outsider" is not a distant and impersonal stranger.[28]

As with psychiatrists, though with less detail and far more concealment, we did become privy in a very limited sense to a kind of private information which would not have come our way except for the study. We trust that this amount of exposure is not resented by our (voluntary and involuntary) subjects and that what we gain from the study is worth the slight risk involved. We say this despite our conviction that social-science research can be, and often is, purchased at too high a price, whether in damage to subjects or to the consciences and sincerities of researchers.

Exposure of Self

Exposure to strangers with whom one probably will never have any personal contact is, of course, much less threatening than exposure to moderately close associates. Before any of us could write unconcealed reports of the sociability of ourselves—not to mention our husbands, wives, and dating partners—we had to be able to *like* and *trust* our colleagues who would read the reports. During the first year, Foote and Riesman were seen as distant and critical. As Andi writes in her retrospective memorandum:

> During the previous meetings and discussions about the observations we brought in, Riesman's interest in sociability as an art product—as a place where stimulating people meet to communicate in a creative way—became clear. I thought he experienced us as dull and uninteresting.

Furthermore, the assistants were experiencing growing friction and mistrust among themselves. They could not confide in one another nor expose themselves to one another. During this initial period, we concluded that it would be impossible for the project to consist of studies of our own sociability; it was essential that the observations be about other persons with whom we were not identified. As already indicated, however, this "solution" had its own difficulties. Acceptance of the technique of observing our own sociability came later, when a new core of congenial persons—Riesman, Watson, and Potter—were able to rebuild the project, eliminating from it those persons who did not "fit" and adding to a few persons who did.

Working at Play

It was not always fun. Many of our friends joshed us about our soft jobs and eagerly volunteered to go on the payroll and attend some parties for us. Unfortunately, we succeeded in making our play into work, rather than our work into play. Perhaps this indicates excessive earnestness on our part.[29] But it is true that the observer at a party has extra work to do. At a time and place when he would ordinarily expect to relax and have fun, he must be keyed up to maintain a continuous process of memorization, a work activity which is partly conscious, partly unconscious. In order to avoid "leaking" the information that he is observing, he must often parry and evade questions about his work —which, among young professionals, means cutting off many of the first overtures which he receives from strangers. When he attends a party as a stranger, he may be under special demand to perform, to entertain others, and to introduce himself. (This requirement is reduced for the man or woman who attends a party as a date or spouse; his "connection" is sufficient to introduce and explain him.) If the other guests, by accident or design, learn of his activities as an observer of sociability, he may be in for a long argument. Particularly among the graduate students there was no hesitation about attacking the project and its representatives.[30] One of our more likable faculty colleagues indicated that he was, in effect, blacklisting us socially because he did not want to be observed.

There was one other unanticipated consequence of the axiom that doing work is incompatible with having fun at a party. We noticed early in the party study that observers had a tendency to report "bad parties": when they were having fun they forgot about observing, and when they were bored they took refuge in going to work. As soon as this tendency was recognized, we tried to correct it, renewing our original objective of reporting what is happening when people are having fun.[31]

The Organization—and Disorganization—of Work

We have described some of the difficulties associated with our assignment, the observation of sociability, but there was another side to the picture: the organization of the work itself. This, too, did much to cause trouble.

The Cult of Individualism

The Sociability Project existed within a milieu which stressed the importance of creative achievement by individuals and disparaged the value of what could be achieved by competent mastery of conventional "rules of the game." Students were enjoined to be imaginative and brilliant, to achieve on their own a kind of magical breakthrough which would astonish their teachers. This demand was heightened for those students who became assistants on the Sociability Project by the way in which their assignment was presented: each was to get a thesis; the project was to consist of one core study with related but detachable parts; the assistants were to help design the research and simultaneously work out their thesis proposals. Moreover, the three top staff members, directly or at least by implication, took the position that no study like this had ever been done before; hence, there were no models to follow, and we were all more-or-less equal partners in an exploration of the unknown.

In the climate of Chicago, what was intended, at least in part, as an invitation to a noncompetitive colleagueship turned out to maximize both ego involvement and individual defensiveness. When our research encountered failure or difficulty, the responses had the character of guilt and blame rather than rational problem-solving. When an individual did have the glimmerings of a new idea, he was highly possessive and protective; if it amounted to anything, it was to be "his baby," and he did not want it stolen from him or made over by others —most especially not by the project director. Hence, the project director was to be excluded from all preliminary (formative) discussions of research ideas originating with the assistants. It was perhaps a corollary that ideas originating with her could not be given thoughtful consideration by the assistants; they could not afford to be dependent, that is, to accept leadership from her. (Perhaps, especially, they could not afford to be dependent on a person who was young, female, and more inclined to listen than to command.) On more heavily structured projects, graduate students often complained of being given routinized tasks, not worthy of their intelligence; yet the excessive freedom of the Sociability Project, which exposed both their need for and their defenses against authority, took perhaps an even greater toll in morale.

Mechanisms for Learning from Experience

It has been evident in our report that the project suffered from an inability to convert negative experience into positive learnings. It would have helped if the early months had been defined more explicitly as exploratory and experimental, with adequate time set aside for evaluation of experience and with plenty of leeway for revision of design.

Denney apparently was successful in establishing a climate of open-minded exploration. Andi writes:

. . . that summer introduced me to the great complexity of the problem we were proposing to study. It was not particularly anxiety-producing at that point because Denney was the only guide and he seemed to convey an attitude that it was a matter of work, trial and error, and thoughtfulness but that it was do-able and, in fact, interesting.

Riesman suggests that Denney's success came from the fact that

[his] whole tactic with people is to be as unthreatening and even defenseless as possible. . . . He assumes everybody else must know ten times as much as he himself knows, when, in fact, he's astonishingly erudite. The combination of brilliant, dramatistic understanding of sociable episodes and modesty about what he could say must have been at once stimulating and reassuring.

But Denney was pinch-hitting, holding the fort. When Watson came on the job, she was under pressure to create a research design, exert leadership, and put people to work—immediately if not sooner. She took it as her major task to produce her own statement of what the project should be, scanning the mass of comment and incoming data for whatever help they could offer. She did the evaluation herself, trying to pass on to the assistants whatever comments she had that might be helpful but trying not to involve them in her own processes of thought and evaluation except as they could be communicated through memoranda. She was more interested in a clear statement of where the project was going than in detailed evaluation of current (fairly unsatisfactory) activities.

As the project developed, her habit of summarizing staff discussions and emergent "bright ideas" in the form of memos to the staff became a mechanism of major significance for the growth of the project. Staff discussions which, in themselves, were chaotic and inconclusive provided the material for memoranda which were clear and evocative. Each memo served as a building block in an emerging structure of ideas, summarizing what had gone before, laying the groundwork for what was to come after. The memos served as visible, tangible products of discussions which often seemed vague and tenuous; they provided an objective record of progress and a continuing stimulus for further inquiry.

On the other hand, these memos were essentially one-way communication. This was a definite advantage at first; by talking to her typewriter instead of to persons, she could ignore the unwanted feedback of resistance, resentment, and mistrust. Later, the habit of presenting her own ideas by way of a memo rather than a conference or discussion may have unduly restricted criticism and evaluation. This was partly intentional. Riesman's habit of focusing attention on possible new approaches to the study of sociability and his continuing doubts about whether the coding already completed was adequate led Watson to feel that it was necessary to maintain tight control of the project lest it lose all forward momentum and once again become stalled on dead center.

Staff Organization—the Assistants

The early hiring of research assistants led to a number of difficulties, many of which have already been indicated. The assistants were, first, assimilated to the wrong project—the project on the Study of Interpersonal Competence, which preceded the Sociability Project at the Family Study Center. In a system whereby funds and personnel are budgeted in relatively inflexible time periods which do not correspond perfectly with the needs of research projects, some overlap from one project to another is almost inevitable. In this case, however, the assimilation to the Competence Project led to difficulties later. Andi reports a positive experience which led her to expect in the Sociability Project a continuation of the emphasis on interpersonal competence and on Foote's ideas about identity; this "really got in my way later when the Sociability Project finally developed a separateness from Foote's view of it."

Bob reports a negative experience:

> The Competence Project had been without result or conclusion. . . . I met A and B, immediately finding in A an excellent thinker, but he, like everyone else, when asked to describe his work talked of it as a failure. I felt here was A, who had done one of the best master's theses I knew about, spending three or four years of his life trying to define an impossible concept, "interpersonal creativity." These experiences led me to reject anything associated with the goals, methods, or execution of the Competence Project. I was haunted by the fear that our project would have the same outcome.

Later, when emphasis shifted to the Sociability Project, there was ambiguity about the purpose of the early assignments given to the assistants. In part, this resulted from differences among top staff members with respect to what they wanted done; in part, it arose from inadequate explanation by top staff members to the assistants. At the workshop, the assistants were expected to gain practice in observation of sociability and also to learn something about the operation of training groups, for it was anticipated by Foote, at least, that the assistants would eventually be running their own groups. This was never made clear to the assistants, and they were dismayed to find that participation in the workshop activities absorbed all their time and energy, leaving them virtually no opportunity to pursue what they thought was their primary objective, namely, the observation of sociability. In the fall, they felt that they were rushed without adequate preparation into the field to get interviews; then, when the interviews were done, no one was interested in reading them. They were urged to make contact with subjects, but when contact had been made, they found that they still had no entree into sociable activities. Assignments which were viewed by top staff members as a means of "learning how to do it" were perceived by assistants as unreasonable demands for production of data. Eventually, these early events came to be seen as evidence of betrayal by the top staff, and a barrier of mistrust was established between the assistants and their supervisors. In addition, as we have seen, the assistants were divided from one another by attitudes of defensiveness and mutual suspicion.

The top staff members were unhappy about the growing antagonism from the assistants, but they had different explanations for it and different remedies. Watson felt that the basic fault was that assistants had been hired when there was no work ready for them; they wanted to do a creditable job which would please their supervisors, and they were made anxious when, instead of being given work to do, they were given in-service training and an invitation to help figure out a plan for the project.

As indicated above, Riesman felt from the outset that at least one and perhaps two of the graduate students should not have been hired, since they were either trained insufficiently to be able to initiate worthwhile research or, for whatever reason, too aggressively uninterested in the project to help its recovery. He could see also that Watson's energies were being distracted, and Potter's as well, by the cumulative mood of anxiety. (Potter, from the very start, had been a hit with all three of the senior staff, despite his own feelings of weakness and inadequacy.) Riesman, out of an excessive dependency, also hoped that by bringing in another gifted and creative person the vicious circle could be broken; that people, rather than sitting around griping about lack of leadership, would become more enterprising in finding tasks that might turn out to lead somewhere.

Foote, more tough-minded, took the view that the assistants should be judged by their productivity: persons not able to produce should not stay on the payroll. All these considerations led to the suggestion that perhaps the assistants were more bother than they were worth, and if so they should be fired to give Watson a clear field; or perhaps two or three of the less productive persons should be fired. Watson, however, felt it was unfair to fire graduate students in November, when other jobs were no longer available, especially since the basic fault was in premature hiring, for which they could not be held accountable.

Eventually a compromise was achieved whereby the two less advanced students were informed that their contracts would terminate in June and that, until then, they should work individually to gather data for M.A. theses related to sociability. The two more advanced students were to work with Watson in the formulation of a new project, and when the project design was clear each was to use some aspect of it for his Ph.D. thesis. In the next few months, it developed that these two more advanced students could not work together, and, for this and other reasons, one of them left the project in the spring.

Staff Organization—Senior Staff

There was a peculiar division of responsibility and experience among Foote, Riesman, and Watson which did not serve the project well. Foote was the one who was ready to do research on sociability; he wrote the proposal, hired the staff (with Riesman's more-or-less reluctant consent), and believed that the project was ready to roll. But when Watson came on the job in September, he turned over to her the responsibility for conduct of the project[32] and concentrated his attention on other things. He continued to be interested in the project, taking part in top-level discussions of goals and strategy, remaining open to requests for aid or consultation, and pressing steadily for greater production of work. But he did not participate in the day-to-day efforts to find a way of getting the job done.

Unfortunately, readiness to do research on a particular problem is something which cannot be easily transferred; it must be individually achieved by each research director. In the fall of 1955, Watson was not ready to "go out and collect data," as Foote wished to do; but by the summer of 1956 she was willing to go with Andi to the summer camp with no other objective than to "collect data." The intervening months had brought her to a point of readiness, a maturity of thought about what could be learned from observing sociability, which she did not have when the project began. Watson writes:

> I came to the project believing that my job on the Sociability Project was to implement the ideas of Foote and Riesman. At first I was relatively unfamiliar with these ideas; I did not feel able to assume responsibility for research based on them, and I expected Foote and Riesman to make the final decisions about the shape of our investigation. As the project progressed, I began to assume more responsibility for making choices and setting directions. It soon became apparent that there was no agreement between Foote and Riesman as to what should be done on the project, and Denney, when he joined our discussions, had still other ideas about what we might do. Each of the assistants had been asked to work out a thesis proposal, and in trying to meet this assignment they came up with still different ways of approaching the study of sociability. I felt as if I were being asked to move in seven different directions at once. The last thing in the world which such a project needed was for me, as project director, to come up with still another suggestion as to what we should do. Instead, I concentrated on trying to work out a research plan which would enable each of the various staff members to gain at least some of the information which he wanted. But before I could do this, I had to have time to study and internalize the different views which were being expressed, particularly those of Foote and Riesman, and I had to have time to work out trial solutions to the problems of design and methodology. Later, the forward movement of the project itself created inflexibilities. Each new commitment to sources of data and means of analysis involved the elimination of alternative and competing choices, and I was in the position of trying to keep the project "on course," that is, of trying to maintain cumulative consistency within the project.

As already indicated, Riesman was away from Chicago during most of the summer and fall quarters, visiting colleges for the Academic Freedom Study and working in New York at the Bureau of Applied Social Research. Yet he was close enough to the project to realize that his anxieties about its development were more than justified. He sought for ways to create a climate of aliveness and high morale on the project. When away from Chicago, he wrote memoranda on potentially relevant reading (including such books as Harold Nicolson's *Good Behaviour*[33] or discussions of parties in novels). He sent in a flow of suggestions which were intended to support any indications of positive action or flight into work. But the effect of some of these early comments was not encouraging. Their whole style indicated so much divergence from what others took to be the basic assumptions of the project that reconciliation seemed virtually impossible. Moreover, Riesman's comments were taken as more binding or critical than intended; for example, references to novels or to essayistic discussions of sociability were received by some staff members as an implicit critique of the heavy-handedness of social-science definitions and procedures.

As it worked out, then, there was no one during the first year and a half of project operation who took final responsibility for making decisions and setting policy. Foote started the project off and then withdrew.[34] Watson felt that she had been hired to implement but not to set policy. Riesman had come to the project as a collaborator rather than a leader; when compelled to lead, he acted as adviser, critic, and idea man but not as research director.

Subsequently, as Watson and Riesman came to assume joint responsibility for the project, new problems emerged. The primary requirement was that the project develop in such a fashion that each could contribute to it in his own way, utilizing his own strengths and being guided by his own curiosities. Riesman writes:

> My approach to social science has never been monolithic, either as a teacher or as a researcher. I believe that researchers should be encouraged to pursue their own temperaments and personal styles in research, to feel free from any going model, including my own. (I have said as much in an essay, "Some Observations on Social Science Research," reprinted in *Individualism Reconsidered*.)[35] I feel that students and scholars often underestimate the range of their own resourcefulness and thus do not try out new styles of work. But I believe we do not know enough to say which styles will be productive for which sorts of people and that there should be encouragement of a great diversity of approaches.

Yet there is a difficulty associated with this philosophy. As Watson puts it:

> I agree in principle with what Riesman says, but in practice it is important to have sustained support for the single one or two method-

ologies which are being utilized by a particular project. It requires time, energy, and continuing commitment to carry through any single inquiry to the point where it yields new and reliable information. Diversity as a model for research is more appropriate for a Center, or an Institute, than for a single project.

But in any case, the research styles of Riesman and Watson were so different from each other that a pluralistic approach to the study of sociability was the only possible course for the Sociability Project. Though Riesman was familiar with the methods of tabulation and presentation of data used by Paul F. Lazarsfeld and his coworkers at the Bureau of Applied Social Research, his own forte lay in the more anthropological methods of observation and description. His approach to the study of parties was to ask what a whole party was like: what were its major, often concurrent themes; what were its blockages, frustrations, liquidities, accomplishments? He hoped to use descriptive reports of parties and fragments of sociability to say something about the manner in which—as suggested in published papers—contemporary codes of equality and sincerity can be self-defeating, leading party-givers and party-goers to blame themselves for failures that would seem to be, in large part, structural. He writes:

> Only full description could show how people stymie themselves in sociability by succumbing to a monopolist or a petty tyrant (just as often happens in other aspects of life). And I thought one might be able to show by detailed description some of the sorts of intervention that could turn the tide, shutting off a person who has become domineering and introducing new persons or allowing new resources. In fact, such concerns bring us back by another route to Foote's original hopes that we would be in a position to say something which could be therapeutic for individuals' understanding of themselves and of their situation, although I think it more feasible to say something about incompetence than about competence.

Watson, influenced by Foote's early emphasis on episodic analysis, preferred to view party interaction as a collection of episodes, each of which could be coded separately, independently, and systematically. Her idea was that the episodes from a number of different parties would be pooled in analysis, and attention would be focused on regularities within the episode. She writes:

> I hoped that the coded data about episode characteristics would include descriptive information about the topics and points of view ex-

pressed in different subcultures, as well as information about a number of different factors which might be expected to influence sociable conversation, for example sex, age, status, type of acquaintance, phase position within the party, group size, and so forth. In addition, I wanted to try out some of the concepts which we had developed for describing sociable information to see how well our categories corresponded with observed variations in behavior and to see what relationship they had with other characteristics of interaction.

Riesman did not believe that the analysis of discrete episodes was the most significant approach to the study of parties. Yet he found that the complexity even of a brief small party made it difficult, if not impossible, to capture and reflect the innermost thoughts of participants, not all of whom are having the same experiences, let alone the same thoughts. Riesman and Potter sought for ways in which the *Gestalt* of a party could be grasped in its own terms, less bound by the unitizing of episodes or a priori code categories. But their efforts, working from party protocols, to set down major themes without coding came up against many of the same problems that had plagued work with small groups where scoring or coding devices have not been used.[36] Hence, while wishing in vain to find ways in which to use more anthropological methods of handling parties, Riesman encouraged work of others with the code and did coding himself under Watson's direction. In addition, he was eager to support the capacities and personal qualities he had come to admire in Watson; he respected her ideas and he identified with her in her perplexities. Riesman writes:

> I cannot emphasize enough how willing and eager I was to go along with any plan that appeared to have a chance of working out and discovering something. Thus, had Watson wanted to do some sort of experimental work, using laboratory groups, I would have been quite content and would have done my best to make a contribution in an unfamiliar area. I wrote up party reports whenever opportunity offered (even if it meant attending parties that I would have skipped otherwise) not only for the sake of the data but also for the sake of supporting the activities of the project staff. I enjoyed coding even when I suffered from misgivings as to what the upshot of all the painstaking coding might be. I did hope to supplement the project by finding another research assistant who could bring it the sort of imagination that Bob Potter had shown; what was lacking in our group was a "critical mass."

The research program which finally emerged was partly a compromise and partly an invitation to each to go his own way. The project

came to focus on party interaction rather than on individual participants, thereby by-passing the concern with identity which had seemed
so fruitless to Riesman; the major investment of time and money was
given to the coding of discrete episodes under Watson's direction. As
the coding progressed, Riesman could see that the code offered a way
of storing an enormous amount of information, but he was an agnostic
as to how observations which were stored in this way could be har-
nessed to the more subtle and fugitive judgments which most interested
him. He continued to work in his own way on the collection and evaluation of reports of party interaction. He gave support to the program of
systematic coding and analysis, but he was not content to have it be
the full measure of the Sociability Project.

Personal Crises

Given a difficult research problem and a staff in which each individual was struggling by himself to find a technique for productive work,
it was perhaps not surprising that individual efforts to resolve the work
problems took on the dimensions of severe personal crises. Just as each
person was guided by his own system of anxieties and defenses in seeking ways of relating successfully to subjects, so, too, each was guided
by the same set of anxieties and defenses in seeking a way to get the
work going (or to justify not doing so). Each, in turn, met with failure
or rejection.

In November, Watson completed her first draft of a memorandum
outlining alternative research designs which might be pursued by the
project. She wrote:

> This memo outlines ten research problems which we might tackle and
> which are suggested by our work up to now. I fully expect that this
> list will be modified and enlarged by myself and the rest of you. I also
> expect that we will not tackle all of these problems but will select a
> few which we find most interesting and most susceptible to study.

There was absolutely no response to this memo: no comment by
assistants or supervisors; no indication that it had been read or was in
any way interesting or relevant.

Riesman does not remember the memo and has no explanation for
his lack of response. Presumably he and Foote assumed that Watson
was moving ahead in her own way to define a course for the project
and that their best tactic would be to give her general support without

specific interference. Watson, however, was unable to move beyond the specification of competing alternatives. She wanted others—in this case, the two principal investigators—to make the final choices and decisions. When they failed to respond as requested, she felt lost. It seemed that they had asked her to work out a research design which would carry further their ideas about sociability; but when she did her best to do this, they were not interested enough even to comment. At this point, she despaired of finding a rational, work-based solution to the problems of the project and decided that problems of internal process must be solved first.

Andi and Bob Potter each had similar experiences in trying to carry out their assignments to produce research proposals. Andi, after much effort, proposed an M.A. thesis which would study the relationship between family role, identity, and sociability; she was told that the proposal was a projection of personal experience and not a reasonable plan for research. Bob and Joan tried to work with each other (and to a lesser extent with Watson) to produce a plan for a core project and two Ph.D. theses, but they found themselves so far apart in their thinking about sociability and research that they gave up in despair. Bob resigned from the project, and it was only with difficulty that he was persuaded to return.

Inability to produce a satisfactory research proposal (Bob) or rejection by others of a proposal created after much concentrated effort (Watson, Andi) served in each case to mark a turning point in the individual's relation to the project. It focused the sense of failure, justified succeeding feelings of despair, and slowed down subsequent efforts to get back to work. Joan and Al did not appear to have such critical turning points, but each spent a long time hovering between hope and despair, immobilized by feelings of resistance and inadequacy, yet somehow too much committed to the project to turn away. Each left the project with these problems unresolved.

For Riesman, this first experience with directing a substantial research project was agonizing and at times bewildering. At the minimum, he was anxious that the project justify itself in terms of the careers of the persons involved—especially Watson and Potter, but also of others who had been recruited to the project—and, as already suggested, he was sensitive to the expectations of the grantor. But, at a maximum, what he hoped for from the work was a contribution to the study of relatively unexplored areas of leisure and sociability as these related to larger patterns of American culture. Furthermore, he hoped that the work would be done in such a way as to increase the openness of the research community, disciplined in the social sciences, to approaches that were not always orthodox. He hoped that the experience of the graduate students on the project might set an example for others who, sometimes in despair and sometimes in eagerness, had thrown in their lot with less risky topics, less marginal faculty members, and more proven methods. And beyond these more general concerns, Riesman, along with others on the project, recognized that sociability is an area where everyone feels himself to some degree an expert; it was perhaps not the ideal ground on which to challenge more established social scientists in the hope of turning their attention to new problems.

Such considerations, together with his own temperament, led Riesman at the outset, when people were exploring different possibilities, to try to keep all avenues open. His response to the difficulties encountered was characteristically to propose some other possibility! If identity was not "researchable," study something else; if observing sociability raised too many problems, at once ethical and practical, do not observe but seek informants or gather information in some other way. He was among those of the staff who wanted to create a "Hawthorne Plant" effect of heightened morale and experimentation based on mutual sympathy. However, while quick to forgive and forget earlier remedies that seemed unrewarding and eager to follow up any clue, no matter how tentative, that promised relief and a way out, he failed to appreciate the extent to which his own suggestions, far from increasing the freedom of choice of the staff, added to the strain by threatening the foundations on which others were trying to build or by suggesting a course of action which others did not feel competent or qualified to follow. Each proposed "solution" or course of action might have succeeded if there had been others listening who were inclined in the same direction; but in each case, the "others" who responded (or failed to respond) were those who were antagonized or threatened or bewildered by the proposed course of action. No one could accept what another proposed.

Analysis of Group Process

Watson made some attempt to help the staff work through its various problems and to gain some perspective on the frictions that were causing difficulty. She was not successful. She was too much involved to have the leverage of an outsider, and she was too much "in the middle" to be able to work from within. She shared in the failure of both the assistants to produce work and the senior staff to provide leadership, but she was not held fully responsible for either. The fact that she wanted things to be different was not enough to convince anyone that they could be made different.

Andi's analysis is as follows:

> As we started working on problems of observation on the Sociability Project, we quickly found ourselves in the middle of this issue (the problem of perception and how it is determined by language, social context, emotional dynamics, and so forth) in a very pressing and personal way. Our projections and distortions on our observations were clearly demonstrated every time we compared notes by several people on observations of the same events. As I see it now, this presented one of the major blocks to the development of the team. Personally, I felt challenged to come to terms with this problem, but I am sure that I was just as defensive as the rest. Looking back, I feel that it was most unfortunate that this problem was contemporaneous with the problem of communication. . . . In other words, it was just at the time that we had to sit down and face ourselves and the problems in observation that conflict and disagreement developed (among the senior staff) about the purpose of the study. . . . I can remember discussions among us wondering how these views could ever be merged into one study. I don't know how I would perceive it now, but at the time it looked to me as though Riesman and Foote had reached an impasse in their communication with each other, and then each started talking to Jeanne. It seemed to me that Jeanne was caught in the middle and that they had decided to "settle" the problem between them by saying that after all it was Jeanne's decision to make in the final analysis since she had to supervise the work involved. . . . If this had not been as serious an issue, I think that working on the observation would have "made" the team into a working organization. As it was, it had to be dropped like a hot cake.

Bob's report stressed the personal impact of the whole situation:

> My first exposure to defense mechanisms was not literary; it was the confrontation of them on the project in myself and others. . . . Throughout the project, I had the sense of trying hard to do something I did not know how to do and in which I had little confidence of success. Fools rush in; I was *surprised* by the failure of our attempts to find young adults who would want to be interviewed concerning sociability and observed at it. This surprise was not at the failure but

at the recognition that a defense *in me* and *in others* [namely, the other assistants] was preventing progress in the solution. . . . The project was responsible for much of the social psychology I've learned. The stress and impact of learning experiences in the research were such that I consider myself much changed from an aloof, detached person to an overinvolved, hypersensitive one. The statement overemphasizes the change by absolutism, but what was once recessive . . . has now became dominant. Reading over the interview done with me about my sociability by Andi at the beginning, I realize that my idea of sociability at that time was that it was intellectual work. There is no mention of emotional components or of extending acquaintance through sociability.

The Absence of Guidelines Defining Our Project

We have tried to describe a situation in which traditional patterns of working were of no use. Ordinary assumptions about work roles were violated: assistants were hired when there was no work to do and then asked to help design the job; one leader left the project, and the other preferred not to lead. The project itself was defined only vaguely and was surrounded by booby traps. Each attempt to create a work-oriented solution broke down, and sometimes these breakdowns were experienced as personal crises of considerable intensity.

This situation might have been mitigated if we had been able to agree on some established model of "good research" as being both desirable and possible for us. The models confronting us included at least the following: the quantitative, scrupulous definiteness of demographers at Chicago's Population Research and Training Center and the rigid criteria of research which they transmitted to students in their courses; the psychoanalytic subtleties of Erik H. Erikson's discussion of identity;[37] the anthropological model of trained and repeated observation of social events where the social location of all participants is known (preferably, as Margaret Mead and others have urged, with camera in hand); the survey-research model of standardized depth interviews; the group-dynamics model of experimental manipulation and observation of small groups; and, beyond these, the social historian's and speculative sociologist's concern for what is changing in the ways in which people dramatize themselves on social occasions and for the changing place of sociability in their lives. Each of these models offered an image of excellence which we felt incapable of achieving on our project; each reminded us of our own deficiencies and the primitive stage of research in the areas which interested us, while at the same time reminding us of the high standards of performance achieved by our colleagues and, in large measure, internalized by ourselves. We became committed, not to any one of these models, but to the proposition that what we were doing was "something new"; we could learn much from the work of others, but none of them could set a pattern appropriate for us.

Just as there were many models of research methodology which had some relevance for us, but none which we could follow in full, so, too, there were many theories and concepts which interested us, but none which was specifically relevant to the formulation of questions for investigation by the project. The theoretical work which interested us the most was highly general in character. Simmel, for example, wrote about sociability as an example of a "pure" form of association; he was interested not so much in sociability per se as in an inquiry into the elements of a "pure sociology." Erikson's discussion of identity draws widely on psychoanalytic thought and cross-cultural and historical comparisons. G. H. Mead and his successors developed the thesis that an individual acquires his sense of self through interaction with others;[38] this idea has become basic to much of modern social psychology. But the proliferation of concepts and "small theories" which makes possible the detailed study of interaction has occurred mainly in connection with the study of persons at work. We felt that the study of sociable interaction required both a new theoretical amalgam and a detailed set of definitions and concepts; these we had to create for ourselves.

All of this was aggravated for the graduate students by the tendency of their professors (as is often true of professors at the leading institutions) to submit all prior research by well-known figures in whatever field to the most relentless and often devastating criticism: if the big shots did work that was so slipshod and that did not prove what it was intended to prove, how could anything valuable come from the fumbling and amateurish efforts of the Sociability staff?

The Growth of Alliances

In the absence of work-based solutions or models for action, we turned to personal alliances. Often the personal congenialities which allowed the emergence of cooperation and productive activity were strangely at odds with professional commitments. Thus, for example, Riesman and Watson tended to differ about almost all aspects of the project; yet their growing respect for each other and their desire to find a way of working together which would permit each to do what he could do best provided the foundation for all the work which followed. There is no one road to wisdom in the social sciences, they believed; each must pursue his own routes of discovery. A multifaceted investigation can yield more information and be more exciting than one which is restricted to a single mode of knowing.

A partnership between Andi and Watson was created for purposes of observation at the summer camp, and it laid the basis for a lasting friendship. These two persons differed greatly in their modes of relating to others and, at the time, in their understanding of research. Nevertheless, their friendship enabled each to work out some of her problems with the project and, in the end, to find a more positive relationship to it.

Potter reports, "It was in the fall of 1956, after having studied the party reports that summer and with the hiring of Kotler, that I began to feel a capacity for initiative. Even this was tentative until December 1956." The reorganization of his relationship to the project began when he had the opportunity to work with Kotler, an unscarred and congenial research assistant with a vigorous enthusiasm for the study of parties, highly trained in another field and seeking to learn about social psychology. As Potter regained confidence in himself and in the potential of the project, he was able to move into a more positive relationship with Watson and Riesman. They had each regarded Potter as a valued and congenial working partner long before he could accept this view of himself.

The three residual legatees of the project, then, are Potter, Watson, and Riesman. These three are alike in that they came to the project more conscious of what they wanted to learn from others than of their own (not inconsiderable) strengths. Beyond this similarity is probably a more basic one: a tendency to exaggerate what one can expect from others and then to be excessively disappointed and angry or hurt when it is not forthcoming. Each is, in his own way, a perfectionist, placing high demands on himself and on others. Each is overly sensitive to criticism from others, quickly roused to sympathy for others, yet capable of considerable indignation himself—Potter perhaps less so than Watson and Riesman. Each tries to work with others in such a way as to emphasize the positive elements in what they have done, and Riesman is outstanding in his ability to register enthusiasm and approval.

As Potter wrote, "In working with Dave, I have had to learn to make fine distinctions between degrees of enthusiastic praise." These are the qualities that have made these three persons congenial working partners—these, plus an active intellectual curiosity, an experimental openness to experience, a flexibility of mind, and a habit of inquiry. We do not want to overstate the similarities; there are also tremendous differences; but it seems to us that it is the similarities that have bound us together as a working team. A final similarity, of course, has come to be our conviction that there is much of interest to be learned from our joint enterprise: the study of sociability.

III. WE GO TO WORK

Early Theoretical Explorations

The Sociability Project came into being, as already stated, as a result of a seminar given by Foote and Riesman on play and leisure. However, this seminar had not made any attempt to focus the thinking about play and leisure in the form of questions for research, and, in any case, the new staff members—particularly Watson—had to create for themselves a theoretical framework for the study of sociable play.

It is impossible in a brief chronicle of this kind to give an exposition of the theories and ideas which occupied us during various phases of the Sociability Project. Nor is it satisfactory simply to list the questions with which we were concerned. The following summary is necessarily brief, oversimplified, and at times, perhaps, somewhat cryptic. The main exception which we shall make in the direction of specificity is to indicate any accidental or adventitious sources of inspiration; for this is a chronicle, and our interest centers on the "how" and "why" of theoretical development more than on the ideas themselves.

The Contrast between Work and Play

During the fall of 1955, our discussions took as their theme, "What is sociability, and how does it contrast with work?" (Foote warned us away from this formulation of the question, asserting that the seminar had found it fruitless to contrast work and play, but we had no alternative jumping-off point.) We recognized that good research about work does not take as its problem, "What is work?" but rather focuses on some more specific question. But before such a specific question can be stated, there must be some context of theoretical understanding about the meaning of work itself, and it was this kind of background understanding which we wished to develop for sociable play.

SEPTEMBER: STRUCTURE AND VOLUNTARISM. At first glance, it appears that work has structure and play has none; work has goals and rules of procedure, whereas play is spontaneous interaction without purpose or obligation. The falsity of this dichotomy led us to consider two sets of rules which operate in sociability: those designed to maintain "integrity of the situation" (rules of the game) and those designed to maintain "integrity of the individual" (tact, etiquette).

OCTOBER: CONTRAST AND BALANCE IN WORK AND PLAY. We considered the functions for the individual of sociable interaction, work interaction, and family interaction and the ways in which concerns arising in one area are reflected or avoided in another.

OCTOBER: MATCHING VERSUS "HOST-GUESTING." Foote and Denney both approached the study of interaction with an eye for detail: for the specific processes by which individuals make contact and establish positions vis-à-vis one another. A memo dated June 3 listed twenty-four dimensions or processes which can be identified in sociability, including the process of matching. During the summer, Denney suggested a fundamental dichotomy in sociable style between those who approach others as equals, seeking areas of common interest (matching), and those who approach others in terms of hierarchy and difference (the host and the guest). In October, Watson and Potter carried further the theoretical analysis of matching and the host–guest relation.

NOVEMBER: THE SOCIABLE SYSTEM. A memo of November 18 from Watson suggested ten alternative topics for research on sociability.

(1) The conventional ritualistic practices utilized for dealing with specified "tasks" in sociability.
(2) Analysis of a range of social situations in terms of structure and voluntarism and analysis of the stresses, supports, and frustrations experienced by an individual who participates in them.
(3) Types of reward associated with participation in each of a variety of types of social situation.

(4) The process by which a "social system" (set of reciprocal obligations and expectations; reciprocal roles) gets established.

(5) Explication of a set of "rates" (reflecting differences in phase, task, and organization) which can be used to describe a sociable system, as the Bales categories can be used to describe a work system.

(6) The processes by which sociable systems expand and contract, adding new members and activities, redefining old ones, and dropping old ones.

These preceding problems take the sociable system in isolation. It is also possible to compare sociable interaction with other forms of interaction, with respect to the ways in which the individual can display, cope with, or confront material relevant to his developing identity.

(7) Comparison focusing on projective representation and working out of basic human conflicts (the problems of autonomy; the self; intimacy).

(8) Comparison focusing on representation and resolution of problems posed for the individual by the society in which he lives (aggression, mate selection, identity diffusion versus integration).

(9) Mechanisms for perfection of role behavior; sociability as theater for coaching and rehearsal of behavior for other settings.

(10) Analysis of ways in which sociability relates to other sectors of living in creating a balanced life for the individual; functions of sociability for the individual.

NOVEMBER AND DECEMBER: SOCIABILITY AS THE PROCESS OF DEVELOPING "RESOURCES." The Lewinian view of work is that it is locomotion toward a goal. By contrast, sociability can be seen as movement which originates with a conversational resource: it moves away from that starting point by developing the resource in such a way that it generates interest and excitement among the participants. Watson made a preliminary attempt to analyze reported observations of sociability in terms of the resource governing each episode, with resource taken to be a "source of tension" to act.

JANUARY: THE MULTILAYERED NATURE OF SOCIABLE COMMUNICATION. Riesman was impatient with Watson for having held a "banal" discussion about the weather, in his home, with a visiting psychologist. Watson replied with a memo citing this and other instances in which banal or trivial topics serve as the vehicle for covert communication about matters which are emotionally meaningful to the participants. In most sociable interaction, there are at least two conversations going on simultaneously: the substantive conversation serves as a vehicle for the development of interpersonal relationships.

FEBRUARY: THE INTERPLAY OF FACT AND FANTASY. Sociability often utilizes a resource which is of serious interest to the people involved but deals with it in a manner that is only half serious. Literal accuracy is set aside in favor of fantasy and imagination; there is occasion for sharing of affect and exchange of personal experience as well as for objective appraisal of reality.

FEBRUARY AND MARCH: SOCIABILITY AS A SECULAR RITUAL. Potter prepared a working paper for the Sociability Project in which he reviewed the work done to date and took as his focus a kind of total description of the party. He conceived of the party as a secular ceremony substituting in modern America for kin ties and religious rituals of the past.

Sociability as a Medium for Definition of the Self

During the winter quarter (January–March 1956), Foote and Watson gave a seminar together on identity. Foote took the view that identity was the sum of characteristics attributed to an individual by others and imposed upon him by his positions and memberships.[39] Watson took the view that identity was that which is retained by the individual from past attributions, positions, and memberships: his "internal society." One emphasized an outside view of identity, the other an inside view; but both agreed that "identity" was a concept which could be used to talk about the residues for an individual of his experience with others. Foote and Watson believed that their conceptions of identity were complementary; Potter reports that to the students they seemed opposed and irreconcilable.[40]

A series of staff discussions was held during February, March, and April on the relationship between self and sociability. Among the topics discussed were: sociability as an opportunity for self-definition, the various forms of self-defining processes, the relationship between what an individual affirms or manifests about himself and the response which he gets from others, the distinction between interaction which assumes overlap between the selves of participants (sharing) and that which assumes separateness (presenting), and the difference between development as it occurs over time in an expanding relationship between two persons and development as it occurs within the span of a single sociable event.

There was special attention given to the situation of young adults, for we expected that they would be our subjects. For young adults, sociability can serve as an opportunity to experiment with different ways of completing the self: by performing in different ways oneself and by choosing different partners or membership groups. We discussed our assumptions about individual change, concluding that change occurs only at points where an individual already feels conflict or ambiguity; given this kind of unsatisfactory *status quo*, new experience with others can sometimes be used as the means of finding a new resolution, a new image of self, a new pattern of behavior.

Potter's analysis of the party as a ritual of association helped to focus our attention on the fact that sociable interaction is used by the individual to define his relationships with the persons and environment around him, and these relationships are the building blocks for identity. Analysis of what is affirmed in sociability about how participants are related to one another and about the essential qualities of self, other, and the external world should lead us to an understanding both of sociable interaction and of the way in which sociability contributes to maintenance and/or change in identity.[41]

Learning from Observations of Sociability

Although we were not yet committed to the study of parties, we were gradually accumulating party reports. Some of them had incidents or an over-all character which suggested new ideas about sociability.

(1) Analysis of a short segment of a report on an office party clarified the role of the punch bowl as terminus or place of interchange: trips to the punch bowl were used to shed old partners, acquire new ones, and otherwise redefine one's relationship to the party.

(2) An evening spent with some friends at the home of a young bachelor dramatized the concept of self-surrogate. The entire evening was occupied with presenting and responding to surrogates for the host: his food, ideas, and possessions. It seemed to us that the use of self-surrogates as resources for conversation was more sociable than literal talk about self but could serve the same functions for the individual in terms of self-esteem and maintenance or change in identity.

(3) A party given to bring together large numbers of young un-

married persons suggested a characteristic pattern of "advance-and-retreat"; individuals would alternate between low-key conversation with old friends and acquaintances and more exciting ventures into the unknown.[42]

(4) Analysis of the conversation which occurred when two staff members met for coffee with two SSA students led to a clearer understanding of the interweaving of episodes. In that conversation, several different topics were pursued concurrently: one about the waitress and getting coffee, a second about the college newspaper, which one person was reading and later passed on to another, and a third about the research project and future plans. We could see that, for purposes of analysis, each topic defined a single episode, though, in occurrence, the three were all mixed up together.

(5) One of the key concepts in our subsequent analysis of sociability was that of axis: a unit consisting of two or more persons who, for the duration of an episode, carry on a conversation.[43] The idea of taking an axis as our unit of observation (rather than, for example, an individual or a party) jelled for us as a result of an incident which we observed at a party in the spring, when two relatively young persons were able to maintain an interesting conversation between themselves but faltered and came to a stop as more and more of the older persons present tried to listen in.

In all probability, none of the five items just mentioned (nor the earlier material abstracted from memos) will seem novel or out of the ordinary to readers with experience of parties and of other sociable markets and exchanges. Indeed, throughout the life of the project, work was often impeded by our sense that we were painfully discovering "what every hostess knows," and the step from common sense to coding and systematization was made harder because to some degree we had internalized the Chicago "aristocratic" pattern and were not satisfied to prove or rediscover the obvious, even if we managed to do so through a "methodology" framed by special usage of familiar terms and sometimes recondite (and correspondingly debatable) judgments in observation and coding. Only when, despite our perfectionism, we became satisfied to make a modest initial contribution to the study of sociability could we be ready to explore, for example, the concept of axis with adequate intensity and commitment.

Two Pilot Studies

Luncheon Conversation among Law Students

Al joined the luncheon group about twice a week during November and December, collecting eleven observations in the period between November 7 and December 21. During the three months of the winter quarter, he collected nine more observations. When he left the project in June, these data and the related interview data had not yet been analyzed.

A year later, after working as a coder on other material, Kenneth D. Feigenbaum undertook to complete the study of luncheon conversation. The result was an M.A. thesis titled, "The Limited Hour: A Situational Study of Sociable Interaction." Feigenbaum conceived of the luncheon group as composed of "familiars," that is, persons related to one another by a common tie, who have an opportunity to express this tie through meetings at frequent intervals. He was able to show that conversation in a group situation toward the end of the year differed both from conversation in a group setting in the fall and from conversation among isolated pairs at the end of the year; in other words, he was able to demonstrate the formation of a "group culture." Also, he was able to show how the substantive themes emphasized in this group culture reflected the situation of the participants: a relatively deprived group within the institutional framework of the law school, in the process of moving away from lower-middle-class origins (where identity was organized around symbols of masculinity) to professional status and identification with the law as a career.[44]

Friendship and Identity Change among Young Adults

As we thought about identity change among young adults, we came to the conclusion that it would be more closely associated with intimate friendship than with casual sociability. Accordingly, Andi undertook to study two trios: one girl with her closest girl friend and current boy friend, and one young man with his closest male friend and his current girl friend. Both romances encountered difficulty during the period of observation, though one eventually resulted in marriage and one of the men moved away from Chicago. Despite these difficulties, all but one of the subjects continued to be available for interviews and observation throughout the period of the study, from November 1955 to May 1956. Andi writes:

During those six months, I was very much involved in the lives of my subjects and all the complexity that went with that. After that, I welcomed the opportunity to observe a large number of people of the same general age group at the summer camp. It gave me perspective on the things I had seen in individuals. The combination of the intensive case studies and the summer at the camp resulted in a view of friendship that is generally applicable and in an understanding of its function in the development of young adults and, particularly, young women. The main finding is that in late adolescence young women from middle-class society engage in all relationships with intense involvement and for the purpose of skill development and identity formation. They are not "casual."

The Camp Study

During the spring of 1956, arrangements were made for two persons to attend an adult summer camp in order to observe the sociable interaction among the more than a hundred college students who served as staff for the camp. The suggestion of observing within an institutional setting, such as a summer camp, came to Watson from her former associate at the University of Michigan, Ronald Lippitt; and Foote helped make the first contacts with the administrative officers of the camp. Thus, most of the initial problems of access to subjects were by-passed: there was no need for individual recruiting by junior members of the staff, and the "subjects" were not ourselves but an outside "tribe" to which we could get access. The expectation of being with the campers all summer made it possible to give a partial explanation of our research interests (studying the relationship between work and play), and the fact that the observers worked half time on regular staff jobs gave them a real (though tenuous) tie with the other staff members. On the other hand, the two persons who went to the camp—Watson and Andi—were somewhat older and more cosmopolitan than the college students whom they were observing, and though they did their best to minimize these differences they achieved only partial entree into a somewhat provincial society dominated by nineteen-year-old college sophomores.

The major activity during the summer was the keeping of research diaries in which Watson and Andi reported all conversations they observed or participated in. Later, these reports were divided into episodes and coded as such; in addition, an individual file was prepared for each camper, which included reports of all episodes in which he had taken part. During the summer, however, no thought was given to the question of how the data would be analyzed; the observers simply tried to remember and report as fully as possible.

In the camp study, observation of sociable interaction was supplemented by collection of data on the campers as individuals. The object was to have some way of relating conversational style to other aspects of an individual's pattern of play and to information about "identity": that is, developmental concerns for summer, past experiences with success and failure, with work and play, with persons of the same and opposite sex.

Accordingly, information was obtained from the application blanks about school grades, field of specialization, and number of years in college, about age and sibling position, and about stated reasons for wanting to come to camp for the summer. From their own knowledge, and with varying degrees of success, the observers coded information about the person's relationship to his family and his readiness for marriage, and they rated him on physical attractiveness. They prepared early- and late-summer ratings on the manner in which each person presented himself to others, distinguishing at the end of the summer among presentations to a group, to individuals of the same sex, to individuals of the opposite sex, and at work. They recorded the ratings which each person received from his employer on his performance at work, and they tried to specify for each what his main concern was for the summer: what was he "busy with" while at camp? A questionnaire was distributed early in the summer, asking what persons did during their leisure time during the first week they were at camp. Finally, after the summer was over, each person's file was reviewed and a summary description prepared of his pattern of sociable interaction. Some follow-up information was obtained from a Christmas reunion.

Feigenbaum prepared a preliminary report on the study of individuals, obtaining positive relationships between the ratings of interactive role and such determinants as physical attractiveness, cosmopolitanism, school grades, and reasons for coming to camp.

Our major concern in analysis, however, was with the record of conversation. Two assistants were hired in the fall of 1956 to do the coding.[45] Watson prepared the code, divided the records of conversation into episodes, and supervised the coding. The record produced 1,638 episodes. Each was coded by both coders, one man and one woman, and differences between the coders were discussed and decided on in conference with Watson. The coding continued through June of 1957.

The first step in the coding process was to divide the reports of conversation into codable units, or episodes. The episode was defined much in terms of the classical unities of the drama, namely, an incident or conversational exchange in which neither the participants nor the topic changed.[46] The problem of defining this unit occupied much of the attention of the preliminary seminar in the spring of 1955, and the choice of the episode as our only unit of analysis aroused doubts in Riesman.[47] But at the time of coding the camp observations, it was a matter-of-fact decision that this should be the unit. Coders were not asked to participate in defining the boundaries of episodes; Watson did this herself, and coders were asked only to code episodes of which boundaries were already given.

The code itself was divided into three sections.

(1) IDENTIFICATION. Episode number, date of interaction, situational context, and number of participants.

(2) EPISODE STRUCTURE. Our notions about episode structure grew out of the staff discussions in the spring: as already suggested, we conceived of each episode as characterized by an axis—two or more persons joined together for the duration of the episode—and by an outside environment to which some relationship is expressed by participants. For each axis, we noted whether the selves of the central participants were treated as overlapping or separate, and we recorded the tie which they used as a basis for interaction. We had a descriptive code to specify what aspect of the environment was discussed, and we also recorded what relationship the speakers expressed toward the environment. With these four codes, we felt we could capture the essence of what was affirmed in the interaction about the relationship of the participants to each other and to the world in which they lived—that is, about these aspects of identity which are rooted in networks of associations. We also noted whether the resource was something common to all (or most) participants or restricted to a single individual or to an individual and his "team."

(3) INDIVIDUAL ROLES ASSOCIATED WITH EPISODE. Because we could follow the same individuals through a number of different interactions, we felt that it was worth while to describe behavior at the level of the individual as well as the episode. For each

salient participant, we coded his role in getting the episode started and his activity within the episode, and we drew upon a number of descriptive codes to describe his personal style.

What the coders had to work with were abbreviated reports of conversation in a variety of settings, including both the peripheral talk that accompanies work and the more sustained and intimate talk that accompanies recreation, the fleeting comments that occupy a walk from one place or occupation to another, and the reiterative comments that characterize talk in the camp dormitories. As can be imagined, the coding was a qualitative and delicate job requiring, on the one hand, a clear understanding of what the code was aiming at and, on the other, an ability to read between the lines of a succinct report and to infer, without stretching matters beyond plausibility, what might have occurred.

Riesman read the camp data in full with perplexity: it was a detailed running account of conversations which exhibited campers griping in stereotyped ways about the guests or the work in general, or asking who was going swimming or to the movies, or discussing the weather and the food or the jewelry and clothes of the girls. The accounts were largely behavioristic, as in Barker and Wright's *One Boy's Day*.[48] Not only did people repeat themselves unendingly, but what they said the first time was little more than stream of consciousness; it had none of the aliveness, subtlety, responsiveness, and humor which can make of conversation a playful and creative activity.[49] Riesman felt that he did not know enough about the group of campers and the worlds of midwestern small colleges out of which most of them came to judge whether or not they were consciously bored or what these conversations, pleasantries, and castings of self and others into stereotypical roles meant either for the group as a whole or for individuals. Riesman's conclusion that he could learn little from these protocols and his impression of their banality were one factor in a shift of emphasis which had begun with Foote's withdrawal from the project in 1955 and was completed with Riesman's return to the project in 1957: we would give up the study of earnest, unsociable young adults and concentrate on more mature persons; we would give up the interest in identity and personal development and concentrate instead on the varieties of sociability and on conversational development.

Accordingly, we did not carry the analysis of camp data beyond the stage of punching IBM cards and obtaining the first information on marginal totals. We planned to return to these data after we had enough information about other populations to permit comparative analysis, but time ran out on us long before this was possible.

Yet the camp study did much more for the project than simply produce data which, in the end, were set aside. It served as a relatively successful experience in the field observation of sociability, thereby easing concern about the ethical and practical problems of relations with subjects. It moved us toward a more manageable research task and a more satisfactory organization of a working team: during the year 1956-1957, Riesman could advise and consult, Watson could direct and supervise, and the research assistants could provide necessary and valuable assistance. Above all, it resulted in the creation of a method for the systematic analysis of reports of sociability: a qualitative code, which, though cumbersome, still was workable. The next phase of the project was to be essentially a transfer of the skills developed in the camp study to a new problem: the study of parties.

The Study of Parties

Early Methodological Explorations

Interest in the study of parties had existed within the Sociability Project from the beginning, but it was not given major attention until the spring of 1956. During the early months of the project, occasional reports of parties were contributed to the files: by the end of February 1956, Foote had completed two reports, Watson had done three, and there was a group report of the party with the SSA students. In March, Potter decided to concentrate his efforts on the study of parties, and in the period of April–June nine more party reports were completed for the project.

Many of these early reports were experimental. We did not know for sure what methods would be best for obtaining a record of party sociability. Watson, Foote, and Riesman were all inclined to reject the method of tape-recording as inefficient and inappropriate: the tape records too much of the noise and confusion and too little of the posture and expression of participants, and the extreme difficulties of transcribing mean that a fantastic amount of time and energy must be invested in the sheer mechanics of making the data available for analysis. Moreover, as compared with clinical analysis of individuals or of therapy groups, our interest was not so much in careful scrutiny of what was said by individuals as in over-all patterns of movement and the qualities generated by use of one or another conversational style or resource; and while, at times, in coding it would have been a great help to have verbatim transcripts of who said what and, especially, recordings giving the tone of voice, we felt that tape-recordings would give us "too much" material for our specific purposes.

Nevertheless, we wanted to give this method a trial, and on two occasions Potter used a tape-recorder at a party in his house. His conclusions were substantially what we expected. He found that it was possible to obtain a record of conversation from tape; losses of comments lasting over one minute occurred at the rate of about seven times per recorded hour, usually at times when two or more conversations occurred simultaneously, but this loss was not intolerable. The critical objection lay, as already suggested, in the tape's sometimes useful lack of selectivity: the record was a long-drawn-out tissue of inanities in which the very diffuseness made analysis more difficult than when one was dealing with the more condensed material of recollection. (Had we used a different scoring or coding system or focused, for example, on content analysis, tape-recordings might have been more appropriate.)

A second experiment was conducted in which four staff members of the Sociability Project attended the same party and all took notes, retiring periodically to a back room set up for this purpose. They found, as might have been expected, that they did not attend the "same party": each missed so much during his "out" periods that differences between observers reflected more a difference in exposure than in recall. Moreover, the business of moving back and forth between participating and observing proved inordinately difficult, if not impossible: Watson and Potter both found that they could recall almost nothing when alone and facing the blank paper, but memories came flooding back when one rejoined the party and was once more a part of the action.

A third experiment focused on the reconstruction of parties by constituents, bringing some or all of the persons who had attended a given party back together in the same place where the party had occurred, setting up a tape-recorder, and asking them to recall and report the party as best they could, starting at the beginning. This method induced good recall on the part of informants, but reporting was inadequate, with statements often taking the form of ". . . and then we talked about housing for a while . . ." and including no adequate report of what was actually said.

In the end, then, we became committed to the technique of reporting from memory: making some notes immediately after the party, before going to bed, and expanding and adding to these throughout the following week. We found that with a little practice we could recall much more than we had expected and that many incidents which were first suppressed, overlooked, or otherwise forgotten were recalled to memory before the report was finished.

It was, of course, more comfortable to rely on memory than to confront participants with the request that they give tolerance or assistance in the matter of recording the party. In instances where the Sociability Project was the topic of conversation, this very fact lent a certain tone, sometimes aggressive and sometimes humorous, to the occasion, and the observer would find himself teased and baited by his fellows about his role at the party and on the project. This was particularly true when the observer was making some demand for assistance. One of our experiments during the spring of 1956 was with a postparty-reaction sheet, similar to the ones used after meetings by some practitioners of group dynamics. This technique proved extraordinarily annoying to participants: they had come to a point of satiation and/or fatigue *before* announcing their decision to leave, and the request that they pause at the door to recapture and reflect on their experience at the party was resented. Some could not mobilize any reactions except their desire to leave. Others told us enough to lend support to the hypothesis that "fun" at a party was associated with new experience and with "sharing," but the postparty-reaction technique itself mobilized so much antagonism that we gave it up.

The Collection of Data

In the fall of 1956, the project acquired, for a short time, the services of Philip Kotler.[50] He worked with Potter and Watson on the party project. During the six months of the fall and winter quarters, these three persons wrote up reports of twenty parties. The two graduate students who were working as research assistants coding the camp data turned in six more party reports, and Anselm Strauss, a faculty colleague who did much to help us in the early days of the party study, reported on two parties which he had attended. Foote sent back one party report from the East Coast.

Potter and Watson brought their major efforts at data collection to an end in March 1957. During the following year, they occasionally wrote up summary descriptions of parties which added some new note of interest, and they gave a detailed report of a large party where observation had been specifically invited. However, they largely ceased their data-collection activities in the spring of 1957 with an eye to the termination of the grant in December of that year and the need to digest and analyze the reports collected during the previous year.

Meantime, Riesman had renewed the practice begun during his Kansas City stay of writing what he termed his "Notes on This and That" for the benefit of the projects with which he was associated. These notes included reports of relevant literature, responses to memoranda written by others, and a kind of field diary which might include first impressions of Kansas City or observations of religious practices there[51] or reports of observations and encounters relevant to the preoccupations of the Sociability Project. He also prepared some party reports specifically for the project, describing events from such distant points as New York City and Hollywood; indeed, his reports were largely written on trips; his wife, with a fiction-writer's eye for the detail of a party, cooperated in these and, in one instance, prepared an independent report. Even after formal data collection ceased, Riesman continued to submit occasional reports to the project, summarizing incidents or events of particular interest.

Theoretical Explorations

The data-collection activities of 1956-1957 were paralleled by theoretical discussions in a staff seminar. In the fall, this seminar consisted most often of Watson, Potter, and Kotler; in the winter and spring, Riesman and Anselm Strauss were more frequent participants, and Kotler eventually had to drop out. In these discussions, Riesman would press ahead with the exploration of ideas, hoping to reach, if not consensus, at least sets of working agreements concerning what might be happening in contemporary sociability. Sometimes he would jump ahead of the data, not waiting to ask whether or not our project would permit the testing of fleeting and fragile ideas. He saw research as a perpetual game of leapfrog between ideas and investigations.

Others in the group found this approach intellectually stimulating but were unwilling to grant the separation between data and ideas: any idea, they said, can be converted into code categories and can help to shape our examination of the party data at hand. Watson, in particular, looked to the systematic coding and analysis of party data as a means of gaining some more solid purchase on the ideas and generalizations which arose in the give-and-take of the seminar.

One of our early ideas about party sociability involved a distinction between what might be called "bass" and "treble" sociability. The former is low-keyed, intimate talk among two or more people; the latter is festive activity in which the fact of being together at a party is used as an occasion for special celebration or exhilaration. We saw these bass and treble themes, not as dividing occasions from each other, but as aspects or accents within the same occasion. They seemed to represent alternative ways of integrating with others, drawing, in the one case, upon personal concerns and, in the other, upon shared group concerns.

One form of group-based festivity which seemed particularly characteristic of party interaction is what we came to call reiterative culture-building. Some word or phrase or idea is detached from its usual context and comes to symbolize the party itself. The twin keys to this process are reiteration and nonsense.

Riesman was quite ready to grant that nonsense and even reiteration could, among intimate friends, serve to intensify bonds, enhance gaiety, and provide a change of pace; he was well aware also that many, especially among high-powered or well-known people, do not want always to talk about "big ideas." Nevertheless, he found it hard to believe that a monotone, serious or frivolous, could actually be refreshing and creative. He was depressed to see that, in our protocols, reiterative culture-building served to prohibit any demonstration of real wit, vision, or excellence.

Such discussions led us to specify alternative models of excellence in party sociability, recognizing that while two or more may be present at different times at the same party, they also reflect differences in the style of different social classes, different ethnic groups, occupational cultures, age grades, and historical periods. We worked out three different standards of excellence and eventually rated each episode on each of the three criteria.

The first we called *artistic quality*. It takes as its standard of excellence the achievement of the kind of dramatic or artistic quality or elegance which might appeal to a noninvolved stranger or to the reader of a novel. The second we called *empathy:* the emotionally moving exposure to the private world of another which, if pursued, lays the basis for intimacy. The third criterion is *group solidarity,* with success measured by the extent to which participants achieve a heightened sense of oneness, of solidarity.

Watson was interested in the process by which excellence is achieved.

If mediocrity is associated with literal and unimaginative reporting of facts—as for Riesman it seems to be—then excellence must be associated with ways of moving away from literal treatment of the topic of conversation. We noted three different styles which participants might use to generate excitement: the development of serious intellectual exchange, the development of collective fantasy or other forms of make-believe, and the development of a kind of fictionalized version of the truth which we tentatively described as legend.

Each of the ideas we have mentioned found its way into the code—the distinction between bass and treble modes of integrating with others; the identification of episodes which are part of a sequence of reiterative culture-building; the evaluation of episodes in terms of artistic quality, empathy, and solidarity; and the description of the manner in which conversation departs (if at all) from the mode of literal accuracy. Many other ideas and insights appeared and disappeared as we tried to make sense of the protocols and incidents brought to our attention in the seminar. Many of the ideas which are reported here in the form they take in the code were explored more fully than can be reported—as, for instance, a careful consideration of how party festivity compares with the "spontaneity" which Polansky, Lippitt, and Redl [52] describe as a condition for "behavioral contagion" among children. But since the code was the major tangible product of this phase of the project, we have focused on the ideas which were reflected in it. An intangible product of considerable significance was, of course, our own renewed sense of excitement about what could be learned from the study of parties.

Formulation of a Thesis Proposal

During the spring of 1957, Potter returned to the task of defining for himself a research problem which could be part of the study of parties and yet sufficiently separate to serve as the basis for a Ph.D. thesis. He reviewed the data on hand, with an eye toward the different definitions of, and intents or motives in sociability and was struck with the difference between the sociability of friends and that of strangers. This difference had been one of four which he had cited in his earlier working paper on sociability. At that time, he had suggested that four explanatory variables affecting expressive behavior at parties are (1) the structure of relationships (degree of acquaintance), (2) the general cultural background (discarded as impossible to study), (3) "shared" predis-

positions and needs (unable to study without being able to collect
background data on individuals), and (4) the party culture created on
the spot (discarded because of difficulty in studying it and conviction
that if anyone could do it well it would be Riesman or Strauss).

He pursued the distinction between acquainted and nonacquainted
partners in sociability in the spring of 1957, preparing a paper which
contrasted two parties at which persons were well acquainted with
each other with two parties at which persons were in process of becoming acquainted. Drawing on this analysis and the findings from the
postparty-reaction sheets, he formulated the hypothesis that the sociability of friends is dull except in instances when they discover some
new resource; whereas the sociability of strangers is or can be exciting.

At this time, he also explored the idea of studying party phases; as
he writes, he was impressed by

> . . . the tremendous difficulties involved (as shown by the work of
> Saul Ben-Zeev with Thelen and Stock).[53] I gave that up, I think, be-
> cause I was sure I lacked both the intuition and technical skill even
> to imitate Ben-Zeev, let alone to advance beyond what he had done.
> Moreover, the study of phases in therapy groups had a sensible appli-
> cation, whereas I couldn't see much practical point in similar knowledge
> of parties. My interest would have to be mainly idle curiosity, but that
> was not strong enough for the kind of work I could see was involved
> in studying phases.

The choice of acquaintance as an explanatory variable worthy of
study, then, he continues,

> . . . just grew, slowly, throughout the project as other ideas became
> unworkable, uninteresting, or stagnated. Looking at the style of total
> parties definitely had something to do with it, but so did a number of
> other things. At the end, it was a question of pinning down a specific
> portion of the code which would provide an interesting analysis, with
> many other possibilities still open.

Potter describes subsequent developments as follows:

> A little later (summer?) I remember distinctly a day when Jeanne,
> coming in to work, stopped by my desk briefly. She said something like,
> "I had a new idea. There is another kind of relation: that of coworkers
> or people who are not friends but see each other routinely." This was
> a striking idea that added a great deal of precision to my thinking

about the relation of acquaintance and interaction. We talked about people who meet occasionally for lunch and what happens when they decide to meet regularly. Joan had described an instance in which this happened to her, reporting that after the lunch became routine people lost interest in each other for conversational purposes. I worked for some time on this distinction and finally wrote a sloppy, two-page memo on three polar types of acquaintance. To each type were coupled a few hypothetical differences in interaction. This was the beginning of my thesis.

The next advance came in the building of the code, when two categories were quickly added to our acquaintance typology: institutionals and expanding acquaintances. . . . At this time I felt ready to write a proposal, but I was blocked in my efforts to do so by a course in research design I was taking where I learned that almost no worth-while research had ever been done in sociology. Faced with this purist view of research, I couldn't freeze a proposal that was within shooting distance of the professor's ideal. . . . And when the time came that I could do so, in fall 1957, the code was already frozen and being applied. Although, in fact, the necessary codes had been built into the code so far as possible, I still had a sense of being trapped and unable to control the research in accordance with the design presented in my proposal. . . . [Members of the committee] insisted on specific hypotheses, though I resisted. Hypotheses in the proposal were hastily constructed to meet this demand by a simple juxtaposition of one set of codes against another. . . . This pressure for the specification of hypotheses has been a block to the interpretation of the data to the present. . . . [They] tended to block synthesis of what the data show to be a much more intelligent conception of the interaction style associated with various degrees of acquaintance. While I see the value of making hypotheses in clarifying thinking, I'm not sure that the timing of this operation in our research didn't create difficulties greater than it solved.

Finally, I think the most important development of this work has been the isolation of the category of "expanding acquaintance"—a category in which interaction reveals clearly the process of getting acquainted. The discovery was . . . unanticipated by me. I thought of the category as a means of separating borderline cases to keep the other categories clean.

Constructing and Using the Code

It was in the summer of 1957 that Riesman, Potter, and Watson sat down to construct a code for use with the party data. As in the camp study, the unit of analysis was to be the episode. Almost all the columns used in the camp study were carried over intact, without serious challenge or discussion; the only ones omitted were those pertaining to the role of individuals in initiating and carrying through the episode. Ideas which had emerged from our previous discussions about party sociability were transformed into code categories. In addition, individuals brought to the coding of sociability certain concerns which originated outside the project.

Riesman was interested in looking for autonomy, adjustment, and *anomie*—both as qualities of substantive meaning, associated with the topic of discussion, and as qualities in the interaction, manifested in the way in which individuals deal with each other to facilitate, thwart, dominate, or otherwise influence the quality of the exchange. (He was convinced that the distinction between inner-direction and other-direction could not possibly be made operational.) Efforts were consequently made to try to codify such vague and elusive concepts as autonomy and *anomie*, as well as other categories involving hardly less difficult judgments.

Potter helped to build into the code various sociological and social-psychological classifications: a report of the number of persons participating in an episode, a description of the structure of the group of participants (dyads, groups, performer-audience), and information about outside attributes which individuals brought with them to the interaction—age, status, sex, ethnic background, and, of course, previous acquaintance. Riesman wanted information about the orbit in which participants had met: graduate school, family, interest group, work. Watson took the lead in developing descriptive codes focusing on the emotional qualities of the interaction. She examined the work of Leary and associates[54] and tried to use their two axes of love–hate and weak–strong in constructing a code for the "vicarious quality of the resource," that is, for the quality which was introduced into conversation simply by virtue of talking about a particular topic. Other codes were constructed to describe the emotional quality generated by the episode for participants and to describe the point of view which they expressed.

Additional codes developed as we tried to fill in the gaps and answer the questions raised by preliminary attempts to use the code. In the end, it became a highly complex and somewhat unwieldy instrument but one, nevertheless, capable of storing considerable information about episodes of sociability.

Watson's conception of the work at that time was that it was still essentially preliminary and exploratory. No one knew which codes would work; no one could say in advance what information we would need in order to give a satisfactory description of an episode of sociability. Watson's solution was to try everything, to suspend doubt and evaluation until the results were in.

And doubts there were! Potter still did not have a thesis proposal, still did not have his hypotheses stated, still did not know what codes he would need to answer the questions posed by his research and his thesis committee. When confronted with the first draft of the code, he felt a distinct sense of shock: many of the columns were totally unfamiliar (having been carried over from the camp code); almost none of them had been submitted to careful, systematic evaluation; and he did not know where the codes came from or why they were there. He did not feel that the code was ready to use. Several years later he was to report, "I remember when I first thought of Jeanne as a research director. It was when she said, 'This is the party code; now let's use it!'"

Riesman felt that the coding, like the preparation of party reports, was something which could be done only as an act of blind faith. He had misgivings about the party reports themselves; persuaded that even the best report could capture no more than a quarter or a third of what happened, aware of all that the observer must miss because of his own position at a party, concerned about the problems of conscious and unconscious distortion which must bias any report written from memory, he felt that he could not place much reliance even on the full party report. How much more arbitrary, then, were the discrete episodes, taken out of context, each to be considered separately and independently.

Similar misgivings haunted the coding of the material after it had been cut into episodes. How could one spend a long period deciding whether a given episode represented adjustment or autonomy if one was not sure of the reliability of the underlying report? Nor was it only a question of the reliability of the report. In subsequent analysis, Riesman was to challenge the adequacy of the episode as a basis for judgment. In one case, a paper was being written about nihilism, and Watson cited similar episodes from a number of different parties as illustrations of what was meant by nihilism. "No," said Riesman, "Mrs. Johnson, in the party which I reported, was incapable of nihilism, and the judgment that her comment was nihilistic was therefore an error." Although recognizing that errors in coding were inevitable, Riesman lacked the comforting belief that such errors would "cancel each other out."

In the coding itself, some of the problems that Riesman had first encountered in the camp data reappeared. His appraisal of a party sometimes differed from that of the graduate students. He would be struck by inanity, lack of artistry, flatness of topic and mood, while they would emphasize sincerity, relaxation, and solidarity. These differences seemed to reflect social class, generation, and sociable experience, and here each had to speak for the world he knew and reported. And when it came to discussing our own sociability in the guise of talking about other people's, we found ourselves as defensive as some of our subjects: Potter might feel he was defending some of his fellow graduate students from Riesman's strictures on nihilism, Riesman might feel that he was defending his own sophistication, such as it was, against strictures on formality or compulsiveness, and Watson might feel the need to defend her preference for quiet, personal talk against strictures on triviality and mediocrity. Yet defensiveness was not the keynote of our discussions. There was a certain edginess, perhaps, and there were occasional rigidities of interpretation which could be resolved, not by consensus, but only by fiat. In the main, we were saved from ourselves by an atmosphere of mutual liking and respect; we could concentrate on trying to understand the significance of the protocols at hand rather than on defending our different points of view.

Faced with these various forms of hesitation and doubt, Watson insisted on the leap of faith which, as already suggested, much quantitative work with qualitative data requires, asking the staff to suspend self-criticism and move ahead with whatever tools and skills it could command. The legitimacy of the written record was to be assumed, much as a court might do on review, and we were to work from it without undue and continual worry as to what underlay it. The episoding was to be done entirely by Watson, with correction offered by coders whenever they found themselves given an episode that was not homogeneous with respect to the basic elements of the code. Two new assistants[55] were hired to do the coding to the best of their ability, and, whenever possible, the observer who reported a particular party would take part in the discussions of coding differences. Thus, the major part of the coding was done by "objective outsiders," but the "insider" was also given his day in court so that he could elaborate on obscure reporting and correct any errors of interpretation.

The coding of parties lasted throughout the three quarters of the academic year 1957-1958. By the end of the year, Watson, Potter, and

Riesman were all doing some of the coding, as well as participating in discussions of coding differences. Twenty-six parties were coded in this fashion, selected from the total collection of available reports because they met such criteria as full reporting of conversation, small enough size for the observer to know much of what happened, and a report long enough to include at least forty episodes. The twenty-six parties yielded a total of 1,873 episodes. In addition, some dozen parties which met the first two criteria but not the last were selected for "schematic coding": here, the reported episodes were taken as illustrative rather than as constituting a full report. The data from these latter parties were set aside and not analyzed; our intention was to use them to check on hypotheses or ideas emerging from analysis of the more fully reported parties.[56]

Analysis and Write-Up

The academic year 1957-1958 was the last year of full-scale operations on the Sociability Project. The coding of party protocols continued through the end of that year, but, in addition, a beginning was made on problems of analysis and write-up.

In the winter quarter of 1958, staff members offered for credit a seminar on "Sociability and Socialization." A thirty-page summary of *Propositions and Assumptions about Sociability* was prepared for discussion in the seminar. Also, the seminar reviewed the empirical materials deriving from the camp study and the study of luncheon sociability, and individual projects carried further the analysis of some of these data.

The end of the academic year 1957-1958 saw the geographical dispersion of members of the Sociability staff. Foote had left the University of Chicago in the summer of 1956. Riesman had stayed on at the University of Chicago on half time during the year 1957-1958—despite Harvard's wish to get him earlier—in large part because he wanted to continue to work on the Sociability Project; but with the coming of summer, he moved to Cambridge. Watson took a leave of absence for the summer of 1958 in order to get married, returned to work alone at the University of Chicago for the year 1958-1959, and then moved to Detroit. In 1960, she retired from professional activity and has since been able to work only sporadically on the Sociability Project. Potter worked in Chicago during the summer of 1958, and then in the fall he

moved to Flint College of the University of Michigan, where he immediately assumed a heavy teaching schedule which has occupied much
of his time since.

The dispersion of project staff members and the accompanying commitment to new interests and obligations tells part of the story of why
the Sociability Project is still unfinished. It is not the whole story.
There was some provision of time and funds for analysis and write-up
before the end of the project, but it was not enough for the job which
had to be done. In addition, differences in emphasis and point of view
which had been glossed over during the period of data collection returned to haunt us as we tried to prepare written reports of our accomplishments. Not all the difficulties which are illustrated by this chronicle have been overcome. For example, Riesman continues to hope that
the study can combine methodical, quantitative, and empirical work
with work in a more essayistic and impressionistic tradition, and, while
Watson supports this, she also wants to see that the empirical findings
are presented in their own right, not just as footnotes to interpretive
essays. Undoubtedly, whatever the project can discover will remain
fragile, tentative, and limited in scope; we can only hope that it adds
to human understanding concerning sociability.

The first two papers to be written after the empirical work of the
project had been completed were initiated by Riesman, with Potter's
assistance. Riesman returned to Chicago for the latter part of the summer of 1958 and, together with Potter, attempted to see whether it was
possible to write up whole parties in any standardized way which
would yet capture their idiosyncratic flavor. In this, they did not draw
on the coded material but on over-all impressions of sociable styles
derived from examination of the full set of party reports. This wholistic
analysis resulted in two papers,[57] but Riesman and Potter still felt that
it was unsatisfactory: the party records were not comparable, the occasions were not comparable, the variables they were looking for were
not clear. Indeed, they found in their shuttling back and forth between
minute analysis of episodes and over-all analysis of total occasions that
they were engaged in a dialectic somewhat similar to that faced by
psychologists when alternating between traits, syndromes, and over-all
"clinical" diagnosis or delineation. The "right" level of generalization
seemed to escape them.

When Watson returned to work in the fall of 1958, she joined with
Riesman and Potter in finishing up the papers which had been started

during the summer. However, her main concern was with the coded data. She arranged for preliminary analysis of the party data, selecting a few columns of special interest to be run against everything else. The columns selected for special examination included four that were relevant to Potter's thesis and eight others: status; sex; time position within the party; participation-structure; an over-all index of quality based on six separate ratings; and the three variables conceived especially for the Sociability Study: axis type (sharing versus presenting), mode of integration (festive versus relational), and treatment of the resource (literal, playful, legendary, intellectual).

During the year 1958-1959, analysis of one of these variables—participation-structure—was completed, and a paper was prepared and read at the fall meetings of the American Sociological Association. This paper was received with considerable interest but was later refused by several editors of professional journals. As one wrote, "The readers suggest two possible articles: (1) a conventional scientific article in which all the data are presented with appropriate statistical tests, etc.; (2) a clear theoretical paper which does not attempt any presentation of the data though it might contain the conclusions." The mixture of qualitative and quantitative modes of understanding and learning which we had worked so hard to achieve was puzzling to readers; it did not seem to fit. In addition, the amount of background information needed to make sense of our findings was too large to fit easily into a journal article. Readers asked for more tables and better presentation of data, clarification as to how observers were recruited and trained and how they managed to make their observations, further description of methodology (for example, reliability), and more precise statements of theory. In short, they wanted a report of the entire study and not just a piece of it.

Difficulties encountered with this paper were much intensified in other papers which were in process during 1958-1959, namely, papers devoted to the analysis of typologies developed specifically for the Sociability Project. These papers consisted of the introduction and examination of concepts which were unfamiliar to the readers. As we learned repeatedly, every attempt to discuss our research with colleagues unfamiliar with our work led us back to the very beginning; things which had become so much of a commonplace to us as to be almost forgotten were still new for others.

Such considerations led us to fix our sights on the production of a book rather than a series of journal articles. But this was a difficult task for us, dispersed as we were and with virtually no time available to do the work of rethinking and writing.

In the summer of 1960, Potter was in New England and worked with Riesman on a paper about differences in party style associated with differences in status and social class. This paper in its preliminary form included extensive descriptions of many of the parties we had observed and provoked the Watson-Riesman discussion about ethics which was cited earlier. The paper was set aside until some resolution of these differences could be achieved, and we thought, again, that some of the problems might be solved if we considered publication of this paper (in any case, ninety pages is too long for a journal article) as part of a book rather than as a separate article.

Watson and Potter worked in 1960-1961 to produce a paper which they called "An Analytic Unit for the Study of Interaction." [58] In it they summarize the basic assumptions made on the project and in the code about the structure of an episode and about how these elements of episode structure are related to identity.

In the spring of 1961, we began work on the research chronicle, and this occupied much of our time during the summers of 1961 and 1962. We had the dual encouragement, here, of the request from the editor of this volume for something which we felt able to provide: a chronicle of a research project and also the conviction that such a chronicle would move us a bit closer to our own book.

In the fall of 1961, Watson took up the two papers which had been completed and set aside—one on participation-structure and one on status—and began to rework the material for a new and more quantitative presentation of findings. She found that a new variable had to be created, new IBM runs made, and, of course, new statistical analyses performed. A first draft of the new paper was completed in the spring of 1963. Meantime, Potter moves ahead with his analysis of acquaintance and interaction.

CODA

Reviewing at this point what our study set out to do, what it has accomplished, and what it may accomplish with the data we have gathered, we find that our major questions remain unanswered. We hope in the months ahead to pursue these questions further. Beyond that, however, we should like to see the study of sociability continued on a quite different scale.

One would begin with the same general suppositions with which we started in 1955, namely, that Americans have not been prepared for the abyss of leisure that has descended on them nor do they know how to manage the resources of sociability to create ease and amenity and enlightenment rather than traffic jams and sterility, which means that much of our sociability is self-defeating. To go into such questions more deeply, one would need a very large study, indeed, of the ecology of sociability: a census of sociable styles, patterns, and preferences, not only in terms of different social groups and regions but also in terms of individual histories of sociable experience. For such a survey, one would need once more to explore what can be discovered about sociability through interviews. Those which we did largely drew a blank: people had no language for discussing sociable encounters, no vocabulary for describing parties except to say that they were "good" or "bad," no way of answering the question, "What do you do for fun?" But this very fact—if indeed it be a fact—is an important finding which needs to be understood better. Is it true that most people have very little discrimination about their sociable experiences? And if there are exceptions to this, who are they and why are they different? Is there any relationship between inadequacy in describing a sociable event and inadequacy in taking part in it in any inventive or enlightening way? Or is it that people fear to become too self-conscious about their sociability, as was, in fact, true of many people in the orbit of our own project?

From the study done at the camp and the early interviews on the project, we gained the impression that many people saw sociability as a difficult area in which they were "shy" or "inhibited," although without much specificity or detail about this. It would be revealing again to see what this involves in practice. Does it imply allergies or fears of particular kinds of people or particular kinds of setting? To what extent do childhood experiences of being ridiculed control and limit later experience, so that people almost automatically avoid situations in which they might be exposed—or in some cases "flee forward" precisely into such situations? And in more sociological terms, where are the people located who have difficulties in talking with others—and which others? How in different subcultural groupings do they learn to defend themselves—to tell stories, engage in flirtation, sing folk songs, or otherwise avoid or postpone the burden of conversation?

But the kinds of questions we would want to ask about sociability would have to include efforts not only to discuss the subject on a conscious level with respondents but also to get some sense of underlying motives influencing attitudes toward others. In the parties we did observe, we were struck by the role that narcissism played in destroying genuine communication. This was true for many of the young people at the summer camp; it was also true for some of the more eminent who as performers may have been entertaining (or intimidating) but for whom sociability was simply an arena in which to seek admiration or, at any rate, attention. Among less established people, for whom life was more stressful, sociability sometimes appeared to serve a desperate desire for companionship; here sociability tended to become a kind of untherapeutic group therapy which emphasized shared antagonism but did little to promote enlightenment, let alone vitality and *esprit*. Meditating on such occasions, we wish we could have asked participants whether they had ever experienced a really exciting and meaningful conversation based not on flirtation, on the desire to impress, or on an exchange of mutual hypochondrias about the world. Even now it is not clear to us what sort of interviews or observations might get at these phenomena within a group of people who are constantly in one another's company, not to speak of gaining a sense of them as they operate in the society at large.

Still, we think more could be done to test hypotheses as to the interplay between the quality of individuals and the quality of arrangements which lead to good and bad sociability. In our own observations, we were able to see a number of occasions where relatively secure people were not in control of their sociable fates but the prey of mischief and malice; in fact, the helplessness of capable people when put in a taxing or "impossible" situation, the inability of many people to communicate with each other often even in benign situations, is one of the reiterated themes of many of the dramatists and novelists of our time.

The experience of the project has reinforced our sense of the limitations of much contemporary social science in gaining understanding of these central aspirations and impasses in contemporary society.[59] To study sociability as we would now like to do it would be at least as difficult as to study prejudice and authoritarianism. One would need to develop a psychological approach which was at the same time sociological. The damping effect of individual insensitivities or needs for domination on social gatherings follows on the abdication of the host, which cannot be sufficiently understood through individual clinical work or through experimental studies of small groups but also requires a grasp of the egalitarian tendencies endemic to American society over a long historical period.

The gap between the styles of the social anthropologist and the social historian is not easy to bridge. The social historian can talk about whole strata and epochs on the basis of the memorabilia that have remained, but the social anthropologist wants to talk about concrete people and how, in our own society, he can define "representative" ones and gain access to them, with a sufficient span of control to understand even a single party, let alone the part that sociability in general plays in the lives of participants. It would seem plain that one needs an interdisciplinary team; but as soon as one uses the word *interdisciplinary*, one runs into trouble, for the word assumes the givenness of the various disciplines, whereas in fact what is necessary here is some reshaping of the disciplines themselves. One needs a psychology that is concerned with social character in its stratified and historical forms, an anthropology that is likewise historical, a sociology that is clinical in psychological outlook and in some of its methods, and a social and intellectual history not confined to elites.

The diversity of background, training, and experience among us on the Sociability Project has turned out, we think, to be fruitful in the long run, even though a mixed blessing in the short run. We do not yet know how to translate diversity into a program for research. Speaking now just of the two authors of this chronicle, each of us was put in the position, in the absence of a common strategy, of supporting approaches taken by the other as much from desperation combined with personal respect as from conviction that this was the right procedure. Each of us brought to the project not only different values and competencies but also different constituencies to whose axioms or fields of inquiry we were sensitive. It took us a long time to understand each other well enough to create a climate in which we could respond energetically and with understanding to the ideas each of us had. Even at best, neither of us could go beyond support and encouragement to commitment: neither could accept the methods and curiosities of the other as adequate for himself. The creative process, whether in sociability or research about it, is an elusive matter. This report, like much of our work on sociability itself, has been descriptive—a necessary first step. But the eventual aim is to understand some of the conditions making for good research in the field of sociability that could, in turn, lead to a better understanding of contemporary society.

• NOTES

1. Research Grant M-891. The original grant was for three years, 1955-1957. One continuation grant was given for the calendar year 1958, and a supplement extended this until the end of the academic year 1958-1959. These grants made it possible for Watson to spend one year working alone, without assistance, on analysis and write-up. Riesman and Foote never received salary from the Sociability Project, and the work which has been done since the end of the project has been done on unpaid time.

2. Jeanne Watson, "A Formal Analysis of Sociable Interaction," *Sociometry,* XXI (1958), 269-280.

3. Jeanne Watson and Robert J. Potter, "An Analytic Unit for the Study of Interaction," *Human Relations,* XV (1962), 245-263.

4. Jeanne Watson, David Riesman, and Robert J. Potter, "Emotional Themes in Sociability," paper in process. Preliminary report given at the 1959 meetings of the American Sociological Association.

5. Robert J. Potter, "Friends, Familiars, and Strangers as They Converse at Parties," paper in process. Preliminary report given at a colloquium of the Sociology Department of the University of Michigan (March 1959). See also Kenneth D. Feigenbaum, "The Limited Hour: A Situational Study of Sociable Interaction," M.A. thesis submitted to the Committee on Human Development, University of Chicago (June 1958).

6. David Riesman, Robert J. Potter, and Jeanne Watson, "Sociability, Permissiveness and Equality: A Preliminary Formulation," *Psychiatry,* XXIII (1960), 323-340. David Riesman, Robert J. Potter, and Jeanne Watson, "The Vanishing Host," *Human Organization,* XIX (1960), 17-27.

7. Georg Simmel, *The Sociology of Georg Simmel,* trans. Kurt H. Wolff (Glencoe, Ill.: The Free Press, 1950), pp. 40-57.

8. Nelson N. Foote and Leonard Cottrell, Jr., *Identity and Interpersonal Competence* (Chicago, Ill.: The University of Chicago Press, 1955).

9. George C. Homans, *The Human Group* (New York: Harcourt, Brace and Company, 1950).

10. David Riesman, M. Glazer, and Reuel Denney, *The Lonely Crowd* (New Haven, Conn.: Yale University Press, 1950).

11. Cf. my article, "Some Clinical and Cultural Aspects of the Aging Process," *American Journal of Sociology,* LIX (1954), 379-383; reprinted in

Individualism Reconsidered (Glencoe, Ill.: The Free Press, 1954), pp. 484-491.

12. In the winter of 1959-1960, *Newsweek* telephoned Riesman to ask about the Sociability Project, and it was clear from the dialogue that they meant to ridicule the project as a waste of taxpayers' money, as well as presenting the amusing picture of a "name" social scientist attending cocktail parties in pursuit of his dreary specialty. He sought to make clear that NIMH was not to be blamed for whatever absurdities one may find in or read into the work of its grantees and that in any case social science needed for its development what natural science had long since attained, namely, the freedom to make mistakes and to violate what passes for common sense. Shortly thereafter, *Newsweek* appeared with a "National Affairs" story and a cartoon of a cocktail party with sloshing glasses and sleuthing social scientists.

There followed, for Watson, Potter, and Riesman, many phone calls from newspapermen and broadcasters wanting to know more—as well as a more serious inquiry from a beverage foundation. More depressing were letters from self-styled taxpayers, often carbon copies of letters sent to their senators or congressmen, demanding in outraged terms an investigation of this immoral or inane waste of government funds. So, too, several congressmen with reputations as Treasury watchdogs or as opponents of "snooping" social scientists wrote to inquire why the study had not been finished, why no book or other such product had yet been received by NIMH, and where had all the money gone anyhow. Throughout this whole proceeding NIMH's conduct has been exemplary; they have spared us even from knowing about most of these attacks and never made the slightest suggestion about altering our work so as to ease their own position vis-à-vis congressional critics. However, in publishing this chronicle, we have feared to again stir up these opponents, who either regard mental health as a Communist plot or vindictively pursue alleged waste in the executive departments with a zeal seldom shown for stock piles, obsolete bases, or counterpart funds.

13. NIMH changed this policy one year later, allowing funds from one year to be carried over to the next if this suited the needs of a project.

14. These comments, like those of Riesman and Andi, were written from memory specifically for the chronicle. No diary was kept while the project was in operation.

15. NIMH policy is explicit that a research proposal is not a contract and that grantees have discretion to change it as the work develops. Retroactively, it would appear that the proposal at times assumed undue hegemony over our thinking simply because it was "there," an apparent authority to which appeal could be made. It is probably not the first time that NIMH or other donors will have noticed efforts to escape freedom in this fashion.

16. Riesman writes:

> I have completely forgotten this and it does not sound like me now. It occurs to me that I began to feel grave doubts as to whether we could succeed in showing what necessarily followed from what, and I was willing to resign

myself to more modest aims of describing and interpreting what occurred in sociability. Certainly I was interested in the use of variables to try to explain attitudes and events and was indeed currently working on Paul F. Lazarsfeld's Academic Freedom Study at the Bureau of Applied Social Research.

Potter writes:

I agree with Jeanne that Riesman gave the impression that he was not interested in cause and effect and was interested in description. I interpreted this in terms of the nonfunctional sort of argument I had learned from Edward Rose at Colorado. All behavior is linked to a cultural matrix acquired unconsciously, willy-nilly; though "traits" may be found in association, generalization must be made for one culture only—in another the same traits might be unassociated or associated for different "reasons." One still searched for "gotogethers," but not to make causal inferences about the linkages. . . . The cause might be historical accident or a third overpower variable like "ethos."

Watson writes:

I was inclined to think of research as stimulated by a comparative question: what is the difference between A and B, and what are some of the possible explanations for this difference? Yet this conception of research never became dominant on our project. Riesman's apparent interest in description outweighed any interest in systematic, comparative analysis, and I eventually accepted the descriptive model as the appropriate one for our project.

17. E.g., Bales; Coffey, Freedman, *et al.*; Carter; Snyder and Rogers; Polansky, Lippitt, and Redl; Anderson; Heyns; Steinzor; Bronfenbrenner; Newcomb; W. Henry; Robert White; Barker and Wright; and Riesman.

18. Riesman did not attend the seminar, and the fact that he was not committed to what I took to be the guiding assumptions of the project did not become visible until later.

19. Change was to be made part of the design by obtaining repeated observations of the same subjects over a period of time in both natural and experimental settings.

20. Even when they were well acquainted, it continued to be true that Foote, Riesman, and Watson had different styles in sociability. No two of the three of them were drawn together as sociable partners.

21. Riesman writes:

My recollection is that I did consult one or two friends at Michigan; and I certainly assumed that Jeanne's Michigan training would be relevant, the more so as I respected the workmanlike quality of the Institute of Social Research; however, it is true I gained only later a cogent idea as to the nature of her various projects at Michigan.

22. Denney had a long association with Riesman and with the intellectual tradition which Riesman brought to the project, as well as an acute sensitivity to the nuances of interaction and a strong interest in sociability.

23. In the spring of 1955, after Prof. Paul F. Lazarsfeld's Academic Freedom Study went into the field, he asked Riesman to make a postaudit of the

survey in terms of the quality of the interviewing, the relevance of the questions, and the substantive concerns of the study itself. By its very nature, this work had to be done at once; by the nature of most projects, it took much longer than either Lazarsfeld or Riesman had anticipated; and it pre-occupied Riesman and kept him away from Chicago during much of the early work on the Sociability Project.

24. This is true of interviewers also, even on standardized surveys. But the interviewer on a survey carries a message from the outside world to which, to some extent, he can subordinate his personal qualities. Compare, for example, David Riesman, "Interviewers, Elites, and Academic Freedom," *Social Problems,* VI (1958), 115-126; June Sachar Ehrlich and David Ries-man, "Age and Authority in the Interview," *Public Opinion Quarterly,* XXV (1961), 39-56.

25. In notes of discussions at that time between Watson, Foote, and Ries-man, there is reference to the fall quarter as a pretest period. However, this image was never firmly established, and it seems likely that Watson was the only one who ever thought of it in that way. Bob Potter reports that he felt that the assignment given to the assistants was to deliver a demonstrated relationship between personality and leisure, experimentally obtained; there was no suggestion that the work at this time was exploratory in nature, no hint of a trial-and-error approach to the research task.

26. We can imagine that more favorable physical settings might have had other drawbacks. For instance, a new and deluxe modern building might have made people anxious lest their efforts not live up to the style of the building. If marginal locations foster poor morale, ideal conditions may, by destroying alibis, foster panic.

27. We do not claim that others would also find the assignment impossible; it might be possible to surmount the difficulties; we report only that at that time we were unable to do so.

28. It should, however, be noted that two-thirds of the coded episodes are classified as conversations among people we termed "casuals," that is, among relative strangers. We made use of our contacts with friends, in other words, to gain access to the sociability of groups composed largely or in part of persons who were relative strangers to us.

29. For Reisman this was perhaps especially true, since he not only does not use alcohol to help him get through a party but finds little opportunity for sociability in a life overfilled with competing demands and obligations. This made all the more ironic the later attack on the project as a cocktail-party binge at the taxpayers' expense.

30. Quite apart from the project, both Foote and Riesman were "contro-versial" at Chicago; the observers, unsure about both the value of the project and their own ethical legitimacy, would understandably at times take too personally and seriously attacks not really directed at them.

31. Although we wanted reports of good parties, we—and Riesman par-

ticularly—continued to be interested in reports of isolated incidents which might bear on his interest in social blockages or pathology.

32. This action was perceived by Watson as an effort to give her a free hand and create conditions which would permit her to be autonomous. Foote's withdrawal from active participation in project activities was hastened when it became evident in the fall that he and Watson wanted to proceed differently. They could not continue to urge contradictory courses of action for the project, so Foote decided to leave the management of the project to Watson.

33. Harold G. Nicolson, *Good Behaviour* (London, Eng.: Constable, 1955).

34. Foote left the University of Chicago to work with the General Electric Company in New York City. He has continued to have a critical interest in the project and has not given up the idea that he may be able to make some further contribution to it.

35. Riesman, *Individualism Reconsidered, loc. cit.*

36. For discussion of contrasting styles in small-group analysis, cf. Richard D. Mann, "The Development of the Member-Trainer Relationship in Self-Analytic Groups," *Human Relations,* forthcoming.

37. Erik H. Erikson, *Childhood and Society* (New York: W. W. Norton and Company, 1950).

38. George Herbert Mead, *Mind, Self, and Society* (Chicago, Ill.: The University of Chicago Press, 1934).

39. See Nelson N. Foote, "Identification as the Basis for a Theory of Motivation," *American Sociological Review,* XVI (1951), 14-21.

40. Potter writes: "Instead of sharpening identification concepts for the study of sociability, the seminar appeared to be a battle between Freud and George Herbert Mead, with neither giving an inch." It should perhaps be added that this same battle was also carried on at Chicago by Anselm Strauss and other "interactionist" theorists.

41. See J. Watson and R. J. Potter, "An Analytic Unit for the Study of Interaction," *op. cit.*

42. See also discussion of bass and treble sociability, page 302.

43. For further elaboration, see Watson and Potter, "An Analytic Unit for the Study of Interaction," *op. cit.*

44. See Kenneth D. Feigenbaum, "Sociable Groups as Pre-political Behavior," *PROD,* II (1959), 29-31.

45. Kenneth D. Feigenbaum, a graduate student in human development, and Lucinda Sangree, a graduate student in sociology.

46. Again, for elaboration, see Watson and Potter, "An Analytic Unit for the Study of Interaction," *op. cit.*

47. As already indicated, Riesman was inclined to pursue the question of what could be said about entire sociable occasions; he was interested in the choreographic flow of interaction, in sociability as a ritual, as patterned movement, as an art form. These "ethnographic" concerns with form and

movement could not be captured in an analysis that focused on separate episodes, which divided the party or other sociable encounter into too-microscopic elements.

48. R. G. Barker and H. F. Wright, *One Boy's Day* (New York: Harper and Brothers, 1951).

49. In a sensitive paper on his experience as a participant-observer in a factory working group, Donald Roy discusses the significance of a seemingly endless round of kidding with an unchanging cast of characters and themes and how his own intervention to upset the routine nearly resulted in the group's dissolution. See D. Roy, "Banana Time," *Human Organization,* XVIII (1959-1960), 158-168.

50. He came to the project with a Ph.D. in economics and an interest in learning about social psychology. He was lured away by the offer of a better position, teaching economics.

51. Cf. David Riesman, "Some Informal Notes on American Churches and Sects," *Confluence,* 4 (1955), 127-159.

52. N. Polansky, R. Lippitt, and F. Redl, "An Investigation of Behavioral Contagion in Groups," *Human Relations,* III (1950), 319-348.

53. D. Stock and H. A. Thelen, "Sociometric Choice and Patterns of Number Participation," *Emotional Dynamics and Group Culture* ("The Research Training Series No. 2"; Washington, D.C.: National Laboratories, 1958).

54. T. F. Leary *et al., Interpersonal Diagnosis of Personality* (New York: Ronald Press, 1957). We used the earlier dittoed report, M. G. Freedman, A. G. Ossorio, T. F. Leary, and H. S. Coffey, "The Interpersonal Dimension of Personality."

55. John Hotchkiss, graduate student in anthropology, and Lucille Kohlberg.

56. Since the text above was written, Watson has analyzed the data from these other parties as a rough sort of control in a forthcoming paper on motive-styles, nihilism, and enthusiasm in parties. Over all, the differences between the parties reported and analyzed with relative fullness and those given somewhat more cursory treatment are very small.

57. See David Riesman, Robert J. Potter, and Jeanne Watson, "The Vanishing Host," *Human Organization,* XIX (1960), 17-27, and "Sociability, Permissiveness, and Equality: A Preliminary Formulation," *Psychiatry,* XXIII (1960), 323-340.

58. *Loc. cit.*

59. The gap between large questions and limited methods can be found in many different areas of contemporary work. For example, many psychologists appear to be driven into a kind of schizophrenia when they seek to relate their civic concern with race relations or the prevention of war to their professional concern with current procedures in, for example, small-group studies.

11 First Days in the Field *

BLANCHE GEER

Participant-observation studies, as we have been doing them,[1] follow a pattern of several years in the field, a period of analytic model-building, outlining, and drafting, and the final assembly of data in the form of description, evidence, and argument for the monograph. The beginning of the final stage, when we go back to the field notes to see if there is evidence for what we want to say, is a time of suspense and self-questioning. It brings up a point seldom mentioned in monographs but frequently discussed by field workers among themselves: the relationship of initial field experiences to our thinking before entering the field and after the field work has been completed. There is so much reiteration of findings in our work that we sometimes think we have always known what we find (everybody knows it; why bother to write a book?) and at other times that it all became clear in some magical way on the very first day in the field.

Most field workers include in their notes material which is not narration, quotation, or description, but comment. It occurred to me that I might use, as a partial answer to questions about what happens in initial field work, the comments recorded during my first days in the field on our current research. Do strategies and concepts change? By what mechanisms? How is subsequent field work affected, and how much of the first experience gets into a final monograph?

METHOD

Field work on our present study of undergraduates at the University of Kansas began during the summer orientation period for prospective freshmen in 1959.[2] At the end of a six-hour day in the field, I dictated

* A chronicle of research in progress.

an account of what I had seen and heard, occasionally inserting comments on the material or appending an interpretive summary. In this report, I shall use the thirty-four comments in the first eight days of field notes. They have been edited when in need of grammatical correction, and names have been changed. When comments duplicate each other, I use one but tabulate all of them as a rough measure of my initial interests.

In addition to the comments, I shall draw on letters and memoranda exchanged by project members before entering the field and the original proposal for a one-year pilot study. The full proposal and a memorandum written after the summer field work complete the documentation.

BEGINNINGS OF THE UNDERGRADUATE STUDY— THE PREVIEWS

Tentative arrangements for a study of undergraduates were made with the University of Kansas while we were still engaged on a study of the University's medical school. During the academic year 1958-1959, we met with members of the administration interested in the project. They provided a gradual introduction to the University and indicated what they wished it to become.

In November of 1958, we submitted a proposal to investigate the following question: What are the differing perspectives on academic work characteristic of an undergraduate college, and how are these influenced by participation in student groups characterized by the possession of a student culture? Our method would be participant observation and interviews of a random sample of one or more subgroups of the student body, beginning with premedical students.

On January 22, 1959, at a lunch meeting with members of the administration, the dean[3] mentioned that the University conducted summer orientation sessions, or previews, for prospective freshmen. In notes on the conversation I summarized what he said as follows:

> Throughout the month of July, groups of prefreshmen come to the University for talks by administrators and for a session of filling out forms. . . . Approximately half the prospective freshman class is processed in July to relieve the pressure at fall orientation time. Each preview group contains about two hundred students with an approximately

equal number of boys and girls. The selection is first-come first-served, so that the groups are mixed in respect to the schools of the University (engineering, fine arts, liberal arts, and so forth) they plan to enter and the region from which they come. There is, however, some tendency for students in the same high school to send in their applications together so that sometimes there are groups that know each other at a preview, particularly from the larger high schools in the state.

In June, I talked to the director of the previews, who agreed to supply me with the names and home towns of students attending, banquet tickets, and a schedule of events. There would be two previews a week—the first on Monday and Tuesday, the second on Thursday and Friday—for the first three weeks of July.

I arrived on campus on the morning of the first preview (July 6) and picked up the list of names from the director's office. At the Union, where orientation sessions were being held, I introduced myself to previewers and attended meetings with them to get their comments and make appointments to meet later on. In the afternoon, many students had free time which gave me a chance to talk to them at length in the Union or dormitories or to explore the campus with them. At six o'clock, everybody went to a social hour and banquet, and I received a name tag and table assignment along with the previewers. After dinner, the previewers again had free time (at some previews there was a dance) until the house meeting held after closing hours at ten thirty.

The second morning was taken up for most students by placement and physical examinations, which left me little time for interviewing. My best opportunities were during the afternoon free time, at dinner, and the dance.

This pattern of going to formal sessions and interviewing during free time followed our plans for the field work. It produced enough data on the previewers' backgrounds and their expectations of what college would be like to orient us. We did not anticipate one feature of the previews which materially changed our plans. No one had told us, thinking it unimportant perhaps, that college seniors acted as speakers, entertainers, and hosts at the dinner table. They were also dormitory counselors.

ANALYSIS OF FIELD COMMENTS

Data, no matter how you collect them, are recalcitrant. They will not always answer the questions you put to them. In this report, since I am dealing with my own work, I want to be particularly careful to stick to the data, to avoid reading more into the comments than is in them by interpreting them in the light of our subsequent findings. I have, therefore, subjected the comments, somewhat grandiosely, to the kind of analysis we use in regular studies, collecting all data in the notes that bear on a topic (in this case all comments), grouping them by subtopic, and then looking to see what I have. The thirty-four comments fall into five categories: the field worker's role, the problem of empathy, the solution of anticipated problems, the nature of working hypotheses, and the recognition of a major theme. I shall consider the categories in order, noting for each of them the number of comments included. Discussion of the relationship of the comments to our original planning of the study and the monograph we are now writing follows the analysis.

The Field Worker's Role

To those who have not done field work, the adoption of a role with informants presents a difficult problem. There is a large literature on the pitfalls of role and failures of rapport, but it is usually written with the interviewer rather than the observer in mind. The interviewer starts from scratch with an informant. He must rapidly establish himself securely enough in a relationship with another person to permit intimate questioning. The participant-observer establishes a relationship with a group. To a single informant, he is much less of a stranger because he moves freely in the informant's own setting and can be seen interacting with other members of the informant's group.

If the setting and the group are reasonably familiar to the observer, his problem in initiating a role is a matter of judicious negatives. He should not have the manner or appearance of any group which his informant group distinguishes sharply from itself. This does not mean forcing identity with the informant group; it does mean that the observer of students, if he wishes a good understanding with them, will avoid the manner of teacher and authoritative adult. Selecting a neutral, approachable role in the sense of acting and speaking in ways which are not threatening to informants smooths the first days of participant observation. But the field worker must soon become aware of the role he plays in more detail to assess its effect. The first comment in my field notes on role deals with this problem.

> 7/9/59: I am conscious that part of my role for [the previewers] is an educational one. I am somebody in a profession that they have not run up against before, and I have a lot of knowledge about universities that they could use. This is quite different from the role you play with medical students who are tolerant of you but not seeking anything.

Occasioned by a previewer who wanted me to tell him whether he must study medicine to be a psychologist, the comment is a warning to myself: Don't wise up your informant! More subtly, the comment suggests the field worker's effort to uncover informants' concepts of who he is and what he wants to know, since such definitions govern what people tell him and how they say it, whether they will volunteer or must be questioned.

After this warning, role in the sense of presentation of self does not turn up in the confessional of the comments, but I become conscious of doing a number of things in the field not planned in our proposal or memoranda and of leaving other things undone. I am concerned about role in the sense of my function as coworker on the larger study of which the previews are a small part. After five days in the field, I evaluate this role as follows.

> 7/13/59: I want briefly to take a look at the kind of questions I have been asking and my general goals during this period of field work. First of all, I've been getting background data [on previewers]; this is necessary if we are to understand the types of students we are dealing with. Second, I've been trying, evidently, to pick up a tremendous body of facts and names, about campus organizations, slang, and customs. This is a natural thing for any field worker, trying to orient himself, to do, and I think it is particularly important in our study if we are to identify groups, as it is by these means that we are able to separate students from each other and recognize them easily and quickly. In other words, if we get all this stuff down we may be able to sort them out without going to such lengthy interviews as this [a three-hour talk with a senior].
>
> I seem to have the feeling that information of this kind is necessary in order to be able to find [the boundaries of] groups. I also seem to be developing a technique for next fall in which I talk with a student

long enough to get his confidence thoroughly and make plans to meet him next fall and attend some class or activity and get to know his friends. This, I think, will be quite a reasonable way of bounding groups, at the same time learning more about the actual activity.

I think I have developed an impatience as a result of our various theoretical and methodological innovations of last year. I apparently want to go right out and put data into perspectives and tables.[4] I am impatient with the mountain of background information that I must first learn before I can interpret any of this. . . . Of course, there are repeats and reiterations coming already, and these are the material of perspectives. At the same time, an outline of the place is forming in my mind, a kind of topographical map in which it becomes more and more easy to locate what the various students say to me.

As I begin to see what we need to learn and ways of learning it, my conception of my role in the study rapidly expands from that of interviewer of prefreshmen to general spadeworker. I do not want to lose the opportunity presented by the previews of laying the groundwork for what we will do in the fall. Getting campus background, learning the lingo, and setting up future meetings with informants are important. Almost in spite of myself, I have embarked on an attempt to apprehend undergraduate culture as a whole.

Our proposal seems forgotten. Of course, there were not enough premedical students at the previews for me to concentrate on them. To limit myself to our broader objective, the liberal-arts college student, was difficult. The previewers did not group themselves according to the school or college of the University they planned to enter. Out of ordinary politeness (at the dinner table, for instance), I found myself talking to prefreshmen planning careers in engineering, pharmacy, business, and fine arts, as well as the liberal arts. Perhaps it is impossible to stick to a narrow objective in the field. If, as will nearly always be the case, there are unanticipated data at hand, the field worker will broaden his operations to get them. Perhaps he includes such data because they will help him to understand his planned objectives, but he may very well go after them simply because, like the mountain, they are there.[5]

Three comments in the field notes deal with role. I have quoted the one dealing with role as presentation of self and one of two extended discussions of my role as coworker in the undergraduate study.

The Problem of Empathy

Developing empathy with informants as a group often presents more of a problem to field workers planning a study, at least in anticipation, than the adoption of an interaction-facilitating role. Role in this sense is a public stance which all of us practice, but empathy is personal. Field workers are not free of prejudice, stereotypes, or other impediments to the understanding of out-groups. But to study, as we proposed to do, the experience of going to college from the point of view of the student necessarily entails at least recognition of personal bias in order to achieve empathy with the informant group.

Throughout the time the undergraduate study was being planned, I was bored by the thought of studying undergraduates. They looked painfully young to me. I considered their concerns childish and unformed. I could not imagine becoming interested in their daily affairs —in classes, study, dating, and bull sessions. I had memories of my own college days in which I appeared as a child: overemotional, limited in understanding, with an incomprehensible taste for milk shakes and convertibles.

Remembering my attitude as I began to sort out the thirty-four comments in the field notes on the prefreshmen, I expected to find evidence of this unfavorable adult bias toward adolescents. But on the third day in the field I am already taking the students' side.

> 7/9/59: Some of these statements by boys like Joe Ropes may sound overdramatic, or self-conscious, or just plain foolish to an adult, and yet . . . when [you] see their serious and thoughtful faces you realize you [are] getting the best fruit of their experience and thoughts and you take them very seriously when they make statements like this.

Humorless seriousness evidently does not fit the stereotype of college students I took with me into the field. In my own person, I might have kidded the students out of it, but as a field worker I must listen and record. My disbelief is reserved for the comment:

> 7/9/59: Here, again, this sounds just too fatuous but I can not do anything but repeat that these kids are speaking seriously from their hearts. They're not trying to be humorous, and they're really not trying to brag. They're explaining the world as they see it to me in [a] kind of man-to-man fashion, and they're not trying to be overearnest about it or anything. It's just apparently a normal conversation with a kid of this age, at least on such an important day in his life as his first day

of a college preview. [This observer's comment comes after a student said, "I think getting into student politics here would be a good thing. You get to be well known and people get to like you. I guess that's an important thing if you want to have your place in the sun."]

I have lost the adult tendency to laugh at such statements, but I think my colleagues will. (They are reading the field notes but are not yet on the scene.) To read what the students say is not as compelling as experiencing it in the flesh, hearing the voices, seeing the gesture and expression. Responding to the students with a solemnity equal to their own, I have fallen into empathy by acting it out.

Later in the field work, I attend a dance in the Trail Room of the Student Union and comment as follows.

> 7/16/59: I am amazed again, in starting the field work, at how quickly . . . you respond in quite other ways than you ordinarily would. I now react with a startled reaction when some dean or other says, as Dean Brown did the other day, "Isn't it boring, talking to all these children?" No, it is not boring when you are doing it, nor is the behavior odd. It has a great deal of inevitability. [What I am referring to here as not being odd is the shyness of the boys at the dance: great numbers of boys walking up and down and standing in groups and some sitting with their heads in their hands, looking at the girls and presumably wondering if they want to dance with them. One or two Romeos leaned nonchalantly against a pillar or two and surveyed the scene without becoming a part of it.]

This comment is part of the notes for the last day of field work on the previews. My reaction to the dean's question suggests not only anger at his failure to understand my job (whatever field work is, it is not boring) but also anger on the student's behalf: why call them children? I am taking up cudgels on their side. Perhaps the rapid development of empathy for a disliked group does not surprise old hands at field work, since it seems to happen again and again. But it surprised me; I comment on it seven times in eight days. Three of the comments have been quoted here.

The Solution of Anticipated Problems

The development of empathy in the field is not the only surprise an observer experiences. Theory, other studies, and common sense make one anticipate difficulties which do not materialize. People one expects to be hostile are not; situations one expects to be incoherent reveal themselves as relatively easily grasped when one is in the midst of them; apparently difficult problems of finding subjects or grouping them in manageable categories are easier in the doing than in theoretical discussion. There are only three comments in the field notes on anticipated problems. One deals with the kind of continuing concern one seldom puts down on paper.

> 7/8/59: Throughout my contacts with administrative officials on this day I express surprise at their friendliness and my ease of access to them. I evidently expect them to ask me personal questions about students.

Absence of one problem provokes anticipation of another. Friendly administrators may expect information about students it would be impossible for us to give while keeping students' confidence. To deal with this problem, we later began efforts to educate members of the administration in field workers' ethics.

In the medical-school study, student fraternity groups were a major variable, meaningfully related to academic perspectives in the freshmen year.[6] But we anticipated serious problems in identifying similar groups among nine thousand undergraduates. We thought and wrote on the formation of groups and the technical problems of observing them. Here is a comment intended to allay some of our fears.

> 7/6/59: I think that this one afternoon and evening has given us some idea of the friend-making process at the beginning of college. When you see it happening under your nose this way, it seems so natural that I wonder why we ever had any questions about it. I think you could make a typology somewhat as follows: There are the students who come from the same high schools or nearby towns and have some acquaintance with each other, and if there is a small number of these they get together and stay together, at least at the beginning of school. It is probable, for instance, that Tom, Dick, Harry . . . from [a small town] will room together in some combination. Those students who come from Kansas City can get together rather easily, although they have come from different schools, by placing each other by means of country clubs, addresses, and boys they have dated. All of these [cues], of course, [are] in addition to the obvious, immediately grasped detail of hair, manner, and clothing. Isolated students from different towns pair up with their assigned roommates for the preview, and they perhaps make a friend to room with the rest of the year or may simply take their chances on being assigned a good person.

Watching people initiate friendships lessens the mystery. Seeing ten students choose a roommate is to begin to structure the activity. The observer now has a list of commonalities to look for: home town, country club, dress, and temporary propinquity.

Getting into the field disposed of another problem.

> 7/16/59: This is really a summary comment. I had the idea before I started this field work that because of the complete turnover in students each time it would be a rather disjointed affair in which the observer did not get a cumulative and continuing sense of growing knowledge of what college is all about. I find that I was quite mistaken in this and that this feeling is going on at a great rate and that I am already at the stage in which I quite falsely pretend to ignorance to get somebody's viewpoint on such things as the fraternity.

Why do field workers so frequently anticipate problems that do not materialize? We do it on each new study. We underestimate people's trust in our neutrality, their lack of interest, perhaps, if we seem to be doing no harm. And we project theoretical problems into the field. Because the process of group formation is difficult to conceptualize, we suppose it will be difficult to observe. We expect ephemeral, unstructured situations like the previews to appear incoherent. Perhaps such mistakes are a necessary part of our efforts to design the study in advance.

The Nature of Working Hypotheses

Although the term *observer* suggests passivity, a participant-observer in the field is at once reporter, interviewer, and scientist. On the scene, he gets the story of an event by questioning participants about what is happening and why. He fills out the story by asking people about their relation to the event, their reactions, opinions, and evaluation of its significance. As interviewer, he encourages an informant to tell his story or supply an expert account of an organization or group. As scientist, he seeks answers to questions, setting up hypotheses and collecting the data with which to test them.

One type of hypothesis is drawn up before entering the field. Essentially a list of variables which theory or common sense suggests may be relevant to what the investigator wishes to study, the hypothesis, for the field worker, takes the practical form of kinds of people to see, places to go, and questions to ask. Although he spends a good part of his time just listening to informants or drifting along with a group to see what will happen, the observer also forms hypotheses during the field-work period. These are called working hypotheses. Some of them are so simple they can be tested immediately by having a look at a group or asking questions of informants. Others, usually based on an accumulation of data, predict an event or state that people will behave in specified ways under certain conditions. These undergo a prolonged process of testing and retesting, preferably by more than one field worker, over a period of months and years. There is no finality about them. They must be refined, expanded, and developed. Checking out may depend on the return of the organizational calendar to its beginning point or the election of a fresh group of officials.

The concept of working hypothesis is not difficult, but field workers often have trouble explaining it to others and sometimes to themselves. The concept is clear as a generality, but its mechanics, the doing of it, smacks of magic. Untrained observers, for instance, can spend a day in a hospital and come back with one page of notes and no hypotheses. It was a hospital, they say; everyone knows what hospitals are like. My comments in the field notes suggest that working hypotheses are a product of the field data itself and of whatever ideas the field worker can summon. The initial stimulus may come from repetition or anomalies in the data which catch the observer's attention so that he searches his mind for explanations. Or he may start from the opposite end— what is in his head—and search the data for evidence of stereotypes from the general culture or notions derived from discussions with colleagues, previous research, and reading.

I do not wish to dignify an impromptu process by formal discussion. My field-note comments are not technical. They are summaries, not of an entire day's work, but of parts of it drawn together into tentative statements of findings, often in quasi-quantitative language. A comment at the end of the first day's field work shows I have been using the general hypothesis: previewers will have concrete knowledge of academic arrangements in college. (We had drawn up a list of questions to test this proposition before I went into the field.) But:

> 7/6/59: The tremendous ignorance of the prefreshmen of college is somewhat startling. They don't know how many hours of classes there will be. I think they are going to get a big surprise when they find that they may have only one class in the morning and another late in the

afternoon on a given day. They have many different notions of how many hours of study they are going to get in.

One day's observation has shaken the prefield hypothesis—students have specific knowledge of the academic side of college before entrance —and substituted the hypothesis that they do not.

The comment continues with a statement about an area of college life we did not plan to study. We had no hypotheses about it, no list of questions prepared.

> 7/6/59 (continued): [The previewers] have very little notion of the tremendous numbers of organizations and activities open to them on the campus. Their ideas are rather limited to the advantages of learning manners, meeting girls, singing or other musical activities, and occasionally sports.

It is not clear why I was struck by the fact that previewers do not talk about college activities. We had on hand University bulletins listing hundreds of organizations. I may have picked up the notion that activities were important from conversation with seniors even on this first day.

It is clear, however, that in the concluding portion of the comment I am drawing on stereotypes in the general culture.

> 7/6/59 (continued): Sports seem, so far, with this group to be a pretty minor undertaking. I wonder if by any chance there is some selective factor acting here and the athletes or even those interested in athletics do not even show up at previews? No freshman has mentioned big football week ends or college rah-rah shindigs in any connection.

Athletes turned up at subsequent previews to lay my sampling doubt at rest, but at the time the hold on me of the tradition of collegiate sports and big week ends was strong. I very tentatively state the working hypothesis that they are not interested.

The first day in the field leaves me with three working hypotheses, each expressed, interestingly enough, in negative form. If I did my work well on subsequent days in the field, I was on the lookout for data to substantiate or disprove these propositions. Reading the educational literature, as we continued to do in the field, probably supplied me with another working hypothesis: college students are not interested in "culture" or religion. There is no comment on these points until July 16, perhaps because I was not getting much data.

7/16/59: They had no notion of "cultural advantages" in the way of concerts, lectures, and so on. I noticed again at the dinner the very solemn and attentive faces that watched the students playing the violin and piano. The faces of the dean's staff and the senior students wore that half-smile of gracious appreciation which particularly in the Middle West, I think, accompanies or encourages a performer—or perhaps what I'm describing is a middle-class phenomenon. All the faces of the freshmen that I could see as I ranged around the room with my eyes, however, were quite unsmiling. It was impossible to tell whether they liked it or didn't like it, but it was clear that they had no social conventions of the proper listener's face.

I haven't heard anybody talk down culture nor have I heard anybody talk it up. It is the same with religion. Those who are religious reveal it by their conversation about something else as very much a part of their lives and connected with their choice of a fraternity or a profession.

The working hypothesis proposed is again negative but more complex than those about sports and knowledge of academic matters, which were relatively simple to test. In a group of x previewers, so many express interest in college sports; so many do not. It is a matter of asking the question and counting. The comment above suggests that previewers use their religion when making important decisions. This is a two-step hypothesis. It directs me to look for religious influences only under certain conditons. It is a statement of the logical type: if this, then that.

If some working hypotheses mercifully narrow the area the field workers must observe by specifying a relationship between variables, others ask him to broaden his perspective, to look at something small and apparently unimportant as a matter of regional history.

7/13/59: I had a conversation the other night with Mary and Dave Newell [sociologists], in which they asked me about the new field work, and I told them as an amusing anecdote about the stress in the fraternities on the teaching of manners. Mary was appalled and said she thought she should immediately leave the Middle West, at least before the great quantities of KU freshmen began to spread finger bowls on the streets of the Plaza. She seemed extremely upset [she is an English girl who has been over here about three years] and seemed to think that this sort of thing was the downfall of education and of America. David said that, on the contrary, he found it very heartening as it was clear the future generations of Kansas Citians would already know what to do with their finger bowls because their fathers, who had been to KU, would be able to teach them as children. And in this way the progress of the race would go on. I concurred in this.

The image of KU as a civilizer taking the grandchildren of the frontier at least one step along the path to a world comparable, at least at the dinner table, all over the world for "educated people," is a good perspective, I think, and one that we must not forget. In this way, it looks quite reasonable for a student whose parents have a grocery store in that distant Kansas town to spend four years acquiring the rudiments of manners that he was not taught at home, and with these he can go on to the real business of life and study in graduate school.

Perhaps because my friends were not midwesterners, they suggested an important relationship to me, putting manners in a historical perspective. My working hypothesis expanded: look for evidence that training in the social graces is a major function of the University.

Inclusive hypotheses, of course, cannot be immediately tested in the field. If something is really important, we should expect to find it cropping out in many sectors of college life. Evidence for and against must be accumulated throughout the study; we must look for it in classes, living groups, and everyday activities. Detailed hypotheses derived from one's own or other people's work on similar institutions may also be useful models for the field worker, whether or not they can be immediately checked. In a long postscript to the summer field notes, I try to relate data on the previewers to a complex model of student development over time which we used for the medical school.[7]

> 7/16/59: No one so far has made any distinction between practical and impractical learning in the fashion of the medical students. I used this as a probe a couple of times and got uncomprehending looks. What there is in college is something to learn, and then in another compartment are their aspirations toward various positions and professions. . . .
>
> Again, [the previewers] have no smartalecky knowledgeableness about rules and regulations or how to beat the dean or the test or anything else. College is a very serious place, and at least at a preview you don't think about these things in connection with it. In this way, they are very similar to the medical-school freshmen, and it will be interesting to see if the conflict between learning for the professor and learning for yourself develops in these students. I know I certainly had an extremely acute case of it as a college student, to the point of wondering whether I should go on.

Working with the hypothesis that previewers may have attitudes toward studying similar to those of medical-school upperclassmen, I find they have instead the idealistic acceptance of college of medical-school freshmen at entrance. Since the latter rapidly become test-wise, I tentatively advance the hypothesis that previewers will also, when they get into college. The basic hypothesis—attitudes toward academic work arise in response to some facet of college experience—is in two steps with a specific time condition. It can be tested only in the fall.

Although our plan was to talk to prefreshmen, I increasingly found myself with older students, present to direct the previews, who told me about the college in great detail. I was acquiring, as I continued to interview freshmen, an idea of what was ahead of them and the contrast in knowledge, interests, and manner between prefreshmen and upperclassmen. The seniors spoke frequently about activities and cultural advantages; previewers did not. The sense of contrast was an unanticipated consequence of what I regarded as a mistake in my field behavior. The next comment reflects information picked up from seniors and staff.

> 7/16/59: Here is an addition to the conversation with Betty Jones [Student Union staff]. She said, "There are some students here who spend practically all their time on activities. I don't know when they do any studying. And really I don't know sometimes what things are coming to. On Monday night they have house meetings in the dorms, and that takes two hours. On Wednesday night there's a dance, and on Thursday night there are all kinds of other meetings."
>
> It is quite clear that the [preview] students are unprepared for this wealth of competition for their time, this wealth of choice, and I would say that probably only the ones with something very specific in mind are going to get over the realization of this variety as a painful and difficult thing when it comes to them after they arrive. This is really a prediction, I suppose, and I feel that it is reinforced by their earnestness and wanting to do well in their studying and have their family be proud of them, which is going to set up a conflict between all these fascinating things and their rather weak direction in themselves. They want to study, but they don't know what and they don't know how much. They are earnest and sincere, but they have no guideposts. They look to me at the moment like . . . a bunch of very sweet lambs being led into a slaughter of decisions to make, pressures to withstand, and moral fiber to reinforce. And yet many of them speak regretfully of their casual high-school years, and you get the strong flavor of missed opportunities and lack of foresight.
>
> This is not true of all students, and I think I am right in saying that the ones from the small schools far out in the state are more apt to have experienced a studious atmosphere in high school.

Provided by seniors with a view of college as full of decisions, pressures, and demands in addition to the academic (to be checked out later) and sufficient acquaintance with previewers to suggest that they do not see it this way, I am possessed of an elaborated hypothesis about change and major learning areas in college.

The five comments quoted in this section (four additional examples are omitted) suggest that the field worker makes constant use of working hypotheses from many sources. My use of hypotheses falls roughly into three sequential types. The first operation consisted of testing a crude yes-or-no proposition. By asking informants or thinking back over volunteered information in the data ("nearly all students today" or "no student"), I stated a working hypothesis in the comments and began the second operation of the sequence: looking for negative cases or setting out deliberately to accumulate positive ones. At the second stage, working with negatively expressed hypotheses gave me a specific goal. One instance that contradicts what I say is enough to force modification of the hypothesis. It is a process of elimination in which I try to build understanding of *what is* by pinning down *what is not*.

The third stage of operating with hypotheses in the field involves two-step formulations and eventually rough models. Hypotheses take the form of predictions about future events which may take place under specific conditions or changes in informants over time in conjunction with events. Needless to say, particularly in the first days in the field, the worker is never at the same stage with all his data; he may be operating at the yes-or-no level in one area and advancing to the model stage with another at any given time.

Recognition of a Major Theme

Participant-observers sometimes say that the major themes of a study appeared very early in the field of work, although they may have been unrecognized. The field comments present an opportunity to investigate the phenomenon—one case in what should be a larger study.

Is there early and sudden insight? If so, what brings it about? On what grounds can an observer predict that a major theme will be central to the study? To get at my reasoning retrospectively, I take the comments related to recognition of a possible major theme in chronological order. After the first day in the field, I tentatively suggest a relationship between two reiterated findings.

7/6/59: I had been probing very gently all day when I asked students about KU as to whether their image of it was a place where you did not study but just had a good time, but I had gotten no flavor of this from any of the students I've talked to. They seem to be very serious about college and, if they were not, were pretending to it or were simply overawed by its social challenge. I think this is part of their saying it is such a big place, certainly a place where you come with some trepidation and with respect.

Since I talked to as many visibly different types of previewers as possible, I was interested that, even after probing, no student provided expected data. As statements on "bigness" were volunteered, I regarded them with added confidence.[8] Considering the comment now, we can say that my tentative statement of the relationship between the academic seriousness of the previewers and their notion that KU is a big place takes the form (following Polya) that A is made somewhat more plausible by B.[9]

The next day I comment on another form of seriousness.

7/7/59: Throughout this day there are small indirect references to the fact that I have found students more serious about academic affairs than I expected. From Dean Brown's conversation, I get the notion that the administration is just as serious about educating them in the social realm as are the students themselves. This is evidently a major change in my orientation to the college.

While the increment is not great, if previewers are serious about social matters (C) as well as academic (A), A becomes more plausible. The dean's concern about the social realm helps still more.

On July 8, I announce "bigness" as a major theme because it is frequent and widespread.[10]

7/8/59: One major theme is evidently already coming out in today's field notes: KU is a very big place. I am already comparing coming to KU as similar to going to New York City [to Eastern kids].

If I think there is a relationship between academic seriousness (A) and "bigness" (B), to feel certain about B, in the primitive analysis I am conducting in the field, increases my confidence in A once more.

Later in the same day's notes, I speculate about the origins of seriousness.

7/8/59: In summary of these two days of field work, I feel frustrated at having talked to a very few students, and yet at the same time I feel that I have the beginning of some sound knowledge of how things are up here and that this will increase if I attend more previews. There already seems to be some repetition in the students' reasons for coming to KU, the amount they expect to study, and the general serious academic outlook that they have. They are more focused on academic matters here at the beginning than you would expect, but I have an idea this may be because these things appear manageable or more manageable to them than the great unspoken questions about whether they will make it socially. They have real fears about this last and very much of a do-or-die attitude, I think, although they did not express this to me directly. Real life has begun, and if you don't make it now perhaps you never will.

The existence of seriousness about academic and social aspects of college (A and C) is made more understandable, hence plausible, if there is evidence of common cause: previewers think college is "real life" (D). We have an explanatory proposition: D leads to $A + C$.

On the third day in the field, taking a subgroup of the preview population—prospective graduate students—I explain their seriousness by the fact that for them college is a step in an irreversible career sequence.

7/9/59: Apparently the recent introduction of psychology into the high schools has had a great effect, and that, plus nuclear physics, is leading many students, who probably would have before been at a loss if they did not want to be doctors, to think of graduate work and academic research careers. These kids are taking a really big jump in coming to college because they have no intention of going back to the small town which they came from, and they are aware that there would be no place for their knowledge there.

Since an irreversible career sequence (D_1) is certainly "real life," we now have a proposition of the form $D_1 = A$. Plausible explanation of seriousness in a subgroup lends greater strength to the proposition $D \to A + C$ for the population as a whole.[11]

At the time of the fourth preview, although I do not use the phrase "major theme," I make a last statement associating the "bigness" of the University, academic seriousness, and seriousness about social life.

7/16/59: I have so far found absolutely nothing to contradict the image of KU students, prefreshmen that is, as very serious and in many

cases academic-minded. Their fears are of its bigness, possible un-
friendliness, and snobbishness and how hard the work will be.

In the primitive analysis carried on during field work, I now seem to
take it for granted that previewers define college as real life and that
their seriousness toward it and reference to it as big express this definition $(A + B + C = D)$.

In what sense can we say that I recognized a major theme in the
eight days of field work? Certainly there is no flash of insight, no sudden revelation. The formulation proceeds slowly along lines of other
working-hypothesis sequences, but the data bear up under elaboration.
Unlike some of the working hypotheses mentioned earlier, the "seriousness" theme does not peter out. According to the comments, the field
notes are full of it. I cannot turn up a negative case. As at least part
of the data is volunteered, I am not afraid that my questioning has put
it in my informants' heads. Insofar as I was able to identify visible
groups or types of previewers, I found seriousness present in each. It
characterized both of the two dimensions of college, academic and
social, that previewers perceived.

Equally important in suggesting status as a major theme, seriousness
has no rivals. It is easy to elaborate[12] and not trivial; if we find in subsequent field work that students consider college real life, it should prove
important for understanding their behavior.

DISCUSSION

The comments make clear that the answer to the question "Do strategies and concepts change in the first days of field work?" is emphati-
cally *yes*. Furthermore, many of the changes are of such a nature as to
affect subsequent field work radically. Table I summarizes the discussion.

Perhaps because of our experience with medical students of the
same University, role as presentation of self to informants changes
least. My role as coworker on the larger study changed rapidly from
interviewer of prefreshmen to general ground-breaker. Largely because of the unanticipated presence of college seniors, several of whom
I interviewed at length, I became aware of students' interests in prestige rankings among living groups, extracurricular activities, and politics. In the fall, we dropped our plan to interview students in order

to map out important areas of interest to them and proceeded on the preview leads.

TABLE I

Changes in Concepts during Early Field Work

CONCEPT BEFORE PREVIEWS	COMMENTS SHOWING		TOTAL
	CHANGE	NO CHANGE	
Role	2	1	3
Empathy	7	0	7
Anticipated problems	3	0	3
Hypotheses	9	0	9
Major theme	9	0	9
Miscellaneous	3	0	3
Total	33	1	34

Three days of field work were enough to change my concept of college students dramatically. Before entering the field, I thought of them as irresponsible children. But as I listened to their voices, learned their language, witnessed gesture and expression, and accumulated the bits of information about them which bring people alive and make their problems real, I achieved a form of empathy with them and became their advocate. The observers who began work in the fall experienced the same change, but not until they got into the field. Reading my field notes did not help.

Parenthetically, one might suppose that empathy for informants, once developed, would become a problem in itself. It often feels like one in the field but drops sharply on leaving it. After a few weeks on analysis, I wondered how I could stand those silly kids. Discussion with coworkers and getting the faculty perspective later in the study also helped to restore a balance.

The problems we anticipated with the administration remained as illusory in the fall as they were during the previews. Identifying student groups, which we thought would be difficult, did not turn out to be a problem in the fall. As at the previews, it was easy to meet individuals who then told us about friends or fellow group-members; we could, over time, get at more of the variables related to formation and maintenance.

Many of the hypotheses I took into the field at the previews did not check out. Prefreshmen did not have concrete enough information about what college would be like to answer our questions about studying, using time, choosing courses or activities. Making the hypotheses and asking questions, however, served to structure initial contacts and produced negative findings which changed our concept of what previewers were like and what they would experience as freshmen. The working hypotheses I used provided a similar entree for work in the fall. The major theme I began to elaborate during the previews—students take college very seriously as part of real life—was a change, not so much from our prefield-work concepts as from the stereotype in the general culture (which the comments suggest I shared) that finds college students, particularly at large state universities, frivolous and sports-minded. This is the way they are in fiction and much of the literature. My efforts to uncover frivolity in the previewers failed. Seniors swamped me with complicated accounts of their activities. In the fall, we started out looking for negative cases and continued to look throughout the field work without much result. The theme should have a place in the monograph in a more differentiated form, along with others we developed later.

CONCLUSIONS

If early field work reaches few conclusions, it may nevertheless have far-reaching effects on the rest of the study. Memoranda written after the previews indicate that the idea of dealing only with premedical students in the fall has gone by the board. We are aware that students in the different schools of the University are so intermingled in living groups and activities that we must deal with all of them. Our proposal to investigate only the academic aspects of college no longer seems feasible. The academic is too closely tied to other aspects of students' lives.

Evidently the preview data—particularly those from the senior leaders—went beyond my comments in giving us a picture of the campus as we might expect to find it in the fall. In a memorandum based on discussion of the summer field notes written before entering the field in September, Becker states:

> The major units of the college in which students belong are based on residence and are contained in the following list: large fraternities,

small fraternities, . . . scholarship halls, other dormitories, married students' residences, and rooms in town . . . each of these groups can be thought as a network of interconnected cliques . . . essentially similar in their views of academic effort.[13]

Becker goes on to hypothesize several student cultures on campus. He later wonders if all may not have the same general goals, but, à la Cohen, differential access to success.[14]

It is clear that while one may reasonably expect initial field work to settle questions of role, empathy, and anticipated problems and to lay the grounds of later work by developing hypotheses and beginning the elaboration of a major theme, the observer may fail to comment on important things. My comments were concerned almost entirely with the previewers. I let the long interviews with seniors speak for themselves. Having more than one person on a study lessens the danger that such leads will be missed. Practically, the early field work provided us with informants to follow up and previewers as foils for college students when we met them. Its most far-reaching result was our broadened objective. We would study all the students, abandon interviews in the formal sense, and take our chances in the field, trusting the students to be as articulate and helpful as previewers.

One must conclude that the first days of field work may transform a study, rightly or wrongly, almost out of recognition.

• NOTES

1. See Howard S. Becker *et al.*, *Boys in White* (Chicago, Ill.: The University of Chicago Press, 1961).

2. The study is financed by the Carnegie Corporation and directed by Everett C. Hughes. The field workers were Howard S. Becker (director), Blanche Geer, and Marsh Ray, then of Community Studies, Inc., Kansas City, Missouri.

3. The "dean" mentioned here and in subsequent comments is a general term for all administrators. There are so few that to give anyone a more precise title might identify him.

4. For a discussion of our use of the term *perspective* and the type of tables we used, see Becker, *op. cit.*, pp. 33-45.

5. I have been accustomed to think of the field worker's tendency to look at wholes as an important theoretical principle, maintaining that those who

344 • *Blanche Geer*

look at limited portions of a community or institution leave themselves open to serious misapprehension. On reflection, the interpretation offered here seems more likely. Compulsive or inelegant origins, however, do not make the "whole" any less necessary and important to understand.

6. There is extended discussion of friendship groups in relation to academic perspectives in Becker, *op. cit.*, chap. 9.

7. *Ibid.*, especially chaps. 8 and 10. Medical-school freshmen often thought they knew better than their teachers what they needed to know to practice medicine.

8. See Howard S. Becker, "Problems of Inference and Proof in Participant Observation," *American Sociological Review*, XXIII (1958) 655-656.

9. George Polya, *Mathematics and Plausible Reasoning* (2 vols.; Princeton, N. J.: Princeton University Press, 1954), vol. 2. *Patterns of Plausible Inference.* On pages 18-54 and 109-141, he discusses a "calculus of plausibility" based on everyday reasoning.

10. See Howard S. Becker and Blanche Geer, "Participant Observation: The Analysis of Quantitative Data," in Richard N. Adams and Jack J. Preiss, eds., *Human Organization Research* (Homewood, Ill.: The Dorsey Press, Inc., 1960), pp. 283-285.

11. Hanan Selvin has called similar reasoning "internal replication." See his discussion in "Durkheim's *Suicide* and Problems of Empirical Research," *American Journal of Sociology*, LXIII (1958), 613-618.

12. The term *elaborate* is used to suggest the general similarity of the reasoning process with that described in Paul F. Lazarsfeld and Morris Rosenberg, eds., *The Language of Social Research* (Glencoe, Ill.: The Free Press, 1955), pp. 121-124.

13. Memorandum entitled "Comments on Theory and Techniques for the Undergraduate Study," September 9, 1959. It begins, "This is written on the eve of going into the field. I am simply recording the more or less tentative conclusions that Ray, Geer, and I arrived at in discussing our preliminary strategy."

14. Albert E. Cohen, *Delinquent Boys* (Glencoe, Ill.: The Free Press, 1955).

12 An American Sociologist

in the Land of Belgian

Medical Research*

RENÉE C. FOX

This is the story of a piece of research[1] that began as the summertime project of an American sociologist who voyaged to a small, reputedly fathomable and friendly land known as Belgium to make a rather specialized study of how social, cultural, and historical factors affect clinical medical research and research careers[2] in a contemporary European society. It is the story of an undertaking that was destined to engage the American sociologist for more than four years in a complex search that by implication touched on many sectors of Belgian life in such a way that the publication of her first findings excited passionate responses from virtually every significant group in Belgian society. The story of this research, then, is distinctive in at least two ways.

(1) Although it is an account of an attempt to make a sociological study of a particular aspect of a particular institution in Belgium— medical research—it is a story that can be deciphered and understood only if it is linked to certain characteristics of the social structure and value system of Belgium that go far beyond its world of medical research.

(2) It is primarily through an analysis of Belgian reactions to the publications of the American sociologist that the largely unanticipated, more general significance of this study of medical research in Belgium emerges.

* A chronicle of research in progress.

345

In the chronicle of research experiences that follows, it is on these aspects of my journey as an American sociologist into the land of Belgain medical research that I shall focus. But first, how did this journey begin?

The idea for a sociological study of medical research and research careers in a present-day European society grew directly out of my special interest in the sociology of medicine, medical education, and science. Since the beginning of my graduate studies in 1949, I have been involved in a series of research projects in this domain, primarily in medical institutions of Boston and New York. These experiences brought me into contact with many European physicians, professionally engaged in academic medical research in their own countries, who had come to the United States for certain kinds of specialized postgraduate training. Some of these physicians were senior men, even professors; most were younger physicians in early stages of their aspired-to careers; all had demonstrated sufficient competence in their fields to be accepted to study and work in medical centers of the topnotch excellence of a Harvard or a Columbia.

I became interested in the fact that such physicians had come to the United States in significant numbers for specialized training. Why had they chosen American medical centers for advanced work, rather than European centers, as most such physicians would have done even as recently as the years preceding World War II? Gradually I also became aware of a problem that many of these European research physicians seemed to share. Especially the younger ones apparently felt uncertain and apprehensive about whether they would be able to pursue the career of their choice successfully in their own countries: Could they expect to find appropriate positions when they returned home? they asked themselves. Would they have anything like the working conditions, equipment, and salaries that would make the career of research physician possible and practicable for them? Could they hope to do work *à leur mesure*—in keeping with their capacity? For the most part, they were skeptical as well as anxious about their future in all these regards; but when their designated training period in the United States was over, one by one they returned to their native countries, as they put it, "to try and to see."

I became interested in learning something about the conditions and problems to which such research physicians were returning in their respective European countries. My professional contact with them raised another kind of question in my mind. I knew from my personal acquaintance with a wide range of medical publications, as well as from formal and informal opinions expressed by members of the American medical profession with whom I worked, that progressively, since World War II, a significant amount of what was considered to be the best current medical research was being produced by investigators in the United States, Canada, England, and the Scandinavian countries rather than in countries of Continental Europe, as had formerly been the case. Why, I wondered, had research on the Continent suffered this relative decline when these countries seemed to have a reservoir of talented, well-trained, committed research physicians like those whom I had been meeting in American medical centers?

I decided to devote the academic summer of 1959 to exploring these questions by visiting some of the medical milieus from which such European research physicians had come and to which most of them eventually returned. My intellectual "baggage" for this voyage consisted of little more than the observations I had made and a conceptual hunch that perhaps there were certain common, possibly traditional, features of the social structures and value systems of the countries of "Old Europe" that created problems for the development of medical research careers. With this very general hypothesis, based largely on intuition, I set sail for Europe in June; my plan was to spend most of the summer visiting medical research groups in Belgium.

"But why Belgium?" This is the question about my research perhaps most frequently asked of me. What this question usually means even when asked by Belgians is: "Why did you pick such a small, relatively inconspicuous country as Belgium?" Though I eventually chose Belgium for what I feel are good sociological reasons, I must admit that the original bases for the decision were a blend of the strictly pragmatic, the sentimental, and even the spurious.

To begin with, since all I envisioned when I left for Europe that first summer was a delimited personal exploration of the problems that had caught my attention, it seemed to me that the most rational and profitable way to proceed would be to try to attain some understanding in depth and detail of social factors that affect medical research by focusing largely on a particular European society rather than making superficial, panoramic observations on a continent-wide geographical scale. A number of elements entered into the choice of Belgium in this connection.

Through my work in the realm of the sociology of medicine and science, I had met quite a few Belgian physicians, temporarily in the United States, who I knew would be willing to help introduce me to the world of medical science in Belgium. The fact that I had made contact with a significant number of Belgian physicians was partly a consequence of something like a tradition among university-educated Belgians to seek a certain amount of their advanced training abroad. More specific to the presence of such physicians in the United States for training was the creation of the Belgian American Educational Foundation after World War I,[3] which has made it possible for two generations of selected Belgians, especially in the fields of engineering and medicine, to come to the United States on fellowships for specialized postgraduate education.

My meeting numerous Belgian physicians was also a result of the fact that I had worked under the aegis of a well-known American professor of medicine who had played such a significant role in the training of Belgian research physicians that he had received an honorary degree from a Belgian university for his contribution to medical science in this respect. He, of course, influenced my choice of Belgium both by his personal attachment to this country and by his offer to write letters of introduction for me to some of the outstanding Belgian professors of medicine. The one foreign language that I could handle with some certainty and skill was French, and this also inclined me to locate my summer-long explorations in Belgium. I was aware that Belgium is a bilingual nation, Flemish- as well as French-speaking; but I was told by persons whom I consulted, Belgians and non-Belgians alike, that with a combination of French and English I could proceed without difficulty to interview Belgian research physicians and that, Flemish nationalism notwithstanding, even the most fervent young Fleming would excuse a foreigner like me for not being able to converse with him in his maternal language.

Given my relative facility with the French language, France perhaps would have seemed a more likely place for my exploratory research, especially because of the historical glory of the "French School of Medicine," on the one hand, and the effort that France is currently making, for the first time since the beginning of the nineteenth century, via the so-called *Réforme Debré,* to bring about major changes in its medical education and the organization and functioning of its hospitals. But because I had only a summer at my disposal, a number of persons discouraged me from locating my inquiry in France and re-inforced me in my tentative decision to work in Belgium. My research would advance much more quickly and get much further in Belgium than in France, they said, for the attitudes of Belgians toward America and Americans were far more receptive than those of the French.

Finally, as important as any other factor in my choice of Belgium was a conception about small countries set forth by a number of the physicians and social scientists whom I consulted, which, at the time, seemed to me both reasonable and inviting. A small country like Belgium, these persons implied, would not only be easier for a lone investigator like me to cover geographically but would also be simpler than a large country to comprehend in a sociological sense. That conception, which I easily accepted, was to prove to be so untrue, at least of the small country known as Belgium, that if I were now asked to formulate a sociological hypothesis about the relationship between the size of a country and the complexity of its social system, I would be tempted to suggest that there is an inverse relationship between the two; that is, the smaller the country, the more complex its social system!

Certain aspects of this sociological complexity and of its possible relevance to the problems of medical research that I wished to explore were manifest almost from the moment of my arrival in Louvain, where I lived that first summer. I had picked Louvain because it was the site of one of the four Belgian universities and hence of one of the medical schools in which a significant amount of medical research is carried out; because it is a town with a medieval Catholic tradition, which I felt would give me a chance, atmospherically and intuitively, to come to know something of "Old Europe"; and because, in this community, via my American contacts, I had excellent introductions to young research physicians and professors of medicine alike.

The first small thing that caught my eye when I debarked from the train was to prove to be connected with a salient characteristic of Belgian life, centrally relevant to my study. The signs on the various *quais* of the station, I noticed, were printed alternately in French and Flemish. "Louvain-Leuven, Louvain, Leuven," they read. These were indicators in miniature, I was soon to learn, of the so-called linguistic problem of Belgium: the vying between French-speaking and Flemish-speaking Belgians of Wallonie, Flanders, and environs for equal recognition and rights in all domains of Belgian life, a struggle symbolized by their disputes over the coexistence and the geographical distribution of their respective maternal languages.

Some of the characteristic sights and sounds of Louvain (Leuven) introduced me from the very first to still another kind of complexity of Belgian life, which I soon realized had consequences for medical research as well as other Belgian activities. On the one hand, there were the ancient, narrow, cobblestoned streets of Louvain, its Gothic town hall and baroque churches, the many priests moving about the town dressed in their cassocks,[4] the lamplighter making his rounds at dusk and at dawn, and the periodic chiming of carillon bells. On the other hand, those same venerable streets were filled with a raucous procession of shiny, fast-driven cars as well as horse-drawn wagons and *vélos;* new buildings were springing up everywhere; and the plate-glass show windows of the town's thriving shops were filled with the newest-model refrigerators, cameras, television sets, transistor radios, ballpoint pens, and typewriters. The deeply rooted traditional and the up-to-the-minute modern lived side by side in these and other ways in the town of Louvain, sometimes in harmony, sometimes in conflict, affecting medical research as well as other factors of Belgian life, I quickly discovered, on a far more than local scale.

During my first week in Louvain, largely through discussions with the physicians and their families to whom I had letters of introduction, I also came to understand more precisely where Louvain "fitted" in the social structure of Belgian medical research. Most such research in Belgium, I learned, is carried out in a department, hospital, or institute affiliated with one of the four Belgian universities (Louvain, Brussels, Ghent, and Liége). Each of these universities, it was explained to me, represents a different combination of some of the social and cultural distinctions institutionalized in many sectors of Belgian life.

Louvain is a "free" (that is, a private, nonstate) Catholic university with a double faculty and student body, the one section of the university being Flemish, the other French. Brussels is also a free university, but Free Thought, rather than Catholic, in its religious-philosophical orientation—Free Thought encompassing a set of attitudes that range from a nonconfessional, humanitarian kind of rationalism to the flagrantly anticonfessional, anti-Catholic attitudes of at least certain chapters of the Free Mason Lodge in Belgium.[5] Originally, all its classes were taught in French, but since World War II various parts of the university (including the medical school) have doubled their faculties, adding professors who give in Flemish the same courses that are given in French. Ghent is a state university, officially neither Catholic nor Free Thought, but with a great many practicing Catholics in the student body and on the faculty. Since the early 1930's, all classes at Ghent have been taught in Flemish. Liége is also a state university, like Ghent, neither Catholic nor Free Thought, but with the greater number of its students and faculty nonpracticing Catholics or non-Catholics. All classes at Liége are taught in French.

But the structure within which Belgian medical research is conducted "is even more complicated than that," my informants assured me; and I subsequently saw this myself, when I began to move from one university research setting to another. For one thing, the two free nonstate universities, as well as the two state universities, receive large subsidies from the Belgian government; for another, though the two state universities are presumably neutral in their over-all political and religious-philosophical orientations, a majority of members of certain faculties of each of these universities, or of particular departments within the faculties, may be affiliated with one of the three major Belgian political parties (Social Christian, Liberal, and Socialist) in an active, partisan way or may be fervently Catholic, non-Catholic, or anti-Catholic.

I carried out my exploratory research that first summer in Belgium in the same general way that I proceeded the three subsequent times I returned to Belgium to go on with the inquiry I had started. What I attempted to do was to divide my time as equally as possible among the various clinical medical research groups within each of the four universities and to function as a participant-observer in these research milieus, as well as to conduct focused interviews with the physicians who worked in each of these settings. Because each of the universities represents a particular constellation of some of the social and cultural differentials that truncate Belgian life, each tends to seal itself off from the others. As numerous Belgians to whom I spoke in the course of this summer put it, the universities are "veritable cloisters," by which they meant that there is a limited interchange of ideas and information between them and an even more restricted exchange of personnel, for the faculty of each one is drawn almost exclusively from its own alumni. Thus, the conviction of numerous research physicians that their professional environment, satisfactions, and problems were distinctly different from those of their colleagues in other Belgian university settings was a consequence of local, particularistic pride on their part, combined with a relative lack of firsthand information about "what it was really like" in Belgian research milieus other than their own. Above and beyond how they affected the reaction to my activities, these restrictions on the exchange of information and personnel among Belgian universities obviously have implications for medical research and research careers in Belgium. Nevertheless, it is methodologically interesting that it was my own personal research experiences that first sensitized me to this important situation.

In contradiction to the sense of uniqueness that each of the groups of research physicians seemed to have regarding their own university situation, as I moved progressively from one research setting to another I was more and more impressed by the fact that certain phenomena affecting medical research and research careers seemed to occur in every one of the four Belgian universities. The differences between the several university worlds of medical research in these respects seemed to be more differences of nuance and degree than differences of kind. Some of the recurrent phenomena by which I was struck include the following.

(1) The paucity of funds available for medical research, partly as a consequence of the relatively small amount of money voted for scientific research each year by the national government, the primary source of research funds in Belgium. Difficulties in obtaining needed equipment and ancillary medical personnel—difficulties primarily, but not exclusively, financial in origin.

(2) Problems of constructing the kind of hospitals and laboratories needed for the most up-to-date, advanced medical research (and care) or, at least, of renovating hospital and laboratory buildings that were constructed years, sometimes even centuries, ago. Problems of completing, opening, and utilizing such new or rehabilitated buildings, once work on them had begun. The sound of hammering could be heard in every one of the four universities, and research physicians eagerly and impatiently awaited the outcome of these building and rebuilding efforts. Yet, in one university, a hospital in construction for twenty-five years was still not completed; in a second, a newly constructed hospital, ready for use, stood idle and closed; in a third, a major piece of research equipment (a betatron), delivered months earlier, remained in the crate in which it had arrived, because the construction of the special

kind of building in which it would have to be housed had only begun.

(3) Problems connected with obtaining an academic position in one of the four universities, primarily on the basis of professional competence, from which a physician could function as a medical researcher with some degree of stability and security. As already noted, Belgium is a country in which practically all medical research is carried out from a position within the university structure; there is virtually no other professional "place to stand," for example, in government or industry, if one wishes to do medical research. Even within the structure of universities there are relatively few formal positions open to a man who aspires to a research career. This was especially true in 1959, when I began my work in Belgium. At that time, more than now, the structure of the staffs of the university departments, hospitals, and institutes where medical research is conducted was, in the words of one physician, "like a building in which the ground and top floors have been constructed, but in which they haven't gotten around yet to putting in the floors in between." Research units were, and still are, typically headed by one full professor, with all the authority and responsibility of a *patron*. Below the rank of professor-*patron*, and greatly subordinate to it, there exist a certain number of university statuses like those of *chargé de cours* or assistant, from which one may carry out medical research (but usually with a heavy load of clinical medical responsibilities as part of one's duties).

A greater number of physicians engaged in medical research hold no such formal university appointment; rather, their positions (often titleless) exist only by virtue of the fact that the *patron* has been able to raise a sufficient sum of money temporarily to pay them a salary and support their research. It was not until 1960 that a law was passed officially creating two new kinds of tenure positions in the structure of Belgian universities in addition to that of full professor: those of *chargé de cours associé* and *professeur associé*. In effect, they represent some of the "floors in between" which were missing in the formal status system of Belgian university faculties; but even so, the number of such positions created and of men appointed to them is still very limited.

As already mentioned, the positions open to a physician who wishes to devote himself primarily to research are further restricted by the relative infrequency with which a graduate from one Belgian university is named to a position in another. Even within his own alma mater, the chances of a physician's being appointed on the basis of scientific competence to one of the few staff or faculty positions in which sustained research is possible are curtailed by the considerable extent to which such particularistic considerations as the political, ethnic, linguistic, philosophical, religious, class, and family affiliations of candidates play a role in determining who is named and by the significant influence that the particularistically or personally based preferences of the *patron* of a given department or institute may exert in choosing the physician to whom a position is accorded. The low salaries typically paid to other than professors engaged in research, combined with this degree of uncertainty about whether one will be appointed to a relatively permanent academic position from which to carry on research, make it difficult, and in many cases impossible, for young physicians with families to support to continue to take the risk and to make the sacrifice of trying to remain primarily in the field of research.

(4) Existing side by side with these problems of medical research and research careers were two phenomena that seemed to be almost paradoxically related to them. Though in each of the universities there was a dearth of adequate positions available to competent research physicians, there were also certain university research positions and even professorial chairs which remained vacant or were not filled for long periods of time. Everywhere medical researchers lacked funds, equipment, and ancillary personnel that would facilitate their work; yet, at the same time, very expensive equipment was duplicated in the four universities for lack of some cooperative arrangement among departments, institutes, and universities for the joint use of such equipment. (For example, I counted sixteen artificial heart machines, four artificial kidney machines, four betatrons, and four cobalt bombs in Belgium, a country smaller than the state of New Jersey.)

(5) Finally, there were certain characteristics of what might be called the psychological atmosphere that seemed to be present in all the university settings. Whatever the specific nature of their problems, medical researchers were all confronted with the general, morale-eroding phenomenon of long delays in trying to deal with them, and sometimes utter paralysis. In the face of these delays and impasses, there was a tendency everywhere to try to find ways around the cumbersome formal procedures and structures that in part caused and perpetuated these blocks. *Petits chemins* (byroads or detours), even *truques* (a Belgian word for somewhat illicit gimmicks or tricks for getting things done more speedily), were invented and used pervasively by medical researchers with frustration-born zest.

At the same time, many research physicians seemed to feel something like a sense of mystery about what they described as the absurd, enigmatic, arbitrary qualities of the conditions under which they had to conduct their work. They expressed apprehension and indignation about what they experienced as the inscrutable or capricious nature of some of the forces that affected their professional activities and destinies. These sentiments often focused on certain strategic persons who, these medical researchers implied, determined often in an invisible and prejudicial way what did and did not happen. Sometimes these presumably powerful individuals were cited by position and name; more often they were referred to, half jokingly, half fearfully, as *messieurs les responsables, Esprits Directeurs,* or *éminences grises.*

As even this brief account of my first Belgian summer makes obvious, what had begun as a June-September voyage of self-education in matters pertinent to my professional interest in the sociology of medicine and science had quickly become far more than that. Almost inadvertently, I had carried out the first stage of a potentially long-term study of medical research and medical-research careers in Belgium; it was already involving me in thinking about certain general attributes and problems of the intricate and diverse little society (in the words of one Belgian physician, the "complex social mosaic") of which these medical scientific institutions are an integral part. For in all the research situations with which I had made contact, in all the universities, the range of problems that confronted medical investigators was significantly related to and affected by rather general traits of Belgian society. These included:

(1) Various forms of particularism—the importance attached to ethnic, linguistic, philosophical, religious, political, community, social-class, special-interest, and family differences in Belgium.

(2) Some of the conflicts and strife that result from this particularism, notably disputes between Walloon and Flemish, Catholic and Free Thought groups.

(3) A very literal "it must be identical and not simply equivalent" conception of equality imposed by one group on the other, which often leads to the replication of resources, facilities, organizations, and so on and, in certain decision-making situations, to delays and impasses because the opposing groups involved are so evenly balanced and so unyielding in their relations with one another.

(4) An elaboration of ministries, agencies, and other divisions of the

central government that enters all domains of life in Belgium—ministries and agencies which have numerous overlapping functions and archaic rules and regulations which are the hard-to-change heritage of a historical past and which have grown more, rather than less, complex by virtue of some of the new functions and positions (created partly in response to pressures for greater representation on the part of vying particularistic groups).

(5) The importance of local communities in Belgian life—of the nine provinces and 2,633 communes that make up the highly autonomous local governments of Belgium—with their tendency to exhibit jealous pride in their own local authority and to regard the central government with suspicion and even a certain disdain.

One or another of these factors, and sometimes virtually all of them, seemed to pervade the stories that lay behind the recurrent problems faced by Belgian medical researchers in the various university milieus. For example, a particularly striking concrete instance was the university hospital begun twenty-five years ago and still not completed. This appeared to be a consequence of a maze of factors, which included the involvement of one of the professors of medicine (a moving spirit behind the building of the hospital) in the Flemish nationalist movement and the disrepute into which the hospital fell when he was accused by persons less "Flamimgant" than himself of having collaborated with the Germans in World War II; conflicts between the local Commission of Public Assistance and the ministries of Public Health and National Education of the central government over various aspects of the work; certain ineptitudes on the part of all three of these bodies in conceiving, planning, and executing the building; and their susceptibility to political, religious-philosophical and ethnic-linguistic pressures which resulted in disputes, delays, and curtailments of funds.

By the time the summer was over, I had committed myself to making a formal study of clinical medical research in a current European society and to locating that study explicitly in Belgium for reasons that were now pre-eminently sociological. The apparent relationship in Belgium between the problems of medical research and numerous other sectors of the society were intriguing. Furthermore, moving in a "Rashomon-like" circle from one university to another, as I would have to do in Belgium to make such a study, had the built-in sociological attraction of enabling me to look systematically at these problems of medical research from the points of view of the different particularistic clusters that each university represented.

The study of clinical medical research in Belgium that began in this way continued for the next three years. My actual physical presence in Belgium consisted all told of three summers (June through September of 1959, 1960, and 1961) and finally a six-month stay extending from July through December of 1962. However, there were a number of respects in which, over the course of those years, I never completely left Belgium behind, even when I was residing and working in the United States. For one thing, I received a large cross section of popular, medical, and scholarly publications of Belgium—newspapers, magazines, journals, books—in French and also in Flemish (which I progressively learned to read with the help of a dictionary and to understand when spoken). Second, I carried on a vast and vigorous correspondence with research physicians—professors as well as younger researchers—in each of the Belgian universities, who kept me closely informed of any happenings while I was away which they thought might be relevant to my study. Third, during those years almost every Belgian connected with medical research or the medical profession in any way who had occasion to be in the United States contacted me. So, too, did a certain number of Belgians outside of medicine, from the worlds of the universities, business, politics, and the Church, who passed through America at that time. This also helped to keep me in constant touch with Belgium and considerably expanded the network of Belgians with whom I was in communication.

Especially because of the light it will eventually shed on the dramatic ways (both positive and negative) in which Belgians reacted to some of the findings of my research when they were first published, it should not be taken for granted that such a sizable number of Belgians actively sought me out and remained conscientiously and enthusiastically in touch with me, even when I was not in Belgium, throughout the entire course of my study. It is easier to see now, ex post facto, than it was at the time that simply by the act of being engaged in this research I was already cast in a rather special role by Belgian medical researchers and by other Belgians, such as the ever-widening group of persons who wrote to me, phoned me, visited with me, both in Belgium and the United States.

In a society like that of Belgium, where particularistic sentiments and attitudes are not only deeply felt but also formally institutionalized, the relations between persons from different groups are sufficiently restricted so that many Belgians found it interesting and valuable to hear what the American sociologist, who in the course of her research was passing more freely than they could from one social universe to another in Belgium, might have to say about "what it was like" in groups other than the ones to which they themselves belonged. The research physicians of each university, for example, asked me whether conditions in medical research units of the other Belgian universities were comparable to their own. Flemings asked about the ideas, opinions, outlook of the Walloons to whom I had spoken; the Walloons about those of the Flemings. Catholics and Free Masons made inquiries about each other. *Patrons* of medicine were interested in knowing how younger research physicians (to whom they referred as *les jeunes*) felt about their professional situations; *les jeunes*, in turn, asked about the *patrons*. It would have been at once methodologically undesirable and morally dubious for me to pass on concrete information of this kind from one group to another. But, within careful and scrupulous limits, I did provide something like a general, analytic overview of the multiple worlds of Belgian medical science—an overview which, as already indicated, suggested as many similarities as differences between these supposedly diverse milieus. Constrained as they had been by their own particularistic attitudes from observing and recognizing such common features, many Belgians seemed to find my overarching perspective original and instructive.

In addition, my continuing and caring interest in things Belgian touched what was apparently a disguised love of country on the part of numerous Belgians. When they spoke about their country, generally what they said was critical, even depreciatory in nature; nevertheless, they were obviously pleased and flattered as well as surprised by the fact that a non-Belgian should be so vitally concerned with what one Belgian physician described as "the absurd and pitiful problems of our small and beautiful country." It also became increasingly apparent that one of the reasons for which Belgian research physicians from each of the universities, and certain other Belgians as well, maintained contact with me and helped with my research was that they saw in me something like a potential *dea ex machina*. As an American, rather than a Belgian, they felt, I could look into some of the problems that beset them, not only with a certain objectivity, but also with relative impunity.

In the tight, invidious, particularistic situations within which they functioned personally and professionally, in their small, tension- and apprehension-ridden country, it was perhaps the better part of wisdom, they reasoned, for them to keep silence about their problems. Partly, this was a matter of pride—personal, local, and national. Partly, it seemed to be a consequence of widespread anxiety, bordering on fear, of the kind of retaliation from a competing individual or group, or even from an all-powerful *monsieur le responsable*, that publicly speaking out might provoke. But if an American sociologist "from outside the System," as a number of persons put it, could be privately encouraged to look into these problems and eventually to make them known, the silence about them might at last be broken in a way that would not involve any specific Belgian or group of Belgians and might help to bring about needed changes. But here we are getting ahead of the story; for this aspect of my role did not really become manifest until my last journey to Belgium, after the publication of two articles.

I returned to Belgium that second summer (1960) as a Special Fellow of the Belgian American Education Foundation and rented an apartment in Brussels. Both of these arrangements were, in part, consequences of what my first year of research there had taught me about the Belgian social system and its implications for my own work. I had applied for the fellowship with the hope of receiving some help in defraying my expenses, of course; but just as important to me was having the moral support of an organization, at once Belgian and American, highly regarded in the Belgian academic world and considered, on the whole, to transcend the divisive conflicts and rivalries of Belgium. I hoped that working with the sponsorship of the foundation would symbolize my commitment to Belgium, on the one hand, and my neutrality, on the other.

The choice of Brussels as my place of residence was also suggested, if not dictated, by the particularism of Belgium. If I had lived in Louvain again I would have been irrevocably classified as a "Louvain-iste" and, as a probable result, not given the kind of access to the other three universities (especially the University of Brussels) that I sought and needed. Brussels, ostensibly, was the least partisan place of residence to choose. Actually, like any community in Belgium, it represents a special particularistic cluster, but I could justify living there in such terms as its being the capital city, centrally located, convenient for obtaining transportation to the other university towns, and so on. The important thing is, of course, that I even had to plan carefully where I should live in order to avoid being classified with any one group in such a way as to disqualify me for working with all the groups pertinent to my research.

So, in 1960 (and again in 1961 and 1962) I located in Brussels. The view from the windows of my apartment there was not so medieval as the one I had in Louvain, but especially one of its landmarks linked my second Belgian summer with my first. On the wrought-iron gate that separated my small, private, nineteenth-century–style street from the bustling modern avenue on which it opened was a shiny red mailbox with two slots. The slot on the right-hand side of the box was labeled with the Flemish word for "letters," *brieven;* on the left-hand side was the equivalent French word, *lettres.*

The planned objectives of my research that summer were several. In addition to my very general aim of acquiring more empirical knowledge and a better analytic understanding of social aspects of clinical medical research in Belgium, one of my specific goals was to equalize my acquaintanceship with the different university settings in which medical research is conducted. Thus, in order to balance out my research experiences of the first summer, I now spent a good deal more time at the universities of Brussels, Ghent, and Liége than at the University of Louvain and, when I was at Louvain, more time in the Flemish section than in the French section of the medical school. Another of my aims that second summer, and in my subsequent research visits to Belgium as well, was to look more deeply into some of the phenomena already mentioned that seemed to be present in many different Belgian research groups and to follow through time the unfolding of certain types of happenings or clusters of events apparently connected with these phenomena. Tracing out in this way what were essentially a series of sociological case histories, I felt, would give me more precise information about how social factors affect medical research and research careers in Belgium and suggest how tenacious or how subject to change were the associated problems.

In this manner, to take one concrete example, I was able to identify and follow some of the social processes involved in the vacating of a particular professorial chair in one of the four universities; the delay of more than two years before that chair was filled (although there were at least four qualified local candidates desirous of that chair); some of the consequences of that delay for these aspiring candidates, for other physicians who worked with them, for the quality and quantity of the research produced by the department during that prolonged waiting period, and for the medical-school faculty at large; the way in which a new professor was finally chosen; and the consequences of that choice, in the context of all that had preceded it, for the new professor, for the candidates not chosen, for their associates, and for their research. Since the stories that I closely observed and followed in this way were chosen because they seemed to be exemplifications of comparable phenomena taking place in many different research groups, these case studies gave me general, as well as particular, insights into the sociodynamics of various facets of Belgian medical research.

Although in a book-length account of my research experiences in Belgium it would merit at least a whole chapter of description and analysis, I can only mention in passing here that my entire second summer of research was carried out against the larger backdrop of Belgium's "Congolese crisis"—the unexpectedly violent way in which Belgium totally lost her former colonial possession, the Congo, one month after she had formally granted its independence. This grave event, so shocking to Belgians, also affected my research in at least two ways. In the discussions I heard about the Congo and Belgium that summer and the reading I progressively did about it, I began to see how some of the same ethnic-linguistic, religious-philosophical, and political divisions of Belgium that I was observing and analyzing in connection with medical research had been transposed to the Congo by the Belgian colonial regime, contributing to the complex pressures for independence in the Congo and the uprisings of July 1960. Being in Belgium during this summer of crisis also gave me a vivid and empathic sense of what it felt like to be a citizen of a small country, suddenly closed in upon itself by the loss of its one open frontier and condemned by the big world powers for its colonial policies. A wave of hurt indignation mounted over the fact that even the United States, with whom there had been especially warm and friendly relations since World War I, had joined those countries that severely criticized Belgium. As a consequence, throughout the summer of 1960 it was no longer especially desirable to be an American sociologist, or an American of any genre, in Belgium. Belgian sensitivity about American criticisms of their policies in the Congo evidently persisted beyond that summer. For, two years later, when my articles about medical research in Belgium were published, among the negative comments about the American sociologist that were verbalized and printed in the lay press were remarks to the effect that she had exhibited the same uncomprehending, unsympathetic attitudes toward Belgian medical research that her country had demonstrated in 1960 when Belgium was faced with its Congo crisis.

During these first two summers, as has been seen, the locus of my research was confined mainly to the four universities and my methods of inquiry were primarily those of participant observation and interviewing of physicians in the medical research groups associated with the universities. Back in the United States, as I reflected on the kind and scope of materials that I had thus far collected and as I prepared myself for a third summer of research in Belgium, I decided that what I now needed to do was to supplement and extend my data in two ways. I should now deliberately go in search of documentary and statistical materials bearing on the status and functioning of medical research in Belgium: for example, figures on how many physicians are engaged full time or part time in clinical medical research, how many different research groups there are, exactly how large or how small the research budget of various of these groups is, from what sources they obtain their funds, how they allocate them; minutes of department or faculty meetings in which matters relevant to medical research and research careers have been discussed; documents connected with the planning and constructing of various hospitals and laboratories, the procuring of equipment, and so forth. In addition, it seemed to me that the time had come to try to make contact with various structures outside the universities that had some bearing on the functioning and financing of medical research, notably particular ministries of the national government, private foundations, and those public-assistance bodies of local communes that own and help to administer university hospitals in which medical research is conducted. Upon my return to Belgium,[6] when I actually set out to enrich and more solidly document my research in these ways, I encountered certain apprehensions and obstacles that once again proved to be related to attitudes and practices deeply rooted in Belgian tradition and social organization rather than just a passing, idiosyncratic response to a querying American sociologist.

The collection of the kinds of documentary and statistical materials that I had in mind was not easy to accomplish. Records of meetings in university, foundation, and government settings that bore on issues or events affecting medical research and researchers, and statistics other than the most gross and general sort on research personnel, grants, budgets, and salaries in some instances did not exist; in others, they existed but through polite evasions were not made available to me. There were even a few occasions on which I was given access to the materials I sought but where I had enough auxiliary information to recognize that the data were inaccurate (for example, the minutes of a medical-school faculty meeting that took such liberties with some of the exchanges between professors as to distort if not falsify them; or the annual budget for the research unit of a professor which significantly underestimated the funds he had available for medical research, since certain private donations he received were not mentioned).

The difficulties of data collection that I experienced, as indicated, were by no means unique but rather were a concrete illustration of the sorts of problems that many other individuals and groups have encountered in trying to assemble and transmit detailed and reliable data bearing on contemporary phenomena in Belgium.[7] To cite just four relevant examples: (1) it has not been possible to conduct a language census in Belgium since 1947; (2) the Sauvy Report, a study of the demographic evolution of the Walloon and Flemish provinces of Belgium and of the *arrondissement* of Brussels, was finally published in 1962, but only after a long prehistory of attempts to discourage first the carrying out of study on which it was based and then the diffusion of its findings; (3) one of the *raisons d'être* of an independent organization known as the Center of Socio-Political Research and Information, or CRISP (*Centre de Recherche et d'Information Socio-Politiques*), is that it enables a group of social scientists, lawyers, journalists, and the like to operate in a personal, anonymous, extraofficial basis, outside of their usual professional organizations and roles, and thereby obtain from individuals and groups data on current social, economic, and political happenings that they would not be granted otherwise; (4) most germane to my own efforts to collect certain data on Belgian medical research are the recent, as yet not very successful, attempts of the National Council of Scientific Policy (*Conseil National de la Politique Scientifique*), an advisory group created by the national government, to do a general survey of the funds, apparatus, and personnel engaged in scientific research of all sorts in Belgium.

The characteristic difficulties that lie behind all these efforts, my own included, to systematically collect and make known facts and figures about various aspects of present-day Belgian society seem to stem largely from some of the same particularistic attitudes already examined. The persons and groups approached for pertinent information tend to be latently, if not always openly, apprehensive and suspicious about one's motives for carrying out the inquiry, about the group by which one is sponsored or employed and its intentions, and about the possible consequences of revealing certain facts to groups other than their own.

As I shall subsequently show in connection with my own research project, there were some very practical, reality-based reasons for which my inquiry evoked this kind of anxiety and wariness in the worlds of medical research and even in a number of milieus outside of them; but, in addition, there were also strong nonrational elements that contributed to it.[8] Nowhere was this more apparent than when I made my first overtures toward accomplishing my other major research goal of that third Belgian summer: direct contact with certain persons in some of the extrauniversity organizations that affect medical research.

Perhaps the interview that most vividly illustrates the kind of anxiety about my research and the potential barriers to it that I progressively began to encounter was the one I made when I first arrived back in Belgium in the summer of 1961. I went to a certain gentleman who had greatly helped with my work during the two prior years and who occupied an administrative position that linked universities, foundations, business, and government. I sought his assistance in meeting the small list of persons I thought essential to my contact with Belgian organizations other than the universities that had some relationship to medical research. This list included a prominent businessman, known to have been a generous donor of funds for scientific research; the Vice President–Director and the Secretary General of the Belgian foundation that plays the most important role in allocating funds for scientific research, the National Fund of Scientific Research (*Fonds National de la Recherche Scientifique*, or FNRS); the Minister and the Secretary General of the Department of Public Health and Family of the Central Government; and the President of the Commission of Public Assistance of Brussels.

Mr. X, the gentleman on whom I called for help in being introduced to these people, reacted to my request with great nervousness. The businessman I wished to interview, he said, would be of no special value to my work; though he was high on the board of administrators of several foundations connected with science, his position was "strictly honorific," and he was "just a businessman" who would not be qualified to discuss research. Then, why did I feel I had to see the Secretary General of the National Fund of Scientific Research if I saw its Director? Would it not be sufficient just to talk with the Director? As for the appointments I had requested at the Ministry of Public Health and Family, did I have to talk with the Minister as well as the Secretary General? Ministers generally do not know as much as their Secretary Generals, Mr. X assured me, for they are always being shifted from one office to another, according to the way the political winds are blowing. Mr. X also ventured the opinion that if I continued on the path that I proposed, I would be moving away from my specific topic of research and from carrying it out "scientifically"; I would be collecting "opinions, rather than facts."

I politely countered each of Mr. X's comments, and then, at a certain point in our conversation when it seemed appropriate, I asked him if the real reason he was warning me that my work was getting unscientific and somewhat off the track was that he was trying to find a way to tell me, as others had begun to, "*Attention!* You had better be careful! There's danger in trying to include some of the people and some of the questions you're now broaching in your research." Mr. X breathed a sigh of relief over the fact that I had so outrightly stated this and admitted that it was so. He told me that I was now moving in a "difficult situation," and he did not want me to have undue trouble or unpleasantness. In the end, Mr. X did decide to help me because he thought my study was necessary and important, he said, and because I had shown a great deal of "courage" in discussing my determination to go on with the research. He made all the appointments for me that I had requested, with the exception of one with the prominent businessman (whom I never did meet). As I left Mr. X's office that afternoon, his last words to me were, "Be careful! What you're doing now is very risky! Move like a snake!"

Despite the admonitions that preceded my meetings with such government and foundation officials, once I was received by them they were not only cordial to me but remarkably open and free in giving me information, some of which might even be considered delicate in nature. For example, from the officers of the National Fund of Scientific Research whom I interviewed I learned the following kinds of things: The fund receives virtually all its money from the national government and allocates it to applying researchers. All research grants awarded are given to persons currently affiliated with one of the four Belgian universities. Who receives these grants is decided primarily by the twenty-five scientific commissions of which the fund is composed, each one organized around a different subgroup of sciences. Each commission consists of four professors, one from each of the four Belgian universities, and a fifth professor who acts as chairman. Chairmanships are also equally distributed among professors of the four universities (with one exception, since there are twenty-five rather than twenty-four commissions). The Vice President of the fund is also its effective Director and the Vice President–Director of a number of other Belgian foundations that distribute funds for scientific research. He himself has no formal training in science. The Secretary General of the National Fund of Scientific Research, a physical scientist, is Secretary General of a number of these other foundations as well.

My interviews with the Minister and the Secretary General of Public Health taught me firsthand that the gentleman occupying the post of Minister at that time, who admittedly was not trained or experienced in medical or scientific matters, had just assumed office; whereas the person who was Secretary General, a physician, had occupied this position for years. The Secretary General of Public Health, I learned from these conversations, was also President of the National Foundation of Medical Scientific Research, a body that receives from the Ministry of Public Health, via the National Fund of Scientific Research, some of the subsidies it distributes to clinical medical researchers.

From my discussions with a number of different officers of the National Council of Scientific Policy came accounts of strong resistance on the part of certain foundation officials and professors to the council's attempts to think through the complex structure within which Belgian research currently operated, to suggest possible modifications in it, and to try to institute a national plan for the development of scientific research in Belgium. My interviews with some foundation officials, on the other hand, as with some professors, had contained apprehensive and sometimes hostile statements about the extent to which they believed the council to be a political instrument of the parties currently in power, hence a far from ideally objective advisory group for national policy concerning medical research.

Certain aspects of the information I gained from meetings with executives in government and foundation positions were potentially delicate in their implications; that is, the simple descriptive facts they supplied about the social structure of the organizations affecting scientific research quite naturally raise certain kinds of questions. For example:

(1) Despite the significant degree of decentralization in the social structure of Belgian medical research suggested by the complex of social organizations within which most of it seemed to take place (four universities, various of their departments, institutes, hospitals, numerous ministries of the national governments, the governments of local communes, a network of voluntary health organizations and of foundations, and so on), is there not, at the same time, an extraordinary amount of authority and power concentrated in the hands of a few agencies and persons, notably in the National Fund of Scientific Research and its Vice President–Director and Secretary General, who occupy these same positions in a number of other foundations as well?

(2) Since, in addition, there is a considerable overlap in the boards of directors of these various foundations that distribute funds for scientific research, and the President of one of them, the National Foundation of Medical Scientific Research, is also Secretary General of the Ministry of Public Health, exactly how independent are these foundations, of each other and from the government, and how independent are they, then, of political and other particularistic influences?

(3) Does the fact that two of the key positions in the social structure of government and foundation organizations concerned with research (those of Director of the National Fund of Scientific Research and of Minister of Public Health) are held by men who are not scientists and do not have advanced scientific training affect the degree to which criteria other than scientific excellence enter into the distribution of research funds?

(4) It is true that the commissions of the National Fund of Scientific Research seem to be the main judges of the merits of the research projects that are submitted to the fund for consideration, and these commissions are composed of professors in various fields of science. But these 125 professors constitute a sizable proportion of the *patrons* who head research groups in the four universities. Does this mean, then, that those who allocate the research grants of the fund are often its recipients either directly or indirectly through the younger members of their various research teams who apply for grants? What implications does this have for the degree of objectivity with which funds are distributed?

(5) Since the commissions are equally divided among representa-

tives of the four universities, might there not be some tendency in these circumstances, in a particularistic society like Belgium's, to divide equally among them the total budget available for research rather than to judge primarily on its scientific merits each project submitted?

These were the questions that my interviews with foundation and government officials suggested to me. It is of some interest to note that I never openly raised them or gave voice to them in any of my interviews or in the two articles about medical research in Belgium that I subsequently published. In the second of these articles, for example, I simply described the social organization of government and foundation agencies connected with scientific research in Belgium largely as they had been explained to me by high-ranking executives within them. Yet, after this article was published, one of the questions most frequently and anxiously raised in many different settings was: Who had provided the American sociologist with her information about the government and foundation structures? Various "sinister others" were implicated even by some of the very persons, foundation and government officials among them, who had supplied me with pertinent information.

This phenomenon of "collective amnesia," and the projecting of the blame for having told the American sociologist things that were described at one and the same time as indiscreet and untrue, was one I experienced frequently during my last year of research in Belgium. In this particular context it is especially interesting, because what I published was no more than a description of the social organization and composition of several government and foundation agencies without any interpretation whatsoever. Here, and in numerous other instances, as will be seen, Belgian reactions to my findings were far more suggestive and significant than the findings themselves.

At the end of the third summer of research in Belgium, I returned to the United States once more, where I now set to work writing the aforementioned two articles based on what I had learned thus far. The first article, entitled "Journal Intime Belge/Intiem Belgisch Dagboek" ("A Belgian Journal"'), appeared in the winter 1962 issue of the *Columbia University Forum*.[9] As the title implies, it was written in the form of a personal diary, through which I tried to evoke an emotional as well as an intellectual understanding of some of the general sociological characteristics of Belgium already described and of certain specifically medical scientific phenomena seen against that backdrop. I used a number of semiliterary devices and images in my attempt to convey this kind of understanding; among them, the gently ironic "doubled" title of the article, evenly balanced between Flemish and French, and a description of the American sociologist standing in front of the two-slotted mailbox on the street where she lived in Brussels, trying to decide whether she should mail a letter she had written in English in the "French slot" or the "Flemish slot." There was an immediate and rather large flurry of responses to this article, both appreciative and critical, first from Belgians in America and then from Belgians at home. These reactions took the form of letters to the editors of *Forum*, to the Belgian Government Information Center in New York, and to me and mention of the article in one of the weekly broadcasts to Belgium from its Information Center in New York.

I shall postpone any real examination and analysis of the comments that were made until the end of this chronicle, for virtually all of them were mild versions of the more heated responses that followed the publication of my second article. The one specific detail that might be noted here is that the image of the bilingual mailbox seems to have elicited more powerful reactions from Belgian readers than any other single aspect of the article. It was referred to in many of the letters I received, and it was the central focus of the Information Center's broadcast to Belgium. Almost a year after the publication of the "Belgian Journal," toward the close of my last research visit to Belgium, a certain professor of medicine devoted a good part of my interview with him to that selfsame mailbox: at first, earnestly denying that such mailboxes even exist in Belgium, and then, after conceding that there might be *some* of this type, chastizing me in all seriousness for not having explained that, no matter which slot one uses to mail a letter, "everything ends up in the same mailbox!" Obviously, even with my first article I had touched on certain very deeply felt Belgian sensitivities. One of the more friendly critics of the "Belgian Journal" who wrote to me after it was published expressed the emotions I aroused in this way:

> In reading it, I felt a blush of shame mount on my face. We are so accustomed to living in our spiderweb, in the unbelievable cloister of our confessional, linguistic, social and scientific quarrels, that we hardly even pay attention to them anymore. They begin to seem normal to us. To see all that, put in black and white, by a foreigner, even by one with much tact and sympathy for us, inevitably gives one a shock.

The publication of my second article, this time in *Science*, seems to have constituted a greater "shock" for the much larger Belgian public who read it. It was called "Medical Scientists in a Château," [10] and its theme, as its subtitle indicated, was: "The traditional social structure creates problems for medical research and researchers in Belgium."

The image of the château, around which I built the first section of the article, was meant in part to symbolize those aspects of traditional social structure in Belgium which tend to curtail medical scientific creativity and productivity and the possibilities for careers in medical research.[11] It also referred more specifically to a medical scientific colloquium held in 1959 in Laeken, at the château of the Belgian royal family. I began the article by describing this colloquium in detail because it seemed in so many ways a ceremonial expression of how social, cultural, and historical factors affect medical research in Belgium. For example, most of the speeches or presentations made at the colloquium were delivered twice, once in Flemish and once in French. The principal guests of honor at the colloquium were three foreign medical scientists (Dutch, French, and American) who had done experimental work that contributed importantly to the field of cardiac surgery, for which they were about to receive the degree of Doctor *honoris causa* from two Belgian universities. The French scientist was scheduled to receive this degree from the Free University of Brussels on the same day that the American and Dutch scientists were to be awarded their degrees from the Catholic University of Louvain.

The other persons invited to the medical scientific gathering in the château included not only physicians, patients who had undergone cardiac surgery and their families, professors of medicine and science, the rectors of the four universities, officials of voluntary health organizations and of foundations, ministers and secretary generals of the national government, and officials of the local governments, but also numerous members of the royal family and of the nobility, university professors from fields other than medical or biological science, directors of various museums and libraries (some with no bearing on science or scientific work), bankers, businessmen (among them several gentlemen known to be especially prominent members of the Free Mason Lodge), Bishops of the Catholic Church, and others.

In short, representatives of a great many of the institutions in Belgian life had been assembled in the château for this presumably medical scientific occasion in a way that suggested the complex interrelationship of these institutions and certain of their *responsables* (as Belgians would say) and, in turn, their influence on medical research in Belgium. Finally, the image of the château in this article was meant to convey the psychological atmosphere in which many Belgian medical researchers felt they were working: an atmosphere with some of the "absurb," "arbitrary," "undecipherable" qualities reminiscent of those evoked by Franz Kafka in his novel, *The Castle*.[12]

The second section of this article described some general characteristics of Belgian society. The third dealt with the social structure of clinical medical research in Belgium, especially the respects in which it shares these general characteristics. The last section of the article traced some of the problematic consequences of this structure and its attributes for medical research and research careers.

Almost immediately after the publication of "Medical Scientists in a Château," a series of remarkable events began to take place—events which rapidly transformed the article and its author into something approaching national *causes célèbres* in Belgium. The first Belgian reaction occurred, only a few days after the article appeared, in the form of a telephone call from an associate of the Belgian Government Information Center in New York. The comments were far from happy or calm. I was accused of having been "unscientific," "erroneous," and "hostile" in what I had said about Belgian medical research and, above all, of having implied that what I had written about it was applicable to other sectors of Belgian society as well. ("You have generalized from a particular case!" the Belgian gentleman on the other end of the line repeated several times.) He went on to say: "What if an American businessman should pick up a copy of this issue of *Science* in his dentist's office and read your article? What will he think? He will think I had better invest my money in Holland, rather than in Belgium!"

A few weeks after this telephone call, five minutes of an Information Center broadcast to Belgium were devoted to my article. The broadcaster referred to it as an indictment of Belgian medical research and said it would be more appropriate for medical researchers than for himself to pass judgment on the accuracy of its contents. He had only two comments to make, he concluded. First, there would be those who would say that its author has a love–hate complex with respect to Belgium; and second, right or wrong, love or hate, "she has courage!"

A swelling stream of personal letters now began arriving from Belgium, mainly from research physicians (more from *les jeunes* at this time than from their *patrons*), telling me that "Medical Scientists in a Château" had already become, as a physician put it, "one of the most read and discussed bundles of papers in this country." Copies of the issue of *Science* in which it had been published, and photocopies of the article itself, I was told, were being passed from person to person. Not only were they circulating among medical researchers, their families, and their friends, but they had also reached various deans of the four universities, their rectors, foundation officials, certain ministers of the central government, members of the Catholic clergy, businessmen, bankers, engineers, museum and library directors, writers, artists, and even some members of the royal family. These reports about the wide circulation of the *Science* article were confirmed by the vast number of requests for reprints that I began to receive from Belgium and by the varied sources from which they came.[13]

The personal mail also indicated, as several of my correspondents put it, that Belgian reactions to the article were "passionate." There are those who consider it a "manifesto," wrote one research physician, and "others who wave it in the air angrily and don't like it, to say the least. . . . I wonder if your planned stay in Belgium will be safe," he added with a flash of humor. "Well, anyway, some of us will protect you!"

The stay in Belgium to which this letter referred was scheduled to begin the third week in July. I thought of it as my definitive research period in Belgium, one that would enable me to complete the data-gathering phase of my study. For I was fortunate enough to have been awarded a Guggenheim Fellowship and to have been granted a semester's leave of absence from my university teaching duties. This made it possible to spend an unbroken period of five months in Belgium from the end of July until the end of December 1962, a time span which, in addition to its duration, had the advantage of including several months of the academic year when university-based medical research activities would be at their height.

But before I set sail for Belgium toward the end of July, a series of reactions of another kind to "Medical Scientists in a Château" began to occur, contributing to the highly charged atmosphere in which I was destined to carry out the last months of my research. On June 15, 1962, an article about "Medical Scientists in a Château" appeared in *Pourquoi Pas*, a weekly French-language magazine published in Brussels. The title of the article (in relatively small print) was "Scientific Research in Belgium," and its subtitle (in much larger print), "Are We a Ridiculous Country?" In rapid succession, over the course of the next month, six more articles appeared in the Belgian press about "the article in *Science* by the American sociologist."

On July 1 and July 11, respectively, there were two articles about it in *Le Soir* (a daily French-language, Free Thought–inclined newspaper of Brussels). The first article was entitled "Is Belgian Particularism Shackling the Progress of Medical Research: A Severe American Indictment," and the second article simply "Particularism." Some of the passage in the *Science* article about the medical scientific colloquium in the royal palace was cited in a July issue of another French-language Belgian magazine called *Europe,* devoted in that instance largely to articles on the Belgian royal family. A bimonthly review called *Tonus,* circulated to Belgian physicians by the Winthrop drug house, carried a summary of my article headed "Progress of Medical Research and Belgian Particularism." A signed editorial was printed in a publication called *Recipe,* a monthly magazine put out by the Medical Circle of the French section of the Medical School of the Catholic University of Louvain. This editorial, written by a Jesuit priest who was editor-in-chief of *Recipe* and chaplain of the Medical Circle, took its title from the *Pourquoi Pas* article which it attacked: "Are We a Ridiculous Country?" Still another article with the same heading subsequently appeared in *Pourquoi Pas,* this one an evaluation of my article by a physician who was a professor of medicine and the head of a clinical medical research unit of the University of Brussels Medical School.

From the beginning of August 1962, after my return to Belgium, until the end of that December, when I finally left Belgium, nine more articles about my own were published in the Belgian press. These included: an article in the August issue of the monthly journal of the College of Physicians of Brussels and environs (*Bullétin du Collège des Médecins de l'Agglomération Bruxelloise*) called "An Odious Attack against Medical Research in Belgium"; an answer to this article in the "Free Tribune" section of the October issue of the same magazine, signed by thirty-five Belgian research physicians, mainly from Brussels, and entitled "Back-Wash (*Remous*) Apropos of the Article of Renée Fox"; an article in the October 26 issue of *De Linie,* a Flemish-Jesuit Christian-Socialism-oriented weekly newspaper, entitled "Medical Research in Belgium: An American Finger on the Wound";

a long discussion of my article, especially the parts of it devoted to the colloquium in the palace, in the context of an article on "What Is Being Done to Promote Scientific Research in Belgium?" that was published in the November 11 issue of *Le Drapeau Rouge*, a French-language Communist newspaper of Brussels; an account in a November issue of *Recipe* of a talk I gave on my research at the "Medical House" (*Maison Médicale*) of the French section of the Medical School of Louvain; a prominent reference to my article in the introductory paragraphs of an article written by a professor of economics of Louvain University whose theme was "Dispersion: Our Economy Is the Victim of It," printed in the September 8 issue of *La Libre Belgique*, a French-language, Catholic, politically conservative newspaper; and still another reference to my article in the November 28 edition of *Le Soir*, in an article called "Magic Skin," [14] which was largely concerned with the low salaries received by members of the faculties and research staffs of the Belgian universities. Finally, just before I left Belgium in December, two small articles appeared, the first in *La Libre Belgique* and the second in *Le Peuple* (a French-language Socialist newspaper published in Brussels), both of which summarized the content of a talk that I had given on Belgian medical research to a Flemish group interested in promoting industry in Flanders, the so-called Union for the Economy (*Vereniging voor Economie*).

There are several general features of this literature that deserve mention here. For one thing, as their titles suggest, most of the articles published were written in the controversial spirit and tone of editorials, either "celebrating" or "denouncing" "Medical Scientists in a Château" and its author, as the case might be. Second, numerous of these articles were written by persons who were not professional journalists: a professor of medicine, for example, a professor of economics, a Jesuit priest. Third, these discussions of my article appeared in publications representing (albeit not in perfect, Belgian-type equality) each of the major particularistic groups of the country: Flemish and Walloon; Catholic and Free Thought; Liberal (Conservative), Christian-Socialist, Socialist, and Communist. Fourth, although the publication of some of these articles about my research was spontaneous, others appeared because they were encouraged or engineered by persons who evidently wished to bring a discussion of problems of Belgian medical research before the larger public. For example, the first "Are

We a Ridiculous Country?" article that appeared in *Pourquoi Pas* was the direct result of the fact that a professor of medicine at one of the Belgian universities went to the editorial office of this magazine with a copy of my article in hand and asked a journalist there to publish what he expressly hoped would be a provocative essay about it.

These aspects of the way in which my article was brought to the attention of Belgian newspapers, magazines, and journals and handled by them are indicative of certain general characteristics of the press in Belgium. The Belgian press not only reports news and attempts to reflect and influence public opinion. It also seems to provide an important channel through which individuals who represent specific social groups in Belgium, but who are not journalists by trade, can speak out, directly or from behind the scenes, on behalf of the interests of those groups. Writing for the popular press, or utilizing it by supplying professional journalists with materials, seems to be a socially acceptable and, to some extent, even prestigious activity in which Belgian physicians, lawyers, professors, priests, and the like engage, as well as politicians and journalists. Apparently it is one of the more important, legitimate, potentially effective ways in Belgium to exert enough social pressure either to foster or to deter social change in various sectors of Belgium life.

This function of the Belgian press[15]—in combination with the zealous way that some persons encouraged me to write another article as soon as possible and the dismayed fashion in which others deplored all the "noise" (*bruit*) that my article had created—suggests that, using journalism as one of their principal means, advocates of change in some of the social conditions surrounding medical research were tourneying with those who were defending the *status quo* in this domain. Another characteristic of many of these articles about "Medical Scientists in a Château"—their tendency to make general references to *all* forms of scientific research in Belgium, to various intra- and interuniversity problems, to certain attributes of the Belgian economy, to the role of the royal family in the political life of Belgium, and even to positive qualities and problems of the country as a whole—also suggests that my article and its press coverage touched off an involved controversy between Belgian groups with vested interests in far more than the status of clinical medical research in Belgium. (I shall return to this point in the last pages of this chronicle.)

At any rate, the kind of coverage that was given to "Mademoiselle/

Juffrouw/Miss Fox" and her "famous (or infamous) article" in the lay press was one of the factors that made it almost impossible for me to carry out the last phases of my research in the discreet and private fashion that had characterized my three prior stays in Belgium. I now found myself moving in an emotional atmosphere in which I was called and considered a "heroine," a "Jeanne d'Arc," by some persons and a "viper" by others.

I was asked to discuss the content of my article and the larger study of which it was a part before a whole series of groups. The audiences ranged in size from a small dinner gathering, like that of the editorial board of a medical review, to an amphitheater filled with almost the entire faculty of one medical school; and they ranged in type from audiences entirely or primarily composed of physicians and medical scientists to those that fell completely outside the sphere of the medical profession—for example, the members of the Civil Service Executive Training Program of one university and the Union for the Economy already mentioned. None of the talks before such groups was conventional, in the sense that first the speaker gave a formal presentation of some length and then a shorter question-and-answer period followed. Instead, most of these occasions were transmuted into heated discussions with the American sociologist and between the members of the audience themselves.

At a particularly animated gathering, for example, a younger research physician, newly appointed to a university position with tenure, bitterly attacked "Medical Scientists in a Château," virtually line by line, whereas an older, full professor of medicine from the same medical school vehemently defended both the article and its author. At a certain point in this discussion, the young physician turned to the *patron* and expressed his surprise and dismay that he, a professor of distinction and accomplishment, should espouse the point of view of the American sociologist and be so severely critical of the Belgian "System," which could not have drastically mistreated him. For, after all, said the young physician, whatever difficulties he may have encountered in the course of his career, the professor had received enough support to do research of quality that was recognized internationally. The response of the professor to these remarks of the *jeune* ringingly ended the evening's debate. Oh, yes, he had received enough money and facilities with which to do research, said the professor. But when? he asked. "At the end of my career," he replied to his own question. "And what did I do before then to have the resources I needed to do research? I'll tell you what I did. I cheated (*J'ai triché*) to have what I needed! I repeat—I cheated!" Again and again, exchanges as emotional as these seemed to be unleashed by my article and by my presence before groups that had invited me to speak about it.

In addition to the public presentations that I was asked, and in a sense required, to give, I was a guest at many more private, social gatherings than during my first three stays in Belgium: luncheons, dinners, after-dinner get-togethers, and so forth. Once again, these occasions were rather out of the ordinary in several significant ways. I was usually the guest of honor or at least a focus of intense interest. In a number of instances, among the other guests who had been invited to meet me were persons prominent in sectors of Belgian life other than the medical: important political, business, or religious figures. Usually these affairs were not purely social but turned into miniature versions of the kinds of discussions and debates in which I was involved when making formal, public presentations of my work. At several of these gatherings, one of the guests privately recommended to me that before my study was finished I ought to discuss some of my observations and impressions with a member of the Belgian royal family. King Baudouin, Queen Mother Elisabeth, Princess Liliane, former King Leopold, were all mentioned in this connection; and in several cases, the persons making the suggestion offered to do what they could to obtain an audience for me—an audience, they said, with joking reference to my article, "in the château."

The excitement generated by the article and by my presence in Belgium produced still another series of extraordinary experiences. I began to receive confidences and data of a kind to which I had not previously had access. Without my actively soliciting them, I was given verbatim transcripts of several meetings highly revelatory of certain attitudes and events that significantly affected Belgian medical research; various documents that medical researchers had drafted at one time or another, individually or in groups, as attempts to analyze their professional situations, criticize or protest against them, or make recommendations for their modification; several reports written by professors of medicine who headed research groups, containing intimate details about the activities of their research teams, the sources of their research funds, and their itemized budgets. Other medical researchers talked to me directly about professional problems they themselves had experienced or about which they knew—problems they had never before discussed with me. And suddenly there also appeared on the scene (once or twice, announced only by their knocking on my front door) individuals with document-accompanied personal histories to tell about how they had been maltreated and professionally blocked by "the System" and sometimes by specific *messieurs* within it. It was hard to ascertain the authenticity of these stories. But from a certain point of view, perhaps as important as the veracity of the reports was the role that such persons imagined that I could play. They had come to tell the American sociologist "all about it," to unburden themselves, to "get the record straight." Above and beyond that, they hoped that through my personal and political influence in Belgium, which they judged to be considerable, and through future articles that I might write, I would be able to help rectify their personal situations and the more general problems of which they conceived them to be a part.

Those whom I did *not* get to see during these eventful six months in Belgium are also significant. Certain persons in the structure of the foundations with whom I tried to renew contact indicated by letter that they felt there was no special reason or value in receiving me. The appointments that people tried to make for me with several high-placed government officials did not materialize. I was received in the royal palace once, by the Chief of the King's Cabinet, but the promised audience with a member of the royal family never took place. And in the medical milieus of each of the four universities there were a certain number of researchers (among them some who had received me with cordiality and helped me with my study during the three previous summers) who now treated me coldly, kept me at a distance, or avoided any contact with me at all.

In the final analysis, how can one explain the degree and kinds of impact that my article in *Science* made on the medical scientific world of Belgium and on other sectors of the society as well? Before attempting an answer to that question, it is enlightening to examine in somewhat more detail the content of the sixteen articles about "Medical Scientists in a Château" published in the Belgian press and of the remarks made about the article at the various formal and informal gatherings to which I was invited during my stay in Belgium. Whether the comments were positive or negative, written or spoken, the majority of them were polemical in nature. On the whole, persons discussing the article were either for it or against it. Rarely were their comments accompanied by objective evidence or counterevidence to support the viewpoints they set forth.

The positive opinions tended to be general. Those well disposed toward the article called it "accurate," "insightful," "well written," and "courageous." By this they meant that they felt that the article evocatively, boldly, and precisely described the problems lived out, largely in silence, by Belgian medical researchers. As one article put it, "No Belgian could or would publish what Miss Fox has written. . . . We should render homage . . . to the foreign writer who dared to touch on these problems, without passion, with a scientific spirit, and, nevertheless, with an enormous amount of sympathy for our country."

The negative criticisms were more copious and specific. The recurrent themes that ran through them are as follows.

(1) How could "Medical Scientists in a Château" have been accepted by *Science,* an outstanding scientific publication? The article is not scientific at all. It is a subjective, impressionistic piece of journalism, a satiric, even cruel indictment of Belgian medical science and scientists. Concrete proof of the fact that the article is not scientific is that it contains no supporting statistics (*"Mais, où sont les chiffres?"*).

(2) This subjective, depreciatory account of medical science in Belgium will be diffused to scientists throughout the world, who will read the article and believe it because it has been published under the distinguished aegis of *Science.* This will adversely affect the way that Belgian medical research and researchers are regarded abroad. It may also jeopardize the financial support for their research that Belgian investigators can hope to receive from foreign sources, especially American ones.

(3) This image of Belgian medical research presented in the article was conceived and produced by an American. She in particular, and Americans in general, cannot hope to penetrate and understand the complexities of European culture and civilization or appreciate the richness and values of its traditions and the high place that it accords to that which is human rather than materialistic or technological. A typical American's perspective on things European is superficial, naïve, upstart, and self-satisfied. It has been perfectly portrayed in the novel *The Ugly American* and recently exemplified by the way that Americans reacted to Belgium in the Congo crisis. In her article, the American sociologist exhibits all these characteristic traits. And let it not be

forgotten, either, that the United States has some of the same problems in the realm of medical research that the sociologist has attributed to Belgium. Why has Belgium been singled out for special scrutiny and criticism?

(4) The account of the medical scientific colloquium held in the royal palace at Laeken and the use of the image of the château to convey certain things about the kind of "house" in which medical research in Belgium is carried out are inappropriate, misleading, and snide. The colloquium in the palace was not so much a scientific gathering as a ceremonial and public-relations occasion, having its American counterpart in certain elaborate receptions held at the White House. Ordinarily, Belgian medical scientists do not spend their time drinking cocktails and tea at four o'clock in the afternoon in palaces and great houses, as the sociologist implies. They are hard-working, dedicated researchers. And the royal family has nothing but a symbolic relationship to scientific research in Belgium. The palace at Laeken in no way controls or affects it.

(5) The American sociologist's description of the social organization of medical research in Belgium is inaccurate and self-contradictory, as well as insinuating. There are not as many research groups and institutes in Belgium as she contends. Universities are not as "cloistered" as she maintains. Relationships between *jeunes* and *patrons* are more colleaguelike than she says. The direction of Belgian foundations is not monopolized by one or two men; and the awarding of research grants by foundations like the National Fund of Scientific Research is determined exclusively by the professors of science who make up its commissions. There is nothing complicated or mysterious about the organizations concerned with medical research in Belgium or the relations between them. This misconception on the part of the American sociologist is due to her being a foreigner and probably to her contact with certain young medical researchers who are still relatively unknowledgeable about these social organizations and with other older researchers who are particularly devious in trying to manipulate "the System." Just as one telling example of her lack of knowledge, consistency, and logic—all three—how can the American sociologist say, as she does, that the social organization of medical research in Belgium is at once highly centralized and highly decentralized?

(6) Certainly none of the observations the author claims to have made are characteristic of *this* university, or of *this* department (which

university, or which department, depending on the affiliation of the critic writing or speaking). What she says may be relevant to other universities or other departments, but not to ours.

(7) As a matter of fact, from whom did the American sociologist get her information? Did she really spend an equal amount of time at each university? She seems to have been more influenced by some milieus than by others. And who were the persons who acted as her informants in the various places she visited? What she has written has the ring of aggrieved statements made by persons who have no special talent for research and yet are discontent because they have been unsuccessful in trying to obtain a university research position. It also sounds as if she were very much influenced by the *jeunes* who, like young people the world over, tend to be critical of their elders and desirous of change. In her remarks about the palace and the foundations, the sociologist must certainly have been skillfully led astray by certain vituperative *messieurs* we could name who constantly and insidiously attack these institutions out of motives of personal, economic, and political aggrandizement.

(8) What was the "real reason" that the American sociologist carried out this inquiry into Belgian medical research and then wrote what amounts to a sensationalistic exposé of it? Her motives surely were not impersonal, objective, and pure. Was she more in sympathy, if not in formal affiliation, with the Flemings or Walloons, with Catholics or Free Masons, with the Liberal, Christian-Socialist, Socialist, or Communist party?

(9) Lurking behind the author's account of medical research in Belgium is the implication that what she has written is applicable to other areas of Belgian life as well. This is grossly untrue. In fact, if there is any validity at all to the picture the American sociologist paints, some of the things she describes might be uniquely true of the world of medical research and the universities.

(10) If the American sociologist's description of the conditions surrounding Belgian medical research were accurate, Belgium would be devoid of any well-trained, competent researchers and would not produce any scientific work of merit. Neither of these things is true. Belgium has even had two Nobel Prize laureates in medicine in the course of the last fifty years. In fact, if the portrait of the problems, confusions, conflicts, and impasses in Belgian society that is suggested by the sociologist's article were authentic, Belgium would be a completely

paralyzed, unproductive, unviable country. We may have our family quarrels, our limitations, our faults. "We are only a small country" geographically and in some of our ways. But we are also a dynamic country, a land of work, achievement, and progress, with the courage to spring back again and again in the face of invasion, occupation, and adversity.

If we consider these reactions to the publication of "Medical Scientists in a Château" not only in relation to the actual nature of the study I conducted, the manner in which I proceeded, and the manifest content of the articles I published but also in the context of some of the characteristics of Belgian society already described, we are led to make certain interpretations about the deeper, more general significance of my explorations in the land of Belgian medical research.

Perhaps most evident of all is that by focusing attention through my study on *problems* of medical research in Belgium and (as numerous Belgian newspaper accounts put it) "diffusing" knowledge of these problems "throughout the world" via my *Science* article, I aroused strong nationalistic sentiments on the part of many Belgians, medical researchers and others. Affirmative pride of Belgians in the hard-working, hard-won, impressive accomplishments of their small country, in its history and some of its traditions, in its physical beauty, and in the kind of bravery for which Belgians have been cited since the days of Caesar all emerged in response to my article, together with a more defensive pride about the sort of denigration to which Belgians felt they had recently been subject by foreign reactions both to their Congo crisis and their linguistic problems. Now, along came an American sociologist who, especially because she wore the respected guise of scientist, could depreciate and damage the public image of Belgium still further. This Belgian *amour-propre* was admixed with certain practical considerations and concerns: namely, how the way Belgium was seen abroad, especially in the United States, might affect the money from foreign sources granted to Belgian scientists for research and, more importantly, the interest of foreign businessmen in Belgium as a site of Common Market investments.

Along with these overarching nationalistic sentiments, various manifestations of Belgian particularism were involved in the reactions to my study and the publication of some of its findings. It was implied by some in the medical research groups of each of the four Belgian universities that whereas my analysis greatly misrepresented the conditions under which they themselves worked, it could possibly be more accurately descriptive of the situations of investigators in "the other" Belgian universities or in a special one of them. This sort of particularistic assumption, as already pointed out, was made all the more probable by that fact that, as a consequence of the relatively limited interchange of ideas and personnel between the universities, Belgian medical researchers do not usually have detailed, firsthand knowledge of professional milieus other than their own.

Some of the particularistic attitudes and sentiments of Belgians partly account for another set of responses to my research activities and articles. It was often implied in private and public gatherings, and several times in print, by friendly as well as by hostile critics that the "real" covert motives of the American sociologist for investigating medical research in Belgium were not scientific, objective, or independent. Rather, she was conceived to be more identified with certain groups in Belgium than with others—possibly even in their employ—and consequently "out to get" the groups and individuals associated with them to whom she and her unseen manipulators and sponsors were opposed. The hidden ("clever," as defined by some, "sinister," as defined by others) ideologically committed role I was regarded as playing in league with certain powerful "others" was to some extent a projection onto my work and my person of the very kinds of suspicion and enmity-ridden attitudes toward members of particularistic groups other than their own and of the sorts of *messieurs les responsables* explanations of complex social events that I had described as being characteristic of many Belgians in my articles.

But the attribution of partisan and subversive motives to the American sociologist also derived from certain realities connected with Belgian social science. For in the research, teaching, and advisory activities in which they have been engaged, and in the planned social action they have initiated both in Belgium and in the Congo, at least two of the major groups of social scientists affiliated with Belgian universities have, in fact, operated primarily as ideologues. Actively and passively encouraged and sponsored by certain influential *messieurs* from outside as well as inside the universities, these Belgian social scientists have worked in an organized, systematic way to promote their particular religious-philosophical, political, and economic points of view. This was one of the implicit, unverbalized referents that certain Belgians had in mind when they speculated on what might be the "real" motives of the American sociologist and the most influential source of her information and support.

In fact, despite all their anxieties and suspicions, Belgians granted me a privilege they would not easily have accorded a compatriot, social scientist or otherwise. Precisely because I was not Belgian, and thus, at least in the earlier phases of my research, assumed to be less implicated in "the System" than any Belgian, I was allowed to pass with remarkable freedom from one university milieu and from one particularistic group to another, gathering sociologically relevant insights and information wherever I went from people who helped me and who confided in me even more than they ordinarily would have in one another. The range and depth of social contacts I was accorded in these ordinarily "boxed up" and "boxed in" Belgian milieus explain certain other features of the ultimate reactions to my work and publications.

For one thing, as already indicated, by virtue of the special opportunities given me I acquired and was able to present a detailed overview of the worlds of Belgian medical science and their relationship to certain more general Belgian characteristics—an overview not ordinarily or easily accessible to Belgians themselves. Thus, my perspective impressed and startled them in both a positive and a negative sense.

Furthermore, it might be said that although none of my research activities would have been possible without at least the tacit consent of Belgians connected with medical research, the observations and interviews I carried out in this connection broke through certain norms of constraint and silence that usually prevailed. Medical researchers everywhere, and even the foundation and government officials whom I got to see, spoke to me with what was evidently far less inhibition than was generally characteristic of them. Not only did they tutor me in the routine things I needed to learn about the organization of medical science in Belgium and point out to me what positive advantages it had for the advancement of research and their own careers, but they were even more vocal about the many social, cultural, psychological, and material conditions under which medical science operates in Belgium and which create problems for research and the investigators who conduct it. They themselves supplied the facts for "Medical Scientists in a Château"; but what seems to have happened when it was published is that virtually all the medical researchers and foundation and government officials reacted with some degree of anxiety, and in some cases shock, when they "saw in black and white" the analysis that was a prod-uct of what each of them had told the American sociologist.

Contributing to this anxiety and shock (along with the nationalistic and particularistic sentiments already considered) were what might be described as certain "guilt feelings" experienced by each of the persons who had helped me with my work, for in so doing they had violated what were apparently their strongly held beliefs that the best and wisest thing to do morally and practically was to keep quiet about one's professional difficulties, or at least discuss them only with the most trusted of intimates among family and friends. Even those persons who were unequivocal and fearless advocates of my work, after as well as before my article was published, admitted in public that until the American sociologist had come along they, too, had "suffered in silence" because they had not had "the courage to speak out" manifested by the sociologist.

Such allusions to the "courage" that it took to "break the silence" sug-gest that Belgian medical researchers felt reluctant to speak out about their problems, not only because of "principles of conscience, loyalty, and pride," as one of them put it, but also because they feared what the consequences might be for their own professional situations and fu-tures if they raised their voices to reveal and to protest, and especially if they were identified by others as the persons who had done so. This widespread fear derived in good part from the very limited professional opportunities that exist in Belgium for a man who wishes to devote himself primarily to medical research—the same situation I described and analyzed in my article.

The apprehensions of medical researchers about their careers signifi-cantly influenced a great many of the reactions to "Medical Scientists in a Château" and to its author. Certain medical researchers, as already indicated, who had done a great deal to advance my work dissociated themselves from my findings and from any contact with me, once my article was published. Others continued to see me and support me on a private basis but publicly indicated that they were not in agreement with what I had written and published. Even those who consistently supported my work and my person, both publicly and privately, were worried to some extent about what this might mean for them profes-sionally.

How individual medical investigators reacted in this regard was de-termined, in part, by their status in the field. For example, in the exchange between a young research physician and an older professor of medicine earlier described, in which the *jeune* attacked the American sociologist and her article and defended the way medical research in Belgium is currently organized and the *patron* defended the article and attacked "the System," it was the firmly established success of the professor in his career which made it just that much more possible for him to speak out with a sense of relative impunity. The younger man, newly appointed to a university position, undoubtedly felt that he would be jeopardizing his career, which was just beginning to unfold in a favorable way, if he were critical of the very organization in which he hoped progressively to succeed.

This is not to suggest that most of the younger Belgian medical researchers defended the *status quo* and attacked my analysis while their seniors generally advocated both my work and certain changes. As might be expected, a greater number of younger medical researchers than older ones regarded my study, the articles I had thus far published, the public discussion they had evoked, and what I could be expected to write in the future as what they called "catalysts" through which certain changes in the social structure of medical research in Belgium might be brought about—changes which could improve their own professional situations and open up new research possibilities to them. As the thirty-five young research physicians who wrote a signed letter to the *Bulletin of the College of Physicians of the Area of Brussels* stated in print: "We wish . . . to remark that medical research, like all human activity, is perfectible; that is why we must rejoice in commentaries, suggestions and even criticisms that are addressed to it, because it is a demonstration of the interest that the powers of public opinion take in it throughout the world. In this regard, the article of Mademoiselle Fox . . . deserves to be read with attention, and ought to be full of lessons for all those, from near and far, who are interested in medical research in our country."

In the case of some of the less easily predictable reactions to my work and publications, certain particularities of a medical researcher's local university situation sometimes exerted an implicit influence. This was one of the elements, for example, that went into the overt and almost perverse dispute that my article seemed to have provoked between the aforementioned young research physician and the professor. It was also true of an even more dramatic occasion when, at the invitation of a relatively young, newly appointed professor of medicine to come and talk with him privately, I arrived at his medical school and, without any advance warning, was whisked before an auditorium filled with members of the medical faculty whom my host had assembled. Presided over by the dean, the faculty engaged me in a two-hour discussion about my work, in the course of which the physician who had engineered all this without telling me about it turned out to be one of the most vocal of my negative critics.

As different as these two incidents were with respect to the concrete details that went into their making (details that I shall not recount here), they had one important element in common. In both these instances, the young research physician in the first case and the young professor of medicine in the second took positions and used means that were essentially what we have already indicated are called *petits chemins* or *truques* by Belgians—that is, calculated, somewhat devious methods to consolidate and further their university research careers and to cover up for other, more illicit techniques they had used in the past to get ahead. In a social system like the Belgian, in which, as we have seen, medical researchers are so often confronted with ambiguity, conflict, delay, and total impasse, the frequency with which nonlegitimate ways are used to get around "the System" is probably very high. (In this connection, we recall the words of the successful professor who, in his heated debate with the young research physician, confessed that he had "cheated" to obtain the funds and equipment he needed to do his research.) Perhaps, then, fear of being detected and exposed in such maneuvers, and guilt and shame about having had to use them, lay behind some of the apprehensive and hostile reactions to my article on the part of certain medical researchers.

Finally, there was a whole series of responses to "Medical Scientists in a Château" that, in one way or another, suggest that the analysis of how social, cultural, and historical factors affect Belgian medical research touched on far more than that. These reactions include the several instances in which my analysis was cited in discussions of characteristics of other Belgian institutions—for example, as illustrative of dispersion in the Belgian economy; the many times that I was accused of having "generalized from a particular case"; the interest that my article aroused among a wide range of persons and groups who were not connected, at least ostensibly, with Belgian medical research; the number of individuals in high-ranking political, economic, and religious positions into whose hands my article was delivered; the royalist and antiroyalist sentiments it evoked; the great houses in which I was received; and the distinguished people who refused me when I asked to see them. These reactions invite one at least to speculate on the extent to which my analysis of medical research in Belgium might be applicable to other institutions as well. They suggest that a number of Belgian institutions have the same general characteristics; that there is an intricate, almost interlocking relationship between many Belgian institutions, the medical scientific among them; and that there may indeed exist a group of *messieurs les responsables* who, in a way that perhaps is not mythical, after all, might wield enough authority and power in many different domains of Belgian life to account for some of their similarities and interconnections.

Two final incidents that occurred before I left Belgium in December 1962, just as my research was officially ending, make a fitting conclusion to this chronicle.

(1) A research physician presented me with a framed reproduction of a primitive Flemish painting, accompanied by a note that said: "Thank you for having done more than anyone ever has for Belgian medical research and researchers, in what may turn out to be our essentially hopeless struggle. . . ."

(2) A Belgian gentleman, indirectly associated with the medical scientific world, told me that the reason I had never quite met some of the important political figures to whom he had volunteered to introduce me was that I had dined too often in the home of a certain Count, regarded by these personages as someone who secretly acted as their enemy. The amusing, mysterious, and perhaps very Belgian thing about this assertion is that in the course of my four consecutive research trips to Belgium I never met the Count cited or even saw him from afar.

Both of these final incidents exemplify some of the attitudes and processes that I tried to study in the land of Belgian medical research and of which I (or, at least, the projected image of the American sociologist as seen by Belgians) inadvertently but not regretfully also became the object.

• NOTES

1. The research project to which this chronicle refers was made possible by special grants from the Council for Research in the Social Sciences of Columbia University (summer 1959) and the Belgian American Educational

Foundation (summers of 1960 and 1961) and by a fellowship from the John Simon Guggenheim Memorial Foundation (1962).

2. I use the term "clinical medical research" in a relatively loose, descriptive sense to refer to medical research which has some ostensible, intended relationship to an understanding of the etiology, treatment, or prognosis of disease, or of the maintenance of health, in human beings.

3. In 1920, the Belgian American Educational Foundation created a number of Commission for Relief in Belgium Special Fellowships as part of the general purposes of this foundation, which are to promote closer relations and an exchange of intellectual ideas between Belgium and the United States and to commemorate the work of the Commission for Relief in Belgium in World War I. According to a memorandum issued by the Foundation on March 18, 1964, in the past 43 years about 1,000 Belgians have studied in the United States for one or more years on such fellowships.

4. It was not until January 1, 1963, that the cardinal and bishops of the Catholic Church in Belgium passed an ordinance which made it officially acceptable for Belgian priests to wear "clergymen suits" in public.

5. The religious structure of Belgian society is a rather special one. Belgium is a predominantly Catholic country, one in which the influence of the Catholic Church in many domains of Belgian life is still strong and pervasive enough for it to be described by some persons as a "clerical" country. It is also a country in which Protestantism is of minor importance, in terms both of its small number of adherents and its relative lack of influence. There are significant groups of Jews only in the communities of Antwerp and Brussels. On the whole, most Belgians who are not practicing Catholics were nonetheless baptized as Catholics. Many such Belgians belong to the religious-philosophical orientation known as Free Thought. Within this category of Free Thought, there is an indeterminate number of persons who are members of the Free Mason Lodge, which, in Belgium, as in France, was historically founded as a secret organization, with one of its primary goals that of combating the doctrine and the influence of the Catholic Church. The influence of the Lodge, in general, in Belgian society, as well as in this particular anticonfessional respect, is thought by many Belgians to be considerable. It is hard to be more precise than this about the beliefs and membership of the Lodge and about its impact on Belgian life since it still remains essentially a secret organization in Belgium.

6. In the summer of 1961, I once again received a Commission for Relief Special Fellowship from the Belgian American Educational Foundation that enabled me to go on with my work, and my place of residence once more was Brussels.

7. There is perhaps a touch of historical irony in these difficulties, since one of the founding fathers of "social statistics" (out of which has been born a tradition of census-taking in many societies) as well as of certain quantitative aspects of the discipline of sociology was Lambert Adolphe Jacques Quetelet

(1796-1874), a Belgian astronomer, meteorologist, and statistician, originally from Ghent.

8. It has been suggested to me by a number of Belgians, among them several former members of the underground that operated in Belgium during World War II when the country was occupied by Germany, that another origin of the lack of documents and data that I encountered—as well as the difficulty in obtaining what records and figures do exist, the falsification of certain materials, and the atmosphere of mystery and tension with which persons like myself who set out to collect such data have been confronted— is the fact that again and again over the centuries Belgium has been besieged and occupied by conquering powers (Spain, Holland, France, Germany, etc.). As a consequence (the individuals who offer this explanation say), Belgians have perfected techniques for making it difficult for powers who have occupied her to gain easy access to information about the country and her inhabitants; and, perhaps in a way that is not completely rational, Belgians have also had a tendency to carry such underground attitudes and sabotaging techniques over to times of relative normalcy and peace.

9. Renée C. Fox, "Journal Intime Belge/Intiem Belgisch Dagboek," *Columbia University Forum*, V (1962), 11-18.

10. Renée C. Fox, "Medical Scientists in a Château," *Science*, CXXXVI (1962), 476-483.

11. The use of the image of the château in this sense was suggested to me by certain comments made by Raoul Kourilsky, professor of clinical medicine at the University of Paris, about the predicament of medical research in present-day France. In the inaugural lesson he gave after being named to his professorial chair, Kourilsky said: "The great sacrifice has been research. We have conserved the old 'château' and its arrangements that belong to another age. . . . We have watched from afar the triumphant ascent of biology. . . ." See Raoul Kourilsky, "Leçon Inaugurale," reprint from *L'Expansion scientifique française*, 1958, pp. 27-28.

12. It is interesting to consider that Kafka, a Czech, was a citizen of another small country of Continental Europe. Perhaps that accounts in part for the striking pertinence to Belgium of certain sociopsychological dimensions of his writings.

13. In addition to all the requests for reprints that I received from Belgium, I received many from other countries of Continental Europe (France, Germany, Italy, Holland), England, the Scandinavian countries of Sweden, Norway, and Denmark, two countries behind the Iron Curtain (Czechoslovakia and Poland), Canada, Latin America (Panama and Venezuela), and many regions of the United States. This attests, of course, to the vast and wide-ranging circulation of *Science*. It also suggests that my analysis of some of the problems of clinical medical research in Belgium might have their counterparts in numerous other societies as well.

14. This is a reference to a novel by Balzac, *La Peau de chagrin*.

15. This is not to imply that I believe these characteristics necessarily to be unique to the Belgian press. It is my impression, for example, that the newspapers and magazines of certain other Continental European countries that I have only visited, and therefore know less well, might have some of the same traits. France and Holland come to mind in this connection.

Index

Abbott, Allan, 223
Aberle, David, 143
  "The Functional Prerequisites of a Society," 228
Agassiz, Alexander, 54
Agrarian Socialism (Lipset), 100, 104–105
American Friends Service Committee, 127
American Sociological Association, 230, 311
American Sociological Review, The, 230
American Soldier, The (Stouffer), 44
"Analytic Unit for the Study of Interaction, An" (Watson and Potter), 312
Antioch College, 226
Applications of Methods of Evaluation (Wright and Hyman), 134
Arensberg, Conrad M., 18
Assistants, in a sociability study, 273–275

Bacon, Francis (*Novum Organum*), 54
Bales, Freed, 225, 228
  Working Papers in the Theory of Action, 157
Barker, R. G. (*One Boy's Day*), 297
Barzun, Jacques, 217
Baudouin, King, 377
Beach, E. F. (*Economic Models*), 230
Beal, Vickie, 224, 226, 229
Beaumont de La Bonninière, Gustave, 1–2
Beck, Ariadne P. ("Andi"), 235–312 *passim*
Becker, Howard S., 342–343
Belgian American Educational Foundation, 348, 359
Belgian Government Information Center (N.Y.), 369, 371

"Belgian Journal, A" (Fox), 368–369
Belgium
  medical research in, 345–391
    reactions to publication of, 369–388
Bell, Daniel, 105
Bellah, Robert N., 5–6, 12
  "Japan: An Example of the Particularistic-Ascriptive Value Pattern," 144
  "Sociology of Religion," 147
Benney, Mark, 242
Ben-Zeev, Saul, 304
Beveridge, W. I. B., 3
Bias, in research, *see* Research, preconceptions in
Blau, Peter, 5, 7, 9, 11, 229–230
Bogoras, Waldemar, 167
Booth, Mary, 226
Bridgman, P. W., 52, 58
Brunner, E. DeS., 125
Bullétin du Collège des Médecins de l'Agglomération Bruxelloise, 373, 386
Bureau of Applied Social Research, 108, 122, 136, 188, 193, 230, 278
  organization and functional framework, 45–48
  consulting colleagues in, 33–37
  problems of field work in, 18–27
  research process in the study of, 16–49
    conceptual refinement, 32–48
    and politics, 31–32
    problems of field work, 18–27
    the role of the observer, 27–32
  statistical records, 39–44
  supervisory authority, 37–39
Bureaucratic Structure and Personality (Merton), 17